D1696240

*Edited by Andre Richardt, Birgit Hülseweh,
Bernd Niemeyer, and Frank Sabath*

CBRN Protection

Related Titles

Voeller, J. G. (ed.)

Wiley Handbook of Science and Technology for Homeland Security

4 Volume Set

2010
ISBN: 978-0-471-76130-3

Burchfield, L. A.

Radiation Safety

Protection and Management for Homeland Security and Emergency Response

2009
ISBN: 978-0-471-79333-5

Richardt, A., Blum, M.-M. (eds.)

Decontamination of Warfare Agents

Enzymatic Methods for the Removal of B/C Weapons

2008
ISBN: 978-3-527-31756-1

Barriot, P., Bismuth, C. (eds.)

Treating Victims of Weapons of Mass Destruction

Medical, Legal and Strategic Aspects

2007
ISBN: 978-0-470-06646-1

Bennett, B. T.

Understanding, Assessing, and Responding to Terrorism

Protecting Critical Infrastructure and Personnel

2007
ISBN: 978-0-471-77152-4

Marrs, T. C., Maynard, R. L., Sidell, F. (eds.)

Chemical Warfare Agents: Toxicology and Treatment

2007
ISBN: 978-0-470-01359-5

Molino, L. N.

Emergency Incident Management Systems

Fundamentals and Applications

2006
ISBN: 978-0-471-45564-6

Khardori, N. (ed.)

Bioterrorism Preparedness

Medicine - Public Health - Policy

2006
ISBN: 978-3-527-31235-1

Langford, R. E.

Introduction to Weapons of Mass Destruction

Radiological, Chemical, and Biological

2004
ISBN: 978-0-471-46560-7

Edited by Andre Richardt, Birgit Hülseweh, Bernd Niemeyer, and Frank Sabath

CBRN Protection

Managing the Threat of Chemical, Biological, Radioactive and Nuclear Weapons

WILEY-VCH

WILEY-VCH Verlag GmbH & Co. KGaA

The Editors

Dr. Andre Richardt
Bundeswehr Institute for
Protection Technologies WIS430
Humboldtstr. 100
29633 Munster
Germany

Dr. Birgit Hülseweh
Bundeswehr Institute for
Protection Technologies WIS430
Humboldtstr. 100
29633 Munster
Germany

Prof. Dr.-Ing. Bernd Niemeyer
Universität der Bundeswehr
Process Engineering
Holstenhofweg 85
22043 Hamburg
Germany

Dr.-Ing. Frank Sabath
Bundeswehr Institute for
Protection Technologies WIS430
Humboldtstr. 100
29633 Munster
Germany

All books published by **Wiley-VCH** are carefully produced. Nevertheless, authors, editors, and publisher do not warrant the information contained in these books, including this book, to be free of errors. Readers are advised to keep in mind that statements, data, illustrations, procedural details or other items may inadvertently be inaccurate.

Library of Congress Card No.: applied for

British Library Cataloguing-in-Publication Data
A catalogue record for this book is available from the British Library.

Bibliographic information published by the Deutsche Nationalbibliothek
The Deutsche Nationalbibliothek lists this publication in the Deutsche Nationalbibliografie; detailed bibliographic data are available on the Internet at <http://dnb.d-nb.de>.

© 2013 Wiley-VCH Verlag GmbH & Co. KGaA, Boschstr. 12, 69469 Weinheim, Germany

All rights reserved (including those of translation into other languages). No part of this book may be reproduced in any form – by photoprinting, microfilm, or any other means – nor transmitted or translated into a machine language without written permission from the publishers. Registered names, trademarks, etc. used in this book, even when not specifically marked as such, are not to be considered unprotected by law.

Composition Laserwords Private Ltd., Chennai, India
Printing and Binding Markono Print Media Pte Ltd, Singapore
Cover Design Adam-Design, Weinheim

Printed in Singapore
Printed on acid-free paper

Print ISBN: 978-3-527-32413-2
ePDF ISBN: 978-3-527-65019-4
ePub ISBN: 978-3-527-65018-7
mobi ISBN: 978-3-527-65017-0
oBook ISBN: 978-3-527-65016-3

Contents

Foreword *XV*

Preface *XVII*

About the Editors *XIX*

List of Contributors *XXIII*

Part I   History and Treaties in CBRN – Warfare and Terrorism *1*

1   A Glance Back – Myths and Facts about CBRN Incidents *3*
Andre Richardt and Frank Sabath
1.1   Introduction *3*
1.2   History of Chemical Warfare *4*
1.2.1   Chemical Warfare Agents in Ancient Times *5*
1.2.2   Birth of Modern Chemical Warfare Agents and Their Use in World War I *5*
1.2.3   Chemical Warfare Agents between the Two World Wars *8*
1.2.3.1   The Italian–Ethiopian War *8*
1.2.3.2   Japanese Invasion of China *9*
1.2.3.3   First Nerve Agents *9*
1.2.4   Chemical Warfare Agents in World War II *9*
1.2.5   Chemical Warfare Agents during the Cold War *10*
1.2.6   Chemical Warfare Agents Used in Terrorism *11*
1.2.7   Conclusions and Outlook *12*
1.3   Introduction to Biological Warfare *13*
1.3.1   Most Harmful Pandemics in History *14*
1.3.2   Biological Warfare Agents in Ancient Times BC *16*
1.3.3   Biological Warfare Agents in the Middle Ages to World War I *18*
1.3.4   From World War I to World War II – the Beginning of Scientifically Based Biological Weapons Research *18*

1.3.5	From the End of World War II to the 1980 – the Great Bioweapons Programs	20
1.3.6	From the 1980 Up Today – the Emerging of Bioterrorism	20
1.3.7	Conclusions and Outlook	20
1.4	Introduction to Radiological and Nuclear Warfare	22
1.4.1	Discovery of Nuclear Fission	23
1.4.2	Manhattan Project – Development of the First Fission Weapons	25
1.4.3	Nuclear Arms Race	29
1.4.4	Status of World Nuclear Forces	35
1.4.5	Radiological Warfare and Nuclear Terrorism	35
1.4.6	Conclusions and Outlook	37
	References	37
2	**International Treaties – Only a Matter for Diplomats?**	*39*
	Martin Schaarschmidt	
2.1	Introduction to the Minefield of Negotiations	39
2.1.1	Arms Reduction and Prohibition of Use	41
2.1.2	Arms Control and International Controlling Bodies	42
2.1.3	Nonproliferation	42
2.2	Why It Is so Difficult to Implement International Regulations?	42
2.2.1	Trust – Devoid of Trust Every Effort Is Useless	43
2.2.2	Negotiation – Special Skills Are Required	43
2.2.3	Dual Use – Good or Bad Technology?	44
2.2.4	Verification – an Instrument for Trust Building	44
2.2.5	Technological Advancement – Gain of Momentum	45
2.3	Historic Development of Treaties – the Link to the Incidents	46
2.4	Today's System of Treaties – a Global Network	47
2.4.1	The Geneva Conventions – the Backbone for Further Treaties	49
2.4.2	Deployment System for Weapons – Control the Carrier Systems	50
2.4.3	Biological and Chemical Weapons	51
2.4.4	Chemical Weapons Convention 1993 and Organization for the Prohibition of Chemical Weapons (OPCW)	52
2.4.5	Implications of the Chemical Weapons Convention (CWC) and the Biological Weapons Convention (BWC)	53
2.5	Nuclear Weapons	54
2.5.1	Nonproliferation	55
2.5.2	Disarmament	56
2.5.2.1	Strategic Arms Limitation Talks/Treaty (SALT)	56
2.5.2.2	Strategic Arms Reduction Treaty (START)	57
2.5.2.3	Strategic Offensive Reductions (SORT) 2003	58
2.5.3	Test-Ban and Civil Use	58
2.5.4	Nuclear-Weapon-Free Zones	60
2.6	Organizations	63
2.7	Conclusions and Where Does the Road Lead?	64
	References	64

Part II CBRN Characteristics – Is There Something Inimitable? 67

3   Chemical Agents – Small Molecules with Deadly Properties 69
Hans-Jürgen Altmann, Silke Oelze, and Bernd Niemeyer
3.1   Are Special Properties Required for Chemical Warfare Agents? 69
3.2   How can we Classify Chemical Warfare Agents? 71
3.2.1   A: Physicochemical Behavior 72
3.2.2   B: Route of Entry into the Body 74
3.2.3   C: Organs to be Affected 75
3.2.4   D: Physiological Effects on Humans 76
3.2.5   E: Identification According to the NATO Code 78
3.3   Properties of Chemical Warfare Agents 78
3.3.1   Blister Agents (Vesicants) 78
3.3.2   Arsenicals 83
3.3.3   Blood Agents 85
3.3.4   Tear Agents (Lachrymators) 89
3.3.5   Vomiting Agents (Sternutators) 92
3.3.6   Nerve Agents 94
3.4   Choking and Irritant Agents 97
3.5   Incapacitating Agents 99
3.6   Dissemination Systems of Chemical Warfare Agents 99
3.7   Conclusions and Outlook 101
        References 101

4   Characteristics of Biological Warfare Agents – Diversity of Biology 103
Birgit Hülseweh
4.1   What Is Special? 104
4.2   Types of Biological Agents 104
4.2.1   Bacteria 105
4.2.2   Viruses 107
4.2.3   Toxins 108
4.2.4   Fungi 109
4.3   Risk Classification of Biological and Biological Warfare Agents 110
4.3.1   Risk Classification of Potential Biological Warfare Agents 111
4.4   Routes of Entry 114
4.5   Origin, Spreading, and Availability 118
4.5.1   Methods of Delivery 120
4.6   The Biological Event – Borderline to Pandemics, Endemics, and Epidemics 121
4.7   The Bane of Biotechnology – Genetically Engineered Pathogens 121
4.8   Conclusions and Outlook 123
        References 123

5	**Characteristics of Nuclear and Radiological Weapons** *125*	
	Ronald Rambousky and Frank Sabath	
5.1	Introduction to Nuclear Explosions *126*	
5.1.1	Nuclear Fission *126*	
5.1.1.1	Critical Mass for a Fission Chain *127*	
5.1.2	Nuclear Fusion *128*	
5.1.3	Weapon Design *129*	
5.1.3.1	Pure Fission Weapon *129*	
5.1.3.2	Fusion-Boosted Fission Weapon *130*	
5.1.3.3	Thermonuclear Weapons *130*	
5.1.4	Effects of a Nuclear Explosion *131*	
5.2	Direct Effects *133*	
5.2.1	Thermal Radiation *133*	
5.2.2	Blast and Shock *137*	
5.2.3	Initial Nuclear Radiation *140*	
5.2.4	Residual Nuclear Radiation *145*	
5.3	Indirect Effects *149*	
5.3.1	Transient Radiation Effects on Electronics (TREE) *149*	
5.3.2	Nuclear Electromagnetic Pulse (NEMP) *152*	
5.3.2.1	Generation of Electric Field *152*	
5.3.2.2	NEMP in High-Altitude Burst *153*	
5.3.2.3	Early Component of NEMP (E1) *154*	
5.3.2.4	Intermediate Component of NEMP (E2) *156*	
5.3.2.5	Late Time Component of NEMP (E3) *157*	
5.4	Radiological Weapons *159*	
5.4.1	Radioactive Material and Radiological Weapons *160*	
5.4.2	Impacts of Radiological Weapons *162*	
5.4.2.1	Radiological Exposure Device *162*	
5.4.2.2	Radiological Dispersal Device *162*	
	References *165*	
	Part III **CBRN Sensors – Key Technology for an Effective CBRN Countermeasure Strategy** *167*	
6	**Why Are Reliable CBRN Detector Technologies Needed?** *169*	
	Birgit Hülseweh, Hans-Jürgen Marschall, Ronald Rambousky, and Andre Richardt	
6.1	Introduction *169*	
6.2	A Concept to Track CBRN Substances *170*	
6.3	Low-Level Exposure and Operational Risk Management *175*	
6.4	Conclusions and Outlook *177*	
	References *178*	

7	**Analysis of Chemical Warfare Agents – Searching for Molecules** 179
	Andre Richardt, Martin Jung, and Bernd Niemeyer
7.1	Analytical Chemistry – the Scientific Basis for Searching Molecules 180
7.2	Standards for Chemical Warfare Agent Sensor Systems and Criteria for Deployment 182
7.2.1	Recommended Chemical Agent Concentration and Requirements for Chemical Warfare Agent Sensors 182
7.2.2	Acute Exposure Guideline Levels (AEGLs) for Chemical Warfare Agents 183
7.3	False Alarm Rate and Limit of Sensitivity 184
7.4	Technologies for Chemical Warfare Agent Sensor Systems 185
7.4.1	Mass Spectrometry 187
7.4.2	Atomic Absorption Spectrometry (AAS) 190
7.4.3	Ion Mobility Spectrometry (IMS) 192
7.4.4	Colorimetric Technology 197
7.4.5	Photoionization Technology (PI) 198
7.4.6	Electrochemical Technologies 199
7.4.7	Infrared (IR) Spectroscopy 200
7.5	Testing of Chemical Warfare Agent Detectors 203
7.6	Conclusions and Future Developments 206
	References 208
8	**Detection and Analysis of Biological Agents** 211
	Birgit Hülseweh and Hans-Jürgen Marschall
8.1	What Makes the Difference? 212
8.2	The Ideal Detection and Identification Platform 215
8.3	Bioaerosols: Particulate and Biological Background 216
8.4	Aerosol Detection – A Tool for Threat Monitoring 217
8.4.1	Cloud Detection 217
8.4.2	Radio Detecting and Ranging (RADAR) and Light Detection and Ranging (LIDAR) 219
8.4.3	Aerosol Particle Sizer (APS), Flame Photometry, and Fluorescence Aerosol Particle Sizer (FLAPS) 220
8.4.4	Detector Layout Topology, Sensitivity, and Response 222
8.5	Sampling of Biological Agents 223
8.5.1	Aerosol Sampling 224
8.5.1.1	Surface Sampling 227
8.6	Identification of Biological Warfare Agents 229
8.6.1	Immunological Methods Based on Enzyme-Linked Immunosorbent Assay (ELISA) 229
8.6.2	Molecular Methods 233
8.6.3	Chemical and Physical Identification 236

8.7	Developing and Upcoming Technologies	238
8.8	Conclusions	239
	References	240

9 Measurement of Ionizing Radiation 243
Ronald Rambousky

9.1	Why Is Detection of Ionizing Radiation So Important?	244
9.2	Physical Quantities used to Describe Radioactivity and Ionizing Radiation	248
9.2.1	Activity	248
9.2.2	Absorbed Dose	249
9.2.3	Equivalent Dose	250
9.2.4	Effective Dose Equivalent	250
9.2.5	Operational Dose Quantities	250
9.3	Different Measuring Tasks Concerning Ionizing Radiation	251
9.3.1	Personal Dosimetry	252
9.3.2	Measuring the Ambient Dose Rate	252
9.3.3	Searching for Gamma- and Neutron-Sources	252
9.3.4	Surface Contamination Measurements	253
9.3.5	Nuclide Identification	253
9.3.6	Measurement of Activity	254
9.3.7	Detection of Radioactive Aerosols	255
9.4	Basics of Radiation Detectors	256
9.4.1	Gas-Filled Detectors	256
9.4.2	Luminescence Detectors	259
9.4.3	Photo-emulsion	260
9.4.4	Scintillators	260
9.4.5	Semiconductor Detectors	262
9.4.6	Neutron Detectors	265
9.5	Gamma Dose Rate and Detection of Gamma Radiation	266
9.5.1	Metrological Dose Rate Measurements	266
9.5.2	Energy Response of a Dose-Rate Detector	267
9.5.3	Quantitative Detection	268
9.6	Conclusions and Outlook	271
	References	272

Part IV Technologies for Physical Protection 273

10 Filter Technology – Clean Air is Required 275
Andre Richardt and Thomas Dawert

10.1	Filters – Needed Technology Equipment for Collective and Individual Protection	275
10.2	General Considerations	276
10.3	What are the Principles for Filtration and Air-Cleaning?	278
10.3.1	Particulate Filtration	279

10.3.2	Gas-Phase Air Cleaning *283*
10.4	Test Methods *286*
10.4.1	Particle filter testing methods *288*
10.4.2	Gasfilter tests *289*
10.5	Selection Process for CBRN Filters *290*
10.6	Conclusions and Outlook *292*
	References *293*
11	**Individual Protective Equipment – Do You Know What to Wear?** *295*
	Karola Hagner and Friedrich Hesse
11.1	Basics of Individual Protection *296*
11.2	Which Challenges for Individual Protection Equipment (IPE) Can Be Identified? *296*
11.3	The Way to Design Individual Protective Equipment *298*
11.4	Function *299*
11.5	Ergonomics – a Key Element for Individual Protection Equipment *301*
11.6	Donning and Doffing – Training Is Required *305*
11.7	Overview of IPE Items – They Have to Act in Concert *306*
11.7.1	Respiratory Protection *307*
11.7.2	Respirator Design *310*
11.7.2.1	Air-Purifying Escape Respirator (APER) with CBRN Protection *313*
11.7.3	Air-Purifying Respirators (APRs) with Canisters for Ambient Air *314*
11.7.3.1	Respirators with Blower Support (Powered Air-Purifying Respirator, PAPR) *315*
11.7.3.2	Respirators with Self-contained Breathing Apparatus (SCBA) *316*
11.7.4	Canisters *317*
11.7.5	Body Protection *317*
11.7.6	Protective Suits *318*
11.7.6.1	Permeable Protective Suits *319*
11.7.6.2	Impermeable Protective Suits *320*
11.7.7	Protective Gloves *322*
11.7.8	Protective Footwear *323*
11.7.9	Pouches *323*
11.7.10	Ponchos *324*
11.7.11	Self-Aid Kit *324*
11.7.12	Casualty Protection *325*
11.8	Quality Assurance *326*
11.9	Workplace Safety *327*
11.10	Future Prospects *327*
	References *328*

12	**Collective Protection – A Secure Area in a Toxic Environment** 331
	Andre Richardt and Bernd Niemeyer
12.1	Why Is Collective Protection of Interest? 332
12.2	Collective Protection Systems – Required for Different Scenarios 337
12.3	Basic Design 341
12.3.1	Air Filtration Unit (AFU) and Auxiliary Equipment 342
12.3.2	Environmental Control Unit (ECU) 344
12.3.3	Contamination Control Area (CCA) 345
12.3.4	Airlock – the Bottleneck for Ingress and Egress 346
12.3.5	Toxic-Free Area (TFA) 348
12.4	Conclusions and Outlook 348
	References 349

Part V Cleanup after a CBRN Event 351

13	**Decontamination of Chemical Warfare Agents – What is Thorough?** 353
	Hans Jürgen Altmann, Martin Jung, and Andre Richardt
13.1	What Is Decontamination? 353
13.2	Dispersal and Fate of Chemical Warfare Agents 354
13.3	Decontamination Media for Chemical Warfare Agents 356
13.3.1	Aqueous-Based Decontaminants 358
13.3.1.1	Water 358
13.3.1.2	Water-Soluble Decontamination Chemicals 359
13.3.2	Non-aqueous Decontaminants 359
13.3.3	Heterogeneous Liquid Media 362
13.3.3.1	Macroemulsions (Emulsions) 362
13.3.3.2	Microemulsions 366
13.3.3.3	Foams and Gels 367
13.4	Selected Chemical Warfare Agents and Decont Reaction Schemes 369
13.4.1	Sulfur Mustard (HD) 370
13.4.1.1	Hydrolysis 370
13.4.2	Sarin (GB) 370
13.5	Soman (GD) 372
13.6	VX 372
13.7	Catalysis in Decontamination 373
13.8	Decont Procedures 375
13.8.1	Generalities 376
13.8.2	Equipment Decontamination 376
13.8.2.1	Wet Procedures 376
13.8.2.2	Dry Procedures 378
13.8.2.3	Clothing and Protective Clothing 378
13.8.2.4	Decontamination of Personnel 379

13.8.2.5	Rapid Decontamination of Personnel and Personal Gear *379*
13.8.2.6	Thorough Decontamination of Personnel *380*
13.9	Conclusions and Outlook *380*
	References *381*

14 Principles and Practice of Disinfection of Biological Warfare Agents – How Clean is Clean Enough? *383*
Andre Richardt and Birgit Hülseweh

14.1	General Principles of Disinfection and Decontamination *384*
14.1.1	Definition of Terms *384*
14.1.2	Physical Methods of Disinfection *385*
14.1.3	Chemical Methods of Disinfection *385*
14.2	Mechanisms of Action of Biocides against Microorganisms *385*
14.2.1	Chemicals for Disinfection *386*
14.2.2	Fumigation – Well-Known for Decontamination of Objects *388*
14.2.2.1	Fumigation with Ethylene Oxide *388*
14.2.2.2	Fumigation with Chlorine Dioxide Gas *388*
14.2.2.3	Fumigation with Formaldehyde Gas *389*
14.2.2.4	Vaporized Hydrogen-Peroxide (VHP) *389*
14.3	Levels of Disinfection *390*
14.4	Biological Target Sites of Selected Biocides *393*
14.4.1	Viral Target Sites *393*
14.4.2	Bacterial Target Sites *394*
14.5	The Spores Problem *395*
14.6	Inactivation as Kinetic Process *399*
14.7	Evaluation of Antimicrobial Efficiency *401*
14.8	Carrier Tests versus Suspension Tests *403*
14.9	Resistance to Biocide Inactivation – a Growing Concern *405*
14.9.1	Resistance of Viruses *406*
14.9.2	Resistance of Bacteria *407*
14.10	New and Emerging Technologies for Disinfection *408*
14.11	"Is Clean Clean Enough" or "How Clean Is Clean Enough"? *408*
	References *409*

15 Radiological/Nuclear Decontamination – Reduce the Risk *411*
Nikolaus Schneider

15.1	Why Is Radiological/Nuclear Decontamination So Special? *412*
15.2	Contamination *414*
15.2.1	Nuclear Weapons *414*
15.2.1.1	Nuclear Fallout Contaminates the Ground *414*
15.2.1.2	Rainout/Washout *415*
15.2.2	Radiological Contaminations – Radiological Dispersal Device (RDD) *416*
15.3	Decontamination *418*
15.3.1	Decontamination Efficiency Calculation *418*

15.3.2	Decontamination Procedures for RN Response	*420*
15.3.3	RN Decontamination Agents	*421*
15.3.3.1	Surfactant	*422*
15.3.3.2	Chelating Agent	*424*
15.3.4	Specific Decon Processes, Alternative Procedures	*424*
15.3.4.1	Spraying/Extraction Systems	*425*
15.3.4.2	Surface Ablation Techniques	*426*
15.4	Conclusions and Outlook	*428*
	References	*429*

Part VI CBRN Risk Management – Are We Prepared to Respond? *431*

16 Preparedness *433*

Marc-Michael Blum, Andre Richardt, and Kai Kehe

16.1	Introduction to Risk Management	*433*
16.2	Key Elements Influencing a Counter-CBRN Strategy	*436*
16.3	A Special Strategy for CBRN	*438*
16.3.1	Chemical Threats	*440*
16.3.2	Biological Threats	*443*
16.3.3	Radiological Threats	*447*
16.3.4	Nuclear Threats	*450*
16.4	Proliferation Prevention	*456*
16.5	Active Countermeasures	*458*
16.6	If Things Get Real: Responding to a CBRN Event	*459*
16.6.1	Fundamentals of Installation of a Response	*459*
16.6.2	Detection, Reconnaissance, and Surveillance	*461*
16.6.3	Risk Assessment	*462*
16.6.4	CBRN Warning and Reporting	*464*
16.6.5	Command and Control, Communication	*465*
16.6.6	Technical Response	*466*
16.6.7	Medical Response	*468*
16.6.8	Risk Communication	*470*
16.6.9	Medical Support and Post-disaster Recovery	*472*
16.7	Research	*473*
16.8	Aftermath Action – Lessons Learned	*474*
16.9	Conclusions and Outlook	*475*
	References	*476*

Index *479*

Foreword

Accidental or deliberate CBRN[1], and events are widely considered as low probability events that might however have a big impact on the citizens and the society. Whenever and wherever they happen, they usually deserve a gradual (regional, national, international) and multi-facetted approach as they tend to provoke severe and unexpected physical, psychological, societal, economical and political effects that might also easily cross the borders.

In that context detection, protection and decontamination against potentially very harmful CBRN agents is of particular importance. It is needed for military staff but also for civilians including a large range of users like firemen, health services, police, civil protection operators who might be involved in such events, whether they are due to terrorism attacks, accidents or natural disasters. What happened in 2011 in Japan around Fukushima is a dramatic example with short-, mid- and probably long term harmful effects.

Many countries have already invested in the CBRN area. At the EU level, as regards CBRN, significant political and technical efforts have been also carried out in recent years: as for example on the development and implementation of the EU CBRN Action Plan (including the concept of Lead States to carry specific priority actions), or the development of CBRN Resilience modules under the EU Civil Protection Mechanism, or the set up of CBRN Centers of Excellence in several sensitive regions in the world.

On the research side, last but not least, considerable investments on CBRN detection, protection and decontamination are carried out under the umbrella of FP7 Security research cooperation theme. They are contributing to provide, to test and to validate complimentary new solutions, tools, equipments, protocols, systems, as well as for example draft standards for CBRN quantified reference materials, reference sampling and analytical method[2]. The EU is indeed currently supporting financially tens of Security Research projects under the FP7[3] which relate to CBRN. This represents, as of 2012, a global budget of 120 million euro. In the next 2 years, 10 to 15 new CBRN activities are expected to start, including the start of a very large scale and unique demonstration programme in 2013, involving

1) Chemical, Biological Radiological and Nuclear
2) E.g. EQUATOX and SLAM projects
3) 7th Framework Programme for Research and Innovation

representative CBRN authorities and end users from EU Member States, as well as Industry and Research representative Organisations.

As regards CBRN CIV-MIL interactions at the EU research level, the recent European Framework Cooperation (EFC) initiative agreed by the EC, EDA and ESA, provides a platform for exchange of informations, ideas, priorities, experts and, to some extent, research results in the area of CBRN, looking for concrete synergies between the different frameworks.

In this evolving and constructive EU context, the release of this textbook on NBC-Protection is a very good opportunity for the CBRN community to reflect on achievements and look forward to the near future.

Tristan Simonart

Preface

Following the public discussion during recent years, it became evident to me that the public fear towards CBRN agents is often based on insufficient scientific information and sometimes leads to unrealistic assumptions. Through numerous discussions with other scientific colleagues my feeling grew that many excellent scientific books and publications dealing with single and isolated CBRN questions are available but get stuck into details without knowing or providing the fundamental principles. The speed of the information about new detection technologies, breakthroughs in nanotechnology and life science tend to result in an overflow of scientific information and the time is missing to sort in the right context to make the right decisions. However, especially for CBRN-questions it is obvious that only to guess is quick – but can de deadly!

Currently, on international level we observe an increasing demand for provision of an all-hazard approach focusing on the prevention, detection, identification, response and preparedness to CBRN threats. New comprehensive concepts to counter a CBRN threat are needed and the changed scenarios call for a unified strategy that covers militarian as well as civilian aspects of defence. In Europe several programs deal with this question and it is obvious that only a close cooperation between scientists and engineers of various affected disciplines will be able to deal with this ambitious stipulation.

The initial intention of the editorial group of this introductive textbook was, to bring together all fundamental technical aspects of CBRN protection based on the knowledge of CBRN-history and existing CBRN treaties. These aspects include physical protection as well as detection and decontamination. Moreover, concepts of preparedness and response should be discussed.

Consequently, in this book we come up with a compromise between clarity, comprehensibility and scientificness. The advanced CBRN reader might miss particular aspects. However, all authors try to provide a scientific and technical understanding that enables the reader to take part in CBRN debates and discussions.

Looking back, it is amazing how many turnarounds happened during the constitution of this book. At the beginning a diverse perspective of CBRN issues enhances our understanding and helps us to figure out the best solution. Then the writing itself – everybody, who ever sat in front of an empty page knows how difficult it is write down something significant.

And last – but not least; sometimes, when we thought, that the project was on the cusp to die a silent death, we got assistance from unanticipated supporters, and we are all very grateful for this support and advice. Also, we would like to thank our friends, colleagues, co-authors as well as the editorial staff at Wiley-VCH for their support, ideas and remarks. Special thanks go to our families for their patience during the endeavour of this book.

Andre Richardt

About the Editors

Andre Richardt has obtained his academic degrees from University Cologne in 1991 (Dipl. Degree in Genetics), Albert-Ludwigs-University, Freiburg (Dr. rer. nat Degree in Microbiology) in 1997 and Helmut-Schmidt-University, Hamburg (Dr. habil Degree in Biotechnology) in 2006. Currently, he is head of Biological and Chemical Decontamination business area at the Bundeswehr Research Institute for Protective Technologies and NBC Protection in Munster, Germany. Most of his career he has been working for the German Armed Forces in the field of CBRN-protection. From 2004 to 2005 he worked at dstl, PortonDown, Great Britain. In the special field catalytic decontamination of biological and chemical warfare agents he has been working for over ten years in national and international working-groups. His current research interests include investigations of non-thermal inactivation of biological and chemical agents as well as the control of the efficiency of a decontamination process. Currently, he is also a lecturer at the Helmut-Schmidt-University, Hamburg and he tutors young officers in the field of CBRN-protection. He is a member in several working groups dealing with fundamental technical and scientific aspects of CBRN protection.

Dr. Birgit Hülseweh studied Biology at the Heinrich-Heine-University of Düsseldorf, Germany with a focus on Microbiology, Molecularbiology and Organic Chemistry. There she received her Diploma in 1990 and did her doctoral thesis (PhD) at the Institute of Microbiology.

From 1994 to 1998 she was as a post-doc at the Max-Planck Institute for Molecular Physiology in Dortmund, Germany and for another 4 year period she worked as a scientific assistant at the University of Essen-Duisburg, Germany. From 2001 to 2002 she was the head of the scientific laboratory of Alpha Technology GmbH in Cologne, Germany, a biotech company, which dealt with the spotting, production and electrical read-out of microarrays for microbial diagnostics. In 2003 she joined as a senior scientist the department of Virology at the Bundeswehr Research Institute for Protective Technologies and NBC Protection in Munster, Germany. Her research focuses on innovative technologies for the identification of microorganisms and her scientific interests include all aspects of real-time-PCR methods, array applications as well as innovative applications of nanotechnology. Dr. Hülseweh has extensive experience in Molecular and Cellular Biology as well as in Immunology and Biochemistry. She is the author of diverse peer reviewed scientific publications and tutors several PhD-students. She has been working as scientific advisor in national and international working-groups and takes care for several international scientific co-operations.

Bernd Niemeyer studied Chemical Engineering and obtained his German Diploma degree (Dipl.-Ing.) at the University Erlangen-Nuremberg, Germany in 1986. His following PhD work focused in the field of bio engineering at the same University. He obtained his PhD-degree in 1990. As Post-doc he visited the Department of Scientific and Industrial Research (DSIR) in Lower Hutt (New Zealand) for one year and researched into separation technologies for health and chemical engineering topics. After his comeback he worked for the Deutsche Aerospace AG (later named DaimlerChrysler Aerospace AG) and the company Thermoselect Suedwest GmbH. He designed, constructed and commissioned new waste treatment plants for ammunition disposal (newly invented process) as well as for municipal waste processing.

Since 1996 he leads the Chair of "Process Engineering with focus on Separation Technology" at the Helmut-Schmidt-University / University of the Federal Armed Forces Hamburg in combination with the Research Group "Molecular Recognition and Separation" at the Helmholtz-ZentrumGeesthacht, Centre for Material Science and Coastal Research.

His research interests are applicable for CBRN-safety (analyses and decontamination of biological warfare agents as well as protection, detection and decontamination of chemical warfare agents), environmental engineering (waste as well as off-gas treatment, like odorous removal), biotechnology (enzyme catalysis and separation of valuables) as well as chemical processing (process design and separation of substances from reaction mixtures). The main methods applied are mainly adsorptive separation technologies, oxidative processes, and development of analytical and sensor systems.

**Frank Sabath** received the Dipl.-Ing. Degree in electrical engineering from the University of Paderborn, Paderborn, Germany, in 1993, and the Dr.-Ing. degree from the Leibniz University of Hannover, Hannover, Germany, in 1998.

From 1993 to 1998, he was with the C-Lab, a Joint Research and Development Institute of the University of Paderborn and the Siemens Nixdorf Informationssysteme AG, Paderborn, Germany, where his responsibilities included research activities on numerical field calculation and the radiation analysis of printed circuit boards. Since 1998, he has been with the Federal Office of Defense Technology and Procurement (BWB). Currently, he is head of the division on *Balanced Nuclear Protection Measures and Nuclear Hardening, Electro-Magnetic Effects, Fire Protection* of the Bundeswehr Research Institute for Protective Technologies and NBC-Protection (WIS), Munster, Germany. He is the author or coauthor of more than 110 papers published in international journals and conference proceedings. His research interests include investigations of electromagnetic field theory, High-Power Electromagnetics, investigations of short pulse interaction on electronics, and impulse radiation.

Dr. Sabath served as Ultra Wide Band (UWB) co-chairman of the EUROEM 2004, Magdeburg, Germany as well of the EUROEM 2008, Lausanne, Switzerland. He has been the Editor-in-chief for several Ultra-Wideband, Short-Pulse electromagnetics books. Currently he is an Associate Editor of the IEEE Transactions on EMC, member of the board of directors of the IEEE EMC Society and chair of the IEEE Germany Section EMC Society Chapter. Due to his outstanding service the EMC Society presented him the Laurence G. Cumming Award in 2009 and the Honored Member Award in 2012. He is a Member of the IEEE Electromagnetic Compatibility (EMC), Antennas and Propagation (AP), Microwaves Theory and Techniques (MTT) societies, and a member of URSI Commission E.

List of Contributors

Hans-Jürgen Altmann
Bundeswehr Research Institute
for Protective Technologies and
NBC Protection (WIS)
Humboldtstr. 100
29633 Munster
Germany

Marc-Michael Blum
Blum-Wissenschaftliche Dienste
Cäcilienstraße 3
22301 Hamburg
Germany

Thomas Dawert
Bundeswehr Research Institute
for Protective Technologies and
NBC Protection (WIS)
Humboldtstr. 100
29633 Munster
Germany

Karola Hagner
Bundeswehr Research Institute
for Protection Technologies –
NBC-Protection
Humboldtstraße 100
29633 Munster
Germany

Friedrich Hesse
Bundeswehr Research Institute
for Protection Technologies –
NBC-Protection
Humboldtstraße 100
29633 Munster
Germany

Birgit Hülseweh
Bundeswehr Research Institute
for Protective Technologies and
NBC Protection (WIS)
Humboldtstr. 100
29633 Munster
Germany

Martin Jung
Bundeswehr Research Institute
for Protective Technologies and
NBC Protection (WIS)
Humboldtstr. 100
29633 Munster
Germany

Kai Kehe
Bundeswehr Medical Office
Dachauer Str. 128
80637 München
Germany

Hans-Jürgen Marschall
Bundeswehr Research Institute
for Protective Technologies and
NBC Protection (WIS)
Humboldtstr. 100
29633 Munster
Germany

Bernd Niemeyer
Helmut-Schmidt-University/
University of the Bundeswehr
Hamburg
Holstenhofweg 85
22043 Hamburg
Germany

Silke Oelze
Bundeswehr Research Institute
for Protective Technologies and
NBC Protection (WIS)
Humboldtstr. 100
29633 Munster
Germany

Ronald Rambousky
Bundeswehr Research Institute
for Protective Technologies and
NBC Protection (WIS)
Humboldtstraße 100
29633 Munster
Germany

Andre Richardt
Bundeswehr Research Institute
for Protective Technologies and
NBC Protection (WIS)
Humboldtstr. 100
29633 Munster
Germany

Frank Sabath
Bundeswehr Research Institute
for Protective Technologies and
NBC Protection (WIS)
Humboldtstr. 100
29633 Munster
Germany

Martin Schaarschmidt
Bundeswehr Research Institute
for Protective Technologies and
NBC Protection
Humboldtstrasse
29633 Munster
Germany

Nikolaus Schneider
Bundeswehr Research Institute
for Protection Technologies and
NBC Protection (WIS)
Humboldtstraße. 100
29633 Munster
Germany

Part I
History and Treaties in CBRN – Warfare and Terrorism

copyright by Jörg Pippirs, http://www.artesartwork.de

CBRN Protection: Managing the Threat of Chemical, Biological, Radioactive and Nuclear Weapons,
First Edition. Edited by A. Richardt, B. Hülseweh, B. Niemeyer, and F. Sabath.
© 2013 Wiley-VCH Verlag GmbH & Co. KGaA. Published 2013 by Wiley-VCH Verlag GmbH & Co. KGaA.

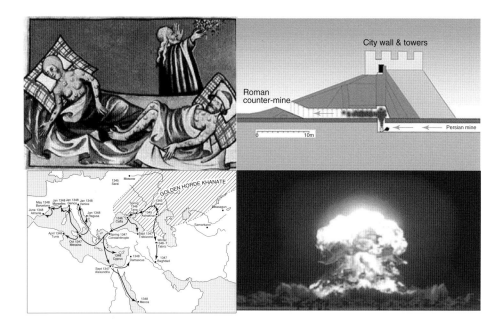

1
A Glance Back – Myths and Facts about CBRN Incidents
Andre Richardt and Frank Sabath

In our human history we can find numerous examples of the application of chemical and biological agents used or proposed as weapons during the course of a campaign or battle. In the twentieth century we saw the rise of a new age in battle field tactics and the abuse of detailed scientific knowledge for the employment of chemicals as warfare agents (CWAs). Another step that crossed a border was the use of nuclear bombs against Nagasaki and Hiroshima in 1945. Although there have been many attempts to ban chemical, biological, radiological, and nuclear warfare agents (CBRN agents), their devastating potential makes them still attractive for regular armies as well as for terrorists. Therefore, it is likely that the emergence of CBRN terrorism is going to be a significant threat in the twenty-first century. However, we need to understand our history if we want to find appropriate answers for current and future threats. For this reason, in this chapter we provide a short history of CBRN, from the beginning of the use of CBRN agents up to the emergence of CBRN terrorism and the attempt to ban the use of this threat by negotiation and treaties.

1.1
Introduction

Why do we fear the use of chemical, biological, and nuclear weapons? What are the reasons behind the obvious? To answer these questions we have to understand that data and facts are only one part of the story. To understand and to be able to lift the veil of myths about CBRN incidents we need a lot more. Therefore, the history section of this part attempts to lay the basis for a deeper understanding of subsequent chapters.

CBRN Protection: Managing the Threat of Chemical, Biological, Radioactive and Nuclear Weapons,
First Edition. Edited by A. Richardt, B. Hülseweh, B. Niemeyer, and F. Sabath.
© 2013 Wiley-VCH Verlag GmbH & Co. KGaA. Published 2013 by Wiley-VCH Verlag GmbH & Co. KGaA.

1.2
History of Chemical Warfare

We can find many examples (Figure 1.1) of how the toxic principle of chemical substances has been used to ambush the enemy, even if the exact mechanism was unknown.

- **Chemical warfare weapon (CWA)**: "... The term *chemical weapon* is applied to any toxic chemical or its precursor that can cause death, injury, temporary incapacitation, or sensory irritation through its chemical action ..." (*http://www.opcw.org/about-chemical-weapons/what-is-a-chemical-weapon/*, accessed 26 January 2011)
- **Chemical warfare (CW) agent**: "... The toxic component of a chemical weapon is called its *"chemical agent."* Based on their mode of action (i.e., the route of penetration and their effect on the human body), chemical agents are commonly divided into several categories: choking, blister, blood, nerve, and riot control agents." (*http://www.opcw.org/about-chemical-weapons/types-of-chemical-agent/*, accessed 26 January 2011).

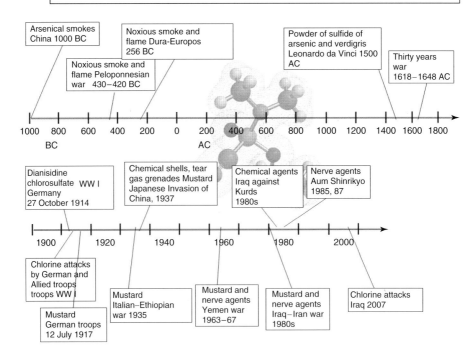

Figure 1.1 Time line of some significant examples of the application of toxic substances as chemical agents.

The deployment of toxic smokes and poisoned fire for advantage in skirmishes and on the battlefield was well known by our ancestors [1, 2]. In addition, we can date some significant changes, where the next level was reached in the discovery and use of toxic chemicals (see Figure 1.4 below).

1.2.1
Chemical Warfare Agents in Ancient Times

We can date the employment of chemicals as chemical warfare agents (CWAs) from at least 1000 BC when the Chinese used arsenical smokes [1]. By the application of noxious smoke and flame the allies of Sparta took an Athenian-held fort in the Peloponnesian War between 420 and 430 BC. Stink bombs of poisonous smoke and shrapnel were designed by the Chinese, along with a chemical mortar that fired cast-iron stink shells. Other conflicts during succeeding centuries saw the use of smoke and flame. However, it is difficult to confirm historical reports about incidents with chemicals by historical facts. One example of the confirmed use of toxic smoke is the siege of the city Dura-Europos by the army from the Sasanian Persian Empire around AD 256, where poisoned smoke was introduced to break the line of Roman defenders [3]. The full range of ancient siege techniques to break into the city, including mining operations to breach the walls, has been discovered by historians [3]. Roman defenders responded with "counter-mines" to thwart the attackers. In one of these narrow, low galleries a pile of bodies, representing about 20 Roman soldiers still with their arms, was found (Figure 1.2). Findings from the Roman tunnel revealed that the Persians used bitumen and sulfur crystals to start it burning. This confirmed application of poison gas in an ancient siege is an example of the inventiveness of our ancestors.

Toxic smoke projectiles were designed and used during the Thirty Years War (1618–1648). Leonardo da Vinci proposed a powder of arsenic sulfide and verdigris in the fifteenth century. Venice employed unspecified poisons in hollow explosive mortar shells during the fifteenth and sixteenth centuries. The Venetians also sent contaminated chests to their enemy to envenom wells, crops, and animals. During the Crimean War (1853–1856), the use of cyanide-filled shells was proposed to break the siege of Sevastopol. However, all these incidents happened without knowing the exact mechanism of poisoning.

1.2.2
Birth of Modern Chemical Warfare Agents and Their Use in World War I

We can date the birth of modern chemical warfare agents (CWAs) to the early twentieth century. Progress in modern inorganic chemistry during the late eighteenth and early nineteenth centuries and the flowering of organic chemistry worldwide during the late-nineteenth and early-twentieth centuries generated renewed interest in chemicals as military weapons. The chemical agents first used in combat during World War I were eighteenth- and nineteenth-century discoveries (Table 1.1).

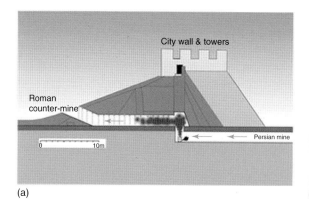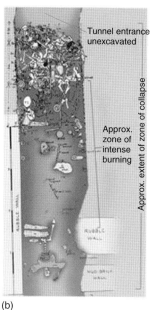

Figure 1.2 Siege of Dura-Europos by an army from the Sasanian Persian Empire (around AD 256 [3]): (a) The Sasanian Persian mine designed to collapse Dura's city wall and adjacent tower. The Roman countermine intended to stop them and the probable location of the inferred Persian smoke-generator thought to have filled the Roman gallery with deadly fumes. (b) A composite plan of the Roman countermine, showing the stock of Roman bodies near its entrance. The area of intense burning marks the gallery's destruction by the Persians and the skeleton of one of the attackers. Credit: images copyright of Simon James.

Table 1.1 Important eighteenth- and nineteenth-century discoveries of toxic chemicals.

Year	Name	Discovery
1774	Carl Scheele, a Swedish chemist	Discovery of chlorine in 1774. He also determined the properties and composition of hydrogen cyanide in 1782
1802	Comte Claude Louis Berthollet, a French chemist	Synthesis of cyanogen chloride
1812	Sir Humphry Davy, a British chemist	Synthesis of phosgene
1822	Victor Meyer, a German chemist	Dichloroethyl sulfide (mustard agent) was synthesized in 1822, again in 1854, and finally fully identified in 1886
1848	John Stenhouse, a Scottish chemist	Synthesis of chloropicrin

In 1887, the use of tear agents (lacrimators) for military purposes was considered in Germany. In addition, a rudimentary chemical warfare program was started by the French with the development of a tear gas grenade containing ethyl bromoacetate. Furthermore, there were some discussions in France about the filling of artillery shells with chloropicrin. The French Gendarmerie had successfully employed riot-control agents for civilian crowd control. These agents were also used in small quantities in minor skirmishes against the Germans, but were largely inefficient. In summary, these riot-control agents were the first chemicals applied on a modern battlefield, and the research for more effective agents continued throughout the war.

In the early stages of World War I, the British examined their own chemical technology for battlefield use. Their first investigations also covered tear agents, but later they put their effort towards more toxic chemicals. Nevertheless, the first large-scale employment of chemicals during World War I was initiated by heavily industrialized Germany. Three thousand 105-mm shells filled with dianisidine chlorosulfate, a lung irritant, were fired by the Germans at British troops near Neuve-Chapelle on the 27 October 1914, but with no visible effect [4]. Nonetheless, the British were the victims of the first large-scale chemical projectile attack. The Germans continued firing modified chemical shells with equally unsuccessful results. This lack of success, and the shortage of artillery shells, led to the concept of creating a toxic gas cloud directly from its storage cylinder. This concept was invented by Fritz Haber in Berlin in 1914.

> The first great attack with CWAs in modern warfare: Ypres in Belgium. Chlorine attack by German troops in April 1915 [5].
>
> German units placed a total of between 2000 and 6000 cylinders opposite the Allied troops defending the city of Ypres in Belgium. The cylinders contained a total of around 160 tons of chlorine. Once the cylinders were in place, and because of the critical importance of the wind, the Germans waited for the winds to shift to a westerly direction toward the trenches of the Allied troops. During the afternoon of the 22 April 1915 the chlorine gas was released with devastating effects. This attack caused between 800 (realistic) and 5000 (mainly propaganda) deaths.

After the first great attack with chemical warfare agents (CWAs) by German troops near Ypres [5], the Allied troops quickly restored a new front line and it took only a short period of time for them to be able to use chlorine themselves. In September 1915, they launched their own chlorine attack against the Germans at Loos. The expansion of the armamentarium with chloropicrin and phosgene was just the beginning of a deadly competition between both sides [6]. We saw the invention and the development of more protective masks, more dangerous chemicals, and improved delivery systems.

Table 1.2 Estimated chemical casualties in World War I.

Country	Nonfatal chemical casualties	Chemical fatalities
Russia	420 000	56 000
Germany	191 000	9000
France	182 000	8000
British Empire	180 000	8100
United States	71 000	1500

A further step in a devastating chemical war was the use of a new kind of chemical agent. On the 12 July 1917, again near Ypres in Belgium, sulfur mustard was spread by the Germans in an artillery attack. Compared to the first agents, mustard was a more persistent vesicant on the ground, and this caused new problems. Not only was the air poisoned, but the ground and equipment was also contaminated. This new agent was effective in low doses and affected the lungs, the eyes, and also the skin. Although fewer than 5% of mustard exposed soldiers died, mustard injuries could easily overwhelm the medical system. Therefore, the need for protective equipment for soldiers and horses that were heavy and bulky at that time led to more difficult and dangerous fighting. In summary, World War I was the dawn of a new military age with devastating effects, with Russia bearing the heaviest burden of chemical casualties (Table 1.2).

1.2.3
Chemical Warfare Agents between the Two World Wars

Throughout the 1920s there was evidence that the military use of chemical agents continued after the end of World War I. Germany worked with Russia, which had suffered nearly half a million chemical casualties during World War I, to improve their chemical agent offensive and defensive programs from the late 1920s to the mid-1930s. During the Russian Civil War and Allied intervention in the early 1920s both sides had chemical weapons, and there were reports of isolated chemical attacks. Later accounts accused the British, French, and Spanish troops of using chemical warfare at various times and places during the 1920s. For example, it is rumored that Great Britain employed chemicals against the Russians and mustard against the Afghans north of the Khyber Pass. Spain has been accused of having deployed mustard shells and bombs against the Riff tribes of Morocco [7].

1.2.3.1 The Italian–Ethiopian War
During the Italian–Ethiopian War the first major employment of chemical weapons after World War I was reported. On 3 October 1935, Mussolini launched an invasion of Ethiopia from its neighbors Eritrea and Italian Somaliland. Italian troops dropped mustard bombs and sprayed it from airplane tanks. Mustard agent was selected

as a "dusty agent" to burn the unprotected feet of the Ethiopians with devastating effects. It was the first time special sprayers were prepared on board of aircrafts to vaporize a fine, deathly rain. This fearful tactic succeeded and by May 1936 the Ethiopian army was completely routed and Italy controlled most of Ethiopia, until 1941 when British and other allied troops re-conquered the country. However, some have concluded that the Italians were a clearly superior force and that the use of chemical agents in the war was nothing more than an experiment. Nonetheless, there were thousands of victims of Italian mustard gas [7].

1.2.3.2 Japanese Invasion of China

The Japanese had an extensive chemical weapons program. They were producing agent and munitions in large numbers by the late 1930s. During the invasion of China in 1937 it was reported that Japanese forces began using chemical shells, tear gas grenades, and lachrymatory candles. In 1939, there was an escalation by the Japanese that led to the application of mustard and lewisite with great effect. The Chinese troops retreated whenever they saw smoke, thinking it was a chemical attack [8].

1.2.3.3 First Nerve Agents

Searching for more potent insecticides the German chemist Schrader of the IG Farben Company discovered an extremely toxic organophosphorus insecticide in 1936. This new compound was reported to the Chemical Weapons Section of the German military prior to patenting, as required by German law of that time. The substance had devastating effects on the nervous system and was therefore classified for further research. The substance was named tabun and after World War II it was designated GA, for "German" agent "A." The research continued and in 1938 a similar agent, sarin (GB), was designated with toxicity five-times higher than that of tabun.

1.2.4
Chemical Warfare Agents in World War II

It is due to common sense that chemical warfare agents (CWAs) were not widely used in World War II. However, pilot plants for production were built on both sides. Germany produced and weaponized approximately 78 000 of tons of CWAs. The key agent was mustard in terms of production. The Germans filled artillery shells, bombs, rockets, and spray tanks with the agent. Why these deadly agents were not used on the battlefield remains a mystery. Thus, the top-secret German nerve agent program remained a secret until its discovery by the Allies after the end of World War II.

Furthermore, there are reports, that Japan produced about 8000 tons of chemical agents during the war. The favored agents were mustard agent, a mustard–Lewisite mixture, phosgene, and hydrogen cyanide. They gained experience during their attacks on China.

The greatest producer of chemical warfare agents (CWAs) during World War II was the United States of America. Ready for retaliation, if Germany had been used chemical warfare agents (CWAs), the United States produced proximately 146 000 tons of chemical agents between 1940 and 1945. With the possible exception of Japan during attacks in China no nation though, employed chemical agents on the battlefield during World War II. However, the positioning of chemical weapons near the front line in case of need resulted in one major disaster in 1943 [9]. In 1943 the Germans bombed the American ship the *John Harvey* in Bari Harbor, Italy. This was a ship loaded with 2000 100-pounds M47A1 mustard bombs. Over 600 military casualties and an unknown number of civilian victims resulted from the raid when they were poisoned by ingestion, skin exposure to mustard-contaminated water, and inhalation of mustard-laden smoke. The harbor clean-up took more than three weeks.

1.2.5
Chemical Warfare Agents during the Cold War

The end of World War II did not stop the development, stockpiling, or use of chemical weapons. During the Yemen War of 1963 through 1967, Egypt in all probability used mustard and nerve agents in support of South Yemen against royalist troops in North Yemen. Attacks occurred on the town of Gahar and on the villages of Gabas, Hofal, Gadr, and Gadafa. Shortly after these attacks, the International Red Cross examined victims, soil samples, and bomb fragments, and officially declared that chemical weapons, identified as mustard agent and possibly nerve agents, had been applied in Yemen [10]. Prior to this, no country had employed nerve agents in combat. The combination of the use of nerve agents by the Egyptians in early 1967 and the outbreak of the war between Egypt and Israel during the Six-Day War in June finally attracted world's attention to the events in Yemen.

The USA, which used napalm, defoliants, and riot-control agents in Vietnam and Laos, finally ratified the Geneva Protocol in 1975, but with the stated reservation that the treaty did not apply either to defoliants or riot-control agents. During the late 1970s and early 1980s, reports of the application of chemical weapons against the Cambodian refugees and against the Hmong tribesmen of central Laos surfaced, and the Soviet Union was accused of using chemical agents in Afghanistan. Widely publicized reports of Iraqi's employment of chemical agents against Iran during the 1980s led to a United Nations investigation that confirmed the attack by the vesicant mustard and the nerve agent tabun. Later during the war, Iraq apparently also began to apply the more volatile nerve agent sarin, and Iran may have used chemical agents to a limited extent in an attempt to retaliate for Iraqi attacks. After the conflict with Iran, Iraq's Saddam Hussein employed chemical weapons to deal with rebellious Iraqi Kurds who had been assisted by the Iranians. The Iraqis used mustard, possibly combined with nerve gases, against the Kurdish town of Halabja in March 1988, killing thousands of people.

Halabja: After two days of conventional artillery attacks gas canisters were dropped on the town on 16 March 1988. The gas aggression began early in this day's evening after a series of napalm and rocket attacks, when a group of up to 20 Iraqi MiG and Mirage aircraft began dropping chemical bombs. The town and surrounding district were further assaulted with conventional bombs and artillery fire. At least 5000 people died as an immediate result of the chemical attack (Figure 1.3) and it is estimated that a further 7000 people were injured or suffered long-term illness. The attack is believed to have included the nerve agents tabun, sarin, and VX, as well as mustard gas.

Figure 1.3 Dead people in Halabja. Image is public domain.

Other countries that have stockpiled chemical agents include countries of the former Soviet Union, Libya (the Rapta chemical plant, part of which may still be operational), and France. Over two dozen other nations may also have the capability to manufacture offensive chemical weapons. The development of chemical warfare programs in these countries is difficult to verify because the substances used in the production of chemical warfare agents (CWAs) are in many cases the same substances that are applied to produce pesticides and other legitimate civilian products.

1.2.6
Chemical Warfare Agents Used in Terrorism

Although terrorism was not unknown in the world through the twentieth century, it was not really widespread until the 1980s. Then the issue began to acquire a higher profile. Aside from some domestic terrorism from the left in the United States during the late 1960s and into the 1970s, most notably in the form of the "Weather Underground" group, by the end of that decade the focus had turned toward the right, first in the form of the "Survivalists" movement and then the rightist/white supremacist "militias" that followed them. In other countries similar organizations, sometimes with a religious background, also presented a potential domestic terroristic threat. One well-known example is the Japanese religious cult,

Table 1.3 Chlorine attacks in Iraq.

Date	Event
21 October 2006	A car bomb carrying 12 120 mm mortar shells and 2 100-pound chlorine tanks detonated in Ramadi
28 January 2007	A suicide bomber drove a dump truck packed with explosives and a 1-ton chlorine tank into an emergency response unit compound in Ramadi
February 2007	Three attacks with chlorine took place in the cities of Ramadi and Baghdad
March 16 2007	Three separate suicide attacks in this month used chlorine in Ramadi and Fallujah
March 28 2007	Suicide bombers detonated a pair of truck bombs, one containing chlorine
April 2007	Three attacks with chlorine where reported in Ramadi and Baghdad
15 and 20 May 2007	Chlorine was used in Abu Sayda and nearby Ramadi
3 June 2007	A car bomb exploded outside a US military base in Diyala, unleashing a noxious cloud of chlorine gas

Aum Shinrikyo. They released nerve agents (sarin) in Matsumoto, Japan, 1994 and in 1995 they used sarin in a crowded Tokyo subway. This is an example of how the employment of chemical warfare agents (CWAs) by terrorists could be a significant threat to the civilian population.

Another actual example is the use of chlorine in Iraq. Chlorine bombings in Iraq began as early as October 2006, when insurgents in the Al Anbar Province started applying chlorine gas in conjunction with conventional vehicle-borne explosive devices (Table 1.3). The inaugural chlorine attacks in Iraq were described as poorly executed, probably because much of the chemical agent was rendered nontoxic by the heat of the accompanying explosives. Subsequent, more refined, attacks resulted in hundreds of injuries, but have proven not to be a viable means of inflicting massive loss of life. Their primary impact has therefore been to cause widespread panic, with large numbers of civilians suffering non life-threatening, but nonetheless highly traumatic, injuries.

1.2.7
Conclusions and Outlook

Parallel to technological developments we have seen the more and more sophisticated use of toxic substances in warfare over the centuries. The understanding of the mechanism of action of chemicals as chemical warfare agents (CWAs) is closely linked to the breakthrough in science and industrial manufacturing over the centuries. From a historical point of view it is important to flag and to understand the periodic leaps by which the use and knowledge of chemicals as chemical warfare agents (CWAs) reached a new phase (Figure 1.4). Therefore it is likely that

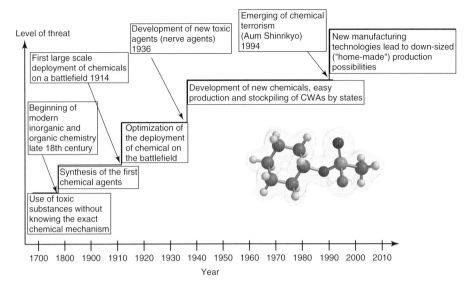

Figure 1.4 Breakthroughs in natural science and industrial development led to important leaps in the development and deployment of toxic substances in human history.

future breakthroughs in natural science and nanotechnology could lead to further leaps in the development of toxic substances designed for chemical warfare.

1.3
Introduction to Biological Warfare

The fear that the possible use of biological agents as weapons could lead to devastating effects for our civilization, economy, and society is still anchored in our minds. Therefore, if we want to estimate the potential of biological agents correctly it is necessary that we understand our history. Natural epidemics of cholera and plague are frightening enough. The notion that rogue states or terrorists could harness these and other diseases as weapons of war is even more chilling. While rare, the use of biological weapons dates back centuries. In this chapter confirmed and unconfirmed examples of biological warfare and bioterrorism are explored, from medieval times to today (Figure 1.5). We will learn more about the reasons why we still have in our memory the devastating effects of illness and a war outside the normal rules of engagement.

- **Biological warfare agents**: the use of disease-producing microorganisms, (bacteria, viruses) and toxic biological products, to cause death or injury to humans, animals, or plants.

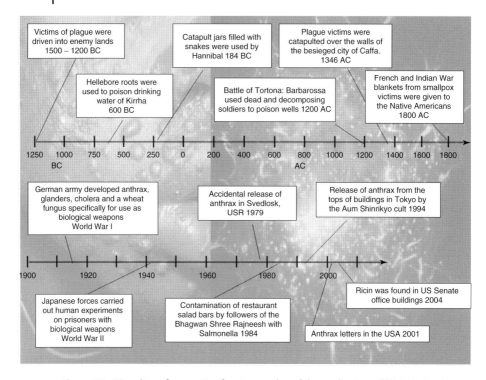

Figure 1.5 Time line of some significant examples of the application of biological substances as biological agents.

1.3.1
Most Harmful Pandemics in History

In our collective human memory the different pandemics of our history are deeply anchored. We can number several different pandemics with a high death toll (Figure 1.6). Even if the death toll is not high, fear can overcome civilizations that a disease will spread across their population.

> - **Pandemic**: A pandemic is an epidemic (an outbreak of an infectious disease) that spreads across international and natural borders, or at least across a large region. A pandemic can start when three conditions have been met: (i) emergence of a new or old disease with a low protection level in a population; (ii) the disease is infectious: agents infect humans, causing serious illness; and (iii) agents spread easily and sustainably among humans.

To get a clue of the heavy death toll of pandemics and epidemics we shed light on plagues and influenzas [11]. The Peloponnesian War Pestilence wiped out over

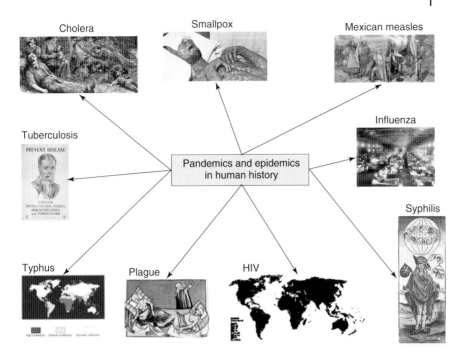

Figure 1.6 Most dangerous pandemics and epidemics in history for human kind. Images are public domain.

30 000 citizens of Athens in 430 BC (roughly one- to two-thirds of the population). Later the Antonine Plague, nowadays thought to be smallpox, was brought to Rome by soldiers returning from Mesopotamia in 165 AD. It was reported that at its peak the disease killed some 5000 people a day in Rome. Finally, 15 years later, a total of 5 million people were dead. The Plague of Justinian (541–542 AD) was recorded as a deadly disease in the Byzantine Empire. At the cumulus of the infection 10 000 people in Constantinople were killed every day. By the end of the outbreak, nearly half of the inhabitants of the city were dead. After the Plague of Justinian, there were many sporadic outbreaks of the plague, but none as severe as the "Black Death" of the fourteenth century. As the origin of the outbreak is unknown this pandemic took a heavy toll on Europe. The fatality level was recorded as over one-fourth of the entire European population. The last epidemic started in 1855 with the initial outbreak in Yunnan Province, China. The disease spread from China to India, Africa, and the American continent. All in all, this pandemic lasted about 100 years (it officially ended in 1959) and claimed over 12 million people in India and China alone.

In human history different influenzas have came across the world [11]. In recent history the highest death toll can be charged to "Spanish Flu." "Spanish Flu" started in March 1918, in the last months of World War I, with an unusually virulent and deadly flu virus. Just six months later the flu had become a worldwide pandemic in all continents, with the result that half of the world's population (1 billion people) had contacted it. From what we know, it is perhaps the most lethal pandemic in the recent history of humankind. Calculations differ between 20 and 100 million dead – more than the number of people killed in the World War I itself. Since then different influenzas have plagued the world. The "Asian Flu Pandemic" (1957–1958) and "Hong Kong Flu" (1968) caused a total of nearly 2 million deaths. The "Swine Flu Threat" (1976), "Russian Flu Threat" (1977), and "Avian Flu Threat" (1997) illustrated that the danger of a potential major pandemic cannot be denied. The ability of influenza viruses to change, become more transmissible among people, and be more difficult to treat is an ongoing concern. Therefore, for illustration, if we have in mind that possible pandemics could wipe out one-fourth of the Asian, European, or American population nowadays, it is more than understandable why we fear the biological nightmare and want to be prepared.

1.3.2
Biological Warfare Agents in Ancient Times BC

Ever since humans brought war upon each other more soldiers have been incapacitated by diseases than by the hand of their human enemies. This observation might have led to the early deployment of poisonous substances during war. The earliest documented incident of the intention to use biological weapons is recorded in Hittite texts of 1500–1200 BC, in which victims of plague were driven into enemy lands. Although the Assyrians knew of ergot, a fungus of rye with effects similar to LSD (lysergic acid diethylamide), there is no evidence that they poisoned enemy wells with ergot, as has often been claimed. In 600 BC Solon of Athens put hellebore roots in the drinking water of Kirrha. About 200 BC, the Carthaginians used mandrake root left in wine to sedate the enemy. To our current knowledge, poisons have been administered to enemy water supplies as early as the sixth century BC. Animal cadavers substituted for poisons during the Greek and Roman eras and Emperor Barbarossa applied human corpses to the same end, although it is likely that there was no intention of spreading disease but rather a simple spoiling of water supply. A more active approach was suggested by Hannibal, who advised the Bithynians to catapult jars filled with snakes toward enemy ships in 184 BC. The panic created rather than poisonous bites likely decided the battle, revealing human psychology as a second important dimension during biological attacks. Catapulting infected human bodies and excrement constituted a further step during the sieges of many towns, although it is not clear if these actions contributed much to the spread of disease.

> **Can We Be Sure about Historical Reports of Biological Incidents?**
>
> There is the legend that during the siege of the Crimean city of Caffa (now: Feodosiya, Ukraine) by the Tartars in 1346 victims of the bubonic plague were catapulted into the city and an outbreak of the Black Death were reported. As the conquered Genoese fled and the victors moved on both were spreading the disease (Figure 1.7). Plague doctors were specifically hired by citizens to treat

nearly everyone – the rich and the poor (Figure 1.8). At the end of the Black Death more than 25% of the European and Chinese population were dead, changing the course of human history.

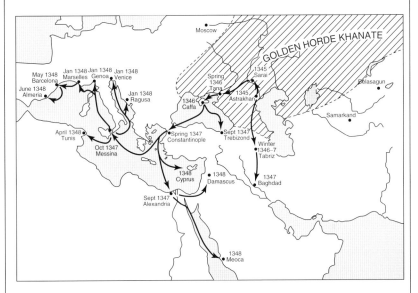

Figure 1.7 Tentative chronology of the initial spread of plague in the mid-fourteenth century. Figure taken from Reference [12].

Figure 1.8 Illustration of a plague doctor ("Doctor Beak from Rome"), engraving Rome 1656). Image is public domain.

However, whether biological warfare or less spectacular hygienic reasons were the beginning of this greatest of the medieval disasters is nearly impossible to prove. Are we able to find simple reasons for the outbreak of the disease?

- the fleas that transmit the disease between humans leave dead bodies rather quickly; this leads to the question of whether the corpses that flew into Caffa were flea infested;
- the rats moving in and out of the city walls – they also could spread the disease much more efficiently;
- it is not clear if the attackers intended to spread the disease among the besieged citizens or simply wanted to get rid of their dead comrades.

These possible reasons illustrate one of the biggest problems in biological warfare history. It is extremely difficult to distinguish between an attack that started the spread of infections and a coincidental natural infection in an "unnaturally" large aggregation of or new encounters between humans [13].

1.3.3
Biological Warfare Agents in the Middle Ages to World War I

During the Middle Ages, victims of the plague and decomposing bodies of humans and animals were used for biological attacks, often by throwing corpses over castle walls by catapults. It is reported that these tactics were not only applied during the siege of Caffa in 1346 [12] but also in the siege of Thun l'Évêque in 1340 during the 100 Year's War and during the siege of Karlštejn Castle in Bohemia in 1422. The last known incident of employing plague corpses for biological warfare took place in 1710. Russian forces catapulted plague-infected corpses over the city walls of Reval (today Tallinn). The Native American population was decimated after contact with the Old World due to the introduction of many different fatal diseases. There is one documented case of alleged germ warfare late in the French and Indian War. The British commander Lord Jeffrey Amherst and Swiss-British officer Colonel Henry Bouquet gave smallpox-infected blankets to Indians as part during the Siege of Fort Pitt (1873).

1.3.4
From World War I to World War II – the Beginning of Scientifically Based Biological Weapons Research

Heinrich Hermann Robert Koch (11 December 1843 to 27 May 1910) (Figure 1.9) was a German physician. He became famous for isolating *Bacillus* *anthracis* (1877), *Tuberculosis bacillus* (1882), and *Vibrio cholera* (1883). Koch's postulates founded modern microbiology. He was awarded a Nobel Prize for his tuberculosis findings in 1905. Robert Koch also inspired such major persons as Paul Ehrlich and Gerhard Domagk.

Figure 1.9 Heinrich Hermann Robert Koch. Image is public domain.

The breakthrough and foundation of modern microbiology by Robert Koch can be dated to the end of the nineteenth century. It was not only a breakthrough for our modern health care system. Unfortunately, this date marked also the beginning of the use of modern science for biological warfare. During World War I, the German Army developed anthrax, glanders, cholera, and a wheat fungus specifically for use as biological weapons. They allegedly spread plague in St. Petersburg, Russia, infected mules with glanders in Mesopotamia, and attempted to do the same with the horses of the French Cavalry. During the Sino-Japanese War (1937–1945) and World War II, Japanese forces operated a secret biological warfare research facility (Unit 731) in Manchuria, China [14]. They exposed more than 3000 prisoners to Yersinia pestis, Bacillus anthracis, Treponema pallidum, and other agents in an attempt to develop and observe the disease. In military campaigns, the Japanese army applied biological weapons on Chinese soldiers and civilians. For example, in 1940, Ningbo was bombed with ceramic bombs full of fleas carrying the bubonic plague. It is estimated that 400 000 Chinese died as a direct result of Japanese field testing of biological weapons. In 1942, the United States of America formed the War Research Service. Anthrax and botulinum toxin initially were investigated for use as weapons. Sufficient quantities of botulinum toxin and anthrax were stockpiled by June 1944 to allow unlimited retaliation if the German forces first applied biological agents. The British also tested anthrax bombs on Gruinard Island off the northwest coast of Scotland in 1942 and 1943. Anthrax-laced cattle cakes were prepared and stockpiled, also for retaliation.

1.3.5
From the End of World War II to the 1980 – the Great Bioweapons Programs

The United States continued research on various offensive biological weapons during the 1950s and 1960s [13, 15]. From 1951 to 1954, harmless organisms were set free at both coasts of the United States to demonstrate the vulnerability of American cities to biological attacks. This weakness was tested again in 1966 when a test substance was released in the New York City subway system.

During the Vietnam War, Viet Cong guerrillas used needle-sharp punji sticks dipped in feces to cause severe infections after an enemy soldier had been stabbed. In 1979, an accidental release of anthrax from a weapons facility in Sverdlovsk, USSR, killed at least 66 people. The Russian government claimed these deaths were due to infected meat, and maintained this position until 1992, when Russian President Boris Yeltsin finally admitted the accident [16].

1.3.6
From the 1980 Up Today – the Emerging of Bioterrorism

Several countries have continued offensive biological weapons research and use [17]. Additionally, since the 1980s, terrorist organizations have become appliers of biological agents [18, 19]. Usually, these cases amount only to hoaxes. However, exceptions have been noted (Table 1.4) and the range of diseases caused by biological agents and their bioterroristic potential is discussed extensively [27].

1.3.7
Conclusions and Outlook

We have discussed why we still fear the use of biological agents. Deadly pandemics and epidemics in the history of human kind have led to devastating effects in the economy and society. With the rise of gene- and biotechnology new scenarios for the possible use of biological agents have plagued the world. The understanding of the mechanism of action of biological agents as possible biological weapons is closely linked to the breakthrough in science and industrial manufacturing over the last 60 years. From a historical point of view it is important to mark/flag and understand the leaps – our understanding of genetics and the possibility of designing new biological warfare agents marks a new phase (Figure 1.10). Therefore, it is likely that future breakthroughs in natural science, especially in gene- and biotechnology, could lead to further leaps in the development of new biological agents tailored for specific missions.

Table 1.4 Examples of the emergence of bioterrorism.

Time	Event
Autumn 1984	Followers of the Bhagwan Shree Rajneesh contaminated restaurant salad bars with *Salmonella* in Oregon; 751 people were intentionally infected with the agent, which causes food poisoning
1985	Iraq began an offensive biological weapons program, producing anthrax, botulinum toxin, and aflatoxin. Iraq disclosed that it had bombs, Scud missiles, 122-mm rockets, and artillery shells armed with the B-agents. They also had spray tanks fitted to aircraft that could distribute agents over a specific target
1994	A Japanese sect of the Aum Shinrikyo cult attempted an aerosolized (sprayed into the air) release of anthrax from the tops of buildings in Tokyo
1995	Two members of a Minnesota militia group were convicted of possession of ricin, which they had produced themselves for use in retaliation against local government officials
2001	Anthrax was delivered by mail to US media and government offices. There were four deaths
2002	Six terrorist suspects were arrested in Manchester, England; their apartment was serving as a "ricin laboratory"
2003	British police raided two residences around London and found traces of ricin, which led to an investigation of a possible Chechen separatist plan to attack the Russian embassy with the toxin; several arrests were made
2004	Three US Senate office buildings were closed after the toxin ricin was found in mailrooms that served the then Senate Majority Leader Bill Frist's office

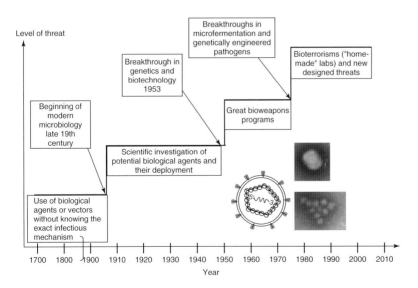

Figure 1.10 Breakthroughs in natural science and industrial development led to important leaps in the development and deployment of biological agents in human history.

22 | *1 A Glance Back – Myths and Facts about CBRN Incidents*

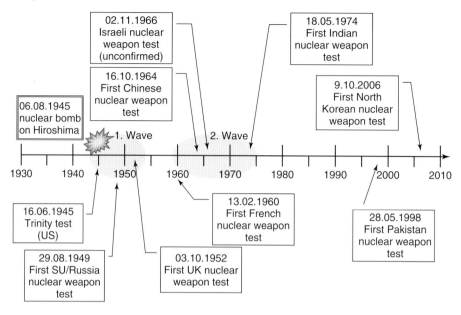

Figure 1.11 Timeline of the nuclear age.

1.4
Introduction to Radiological and Nuclear Warfare

Here we provide a brief survey of the development of radiological and nuclear warfare, from the discovery of nuclear fission, through the development of the first nuclear bomb and the nuclear arms race, to today's nuclear proliferation and threats caused by radiological warfare devices (Figure 1.11). Our objectives are to (i) sum up key developments in nuclear warfare, (ii) provide background information on nuclear armament (e.g., political aspects, nuclear doctrines, and operational aspects), (iii) introduce constitutive stages of nuclear armament, (iv) analyze the historical tendency of nuclear warfare, and (v) explain basic ideas of radiological warfare.

Before we start our tour through the history of nuclear and radiological warfare we need to set the terminology employed by defining often used basic terms:

- **Nuclear warfare**: a military conflict or political strategy in which nuclear weapons are used.
- **Nuclear weapon**: An explosive device that derives its destructive force from nuclear reactions, either fission or a combination of fission and fusion. Nuclear weapons are considered weapons of mass destruction, and their

> use and control has been a major aspect of international policy since their debut.
> - **Thermonuclear weapon**: A nuclear weapon that derives its energy from the fusion of hydrogen. Also known as a *hydrogen weapon*.
> - **Radiological warfare**: any form of warfare involving deliberate radiation poisoning, without relying on nuclear fission or nuclear fusion.
> - **Radiological weapon**: any weapon that is designed to spread radioactive material with the intent to kill and cause disruption upon an area (e.g., city).
> - **Nuclear fission**: A nuclear reaction in which the nucleus of an atom splits into smaller parts, often producing free neutrons and lighter nuclei, which may eventually produce photons (in the form of gamma rays). Fission of heavy elements is an exothermic reaction that can release large amounts of energy both as electromagnetic radiation and as kinetic energy of the fragments
> - **Nuclear fusion**: The process by which multiple like-charged atomic nuclei join together to form a heavier nucleus. It is accompanied by the release or absorption of energy, which allows matter to enter a plasma state. The fusion of two nuclei with lower mass than iron generally releases energy, while the fusion of nuclei heavier than iron absorbs energy.

1.4.1
Discovery of Nuclear Fission

The history of nuclear weapons started with the discovery of its fundamental physical mechanisms, nuclear fission and nuclear fusion. In the first three decades of the twentieth century fundamental developments in our understanding of the nature of atoms, including radioactivity, revolutionized physics. In 1932 John Cockcroft and Ernest Walton "split the atom" for the first time by bombarding lithium with protons. In 1934 Enrico Fermi and his colleagues in Rome studied the results of bombarding uranium with neutrons. Inspired by Fermi's results Lise Meitner, Otto Hahn, and Fritz Strassmann performed similar experiments in Germany. In December 1938 Hahn and Strassmann submitted a manuscript to *Naturwissenschaften* reporting that they detected the element barium after bombarding uranium with neutrons. Lise Meitner and Otto Robert Frisch correctly interpreted these results as being nuclear fission [20].

The news on nuclear fission was spread further during the Fifth Washington Conference on Theoretical Physics in January 1939, which fostered many more experimental demonstrations. Frédéric Joliot-Curie's team in Paris discovered that secondary neutrons are released during uranium fission, thus making a nuclear chain-reaction feasible. With the news of fission neutrons, Leo Szilárd immediately understood the possibility of a nuclear chain reaction using uranium. In the summer of 1939, Fermi and Szilard proposed the idea of a nuclear reactor (pile) to mediate this process. The pile would use natural uranium as fuel, and graphite as the moderator of neutron energy. At that time scientists in America as well as in Europe were well aware of the potential of utilizing nuclear fission as a powerful weapon, but no one was quite sure how it could be done.

Leo Szilárd (11 February 1898 to 30 May 1964) was a Hungarian physicist who conceived the nuclear chain reaction and worked on the Manhattan Project. During 1936, he assigned the chain-reaction patent to the British Admiralty to ensure its secrecy (GB patent 630726). Szilárd was also the co-holder, with Enrico Fermi, of the patent on the nuclear reactor (US Patent 2,708,656).

Hans Albrecht Bethe (2 July 1906 to 6 March 2005) was a German–American physicist, and Nobel laureate in physics for his work on the theory of stellar nucleosynthesis. He was head of the Theoretical Division at the secret Los Alamos laboratory developing the first atomic bombs. There he played a key role in calculating the critical mass of the weapons, and carried out theoretical work on the implosion method. Along with Richard Feynman, he developed a formula for calculating the explosive yield of the bomb. During the early 1950s, Bethe also played an important role in the development of the larger hydrogen bomb.

J. Robert Oppenheimer (22 April 1904 to 18 February 1967) was an American theoretical physicist and professor of physics at the University of California, Berkeley. He is best known for his role as the scientific director of the Manhattan Project, the World War II effort to develop the first nuclear weapons at the secret Los Alamos National Laboratory in New Mexico. After the war Oppenheimer was a chief advisor to the newly created United States Atomic Energy Commission and used that position to lobby for international control of nuclear power and to avert the nuclear arms race with the Soviet Union.

At the Cavendish Laboratory, at the University of Cambridge, Mark Oliphant observed the fusion of hydrogen isotopes for the first time in 1932. In 1939 Hans Bethe theorized about nuclear fusion and worked out the main cycle of nuclear fusion in stars. Bethe suggested that much of the energy output of the Sun and other stars results from reactions in which four hydrogen nuclei unite to form one helium nucleus while releasing a large amount of energy.

Later research proved that fusion reactions can indeed occur, but only at many millions of degrees kelvin when the electrostatic forces of repulsion that result from the presence of positive electric charges in both nuclei can be overcome so that the nuclear forces of attraction can perform a fusion. Such high temperatures, however, only occur in stars or in uncontrolled nuclear chain reactions.

1.4.2
Manhattan Project – Development of the First Fission Weapons

Up to the beginning of World War II many key discoveries in nuclear physics and nuclear chemistry were made by German scientists. Owing to this fact, there was concern among scientists in the Allied nations that Germany might have a project to develop fission-based weapons. In March 1940 Otto Frisch and Rudolf Peierls, two exiled German scientists living in Britain, reported in their famous Frisch–Peierls memorandum that if uranium-235 is completely separated from uranium-238 there is no need to slow the neutrons down, so no moderator was required. Consequently, in Great Britain a top-secret committee of experts (later known as the *MAUD* [1] Committee) was formed to investigate the feasibility of an atomic bomb. The MAUD report led to the "Tube Alloys Project," the first organized research on nuclear weapons. The "Tube Alloys Project" remained as the leading nuclear project until the start of the "Manhattan Project" in 1942.

The USA had started investigations into nuclear weapons with the Uranium Committee in 1939. Owing to MAUD reports and first results of the "Tube Alloys Project," which indicated that a fission weapon could be accomplished within a few years, in 1942 the USA reorganized its nuclear research under the control of the US Army as the "Manhattan Project" (Figure 1.12). In August 1943 Winston

Figure 1.12 Major research sites of the Manhattan Project.

Churchill and Franklin D. Roosevelt agreed on cooperation on nuclear research by signing the Quebec Agreement. The United Kingdom handed over all of its material to the United States and, in return, received all the copies of the American progress reports to the president. The British atomic research was subsumed then into the "Manhattan Project" until after the war, and a large team of British and Canadian scientists moved to the United States.

The Manhattan Project encompassed research activities at over 30 sites across the United States (Figure 1.12), Canada, and the United Kingdom. The three primary research and production sites of the project were the plutonium-production facility at what is now the Hanford Nuclear Reservation, Hanford (WA), the uranium-enrichment facilities at Oak Ridge (TN), and the weapons research and design laboratory now known as Los Alamos National Laboratory, Los Alamos (NM). The Manhattan Project resolved its first key scientific hurdle on 2 December 1942, when a team at the University of Chicago was able to initiate the first artificial self-sustaining nuclear chain reaction. Using these designs, massive reactors was secretly created at the Hanford Site to transform uranium 238 (^{238}U) into plutonium 239 (^{239}Pu). Plutonium-239 is a relatively stable element that does not exist in nature and is also fissible.

Another key scientific hurdle for the Manhattan Project was the production and purification of ^{235}U. Some 99.3% of natural uranium is ^{238}U, which cannot be used for a fission weapon as it absorbs neutrons and does not split. Two methods were developed to overcome this problem, electromagnetic separation and gaseous diffusion. Both methods separate isotopes based on their different weights. For the large-scale production and purification another secret site was erected at Oak ridge (TN).

The highly purified ^{235}U was used to build a gun-type fission weapon, called "*Little Boy.*" The gun-type design consists of a mass of ^{235}U, which is fired down a gun barrel like tube into another mass of ^{235}U. In the moment of the impact both mass rapidly create the critical mass of ^{235}U, resulting in a nuclear explosion. Even though the fundamental assumptions were verified in extensive laboratory work, no system level test was carried out as the design was so certain to work. In addition, the bomb that was dropped on Hiroshima used all the existing highly purified ^{235}U, and so there was no ^{235}U available for such a system level test.

Trinity was the first test of technology for a nuclear weapon (Figure 1.13). It was conducted by the United States on 16 July 1945, on the White Sands Proving Ground. Trinity was a test of an implosion-design plutonium device. The Trinity detonation was equivalent to the explosion of around 20 kt of TNT (trinitrotoluene) and is usually considered the beginning of the Nuclear Age.

Figure 1.13 Trinity explosion 0.016 s after detonation. The fireball is about 200 m wide. Image is public domain.

In 1943–1944, work with regard to a plutonium-based bomb focused on a gun-type design called *"Thin Man."* In April 1944 the research laboratory at Los Alamos received the first sample of Hanford-produced plutonium. Within two weeks, the research team discovered a problem: reactor-bred plutonium was far less isotopically pure than cyclotron-produced plutonium, which made the Hanford plutonium unsuitable for use in a gun-type weapon.

This problem was solved by changing to the idea of "implosion," an alternative detonation scheme that had existed for some time at Los Alamos. The implosion design employed chemical explosives to squeeze a sub-critical sphere of fissile material into a smaller and denser form. When the fissile atoms were packed closer together, the rate of neutron capture would increase, and the mass would become a critical mass. The metal needed to travel only very short distances, so the critical mass would be assembled in much less time than it would take to assemble a mass by a bullet impacting a target. Because of the complexity of an implosion-type weapon, it was decided that, despite the waste of fissile material, an initial test would be required. The first nuclear test took place on 16 July 1945 on White Sands Proving Ground under the code name "Trinity" (Figure 1.13).

> **Use of nuclear weapon**: In the history of warfare, only two nuclear weapons have been detonated as part of military operations. The first was a uranium gun-type device code-named "Little Boy," which was dropped by the United States on the Japanese city of Hiroshima on the morning of 6 August 1945 (Figure 1.14). The second was detonated three days later when the United States dropped a plutonium implosion-type device code-named "Fat Man" on the city of Nagasaki, Japan.

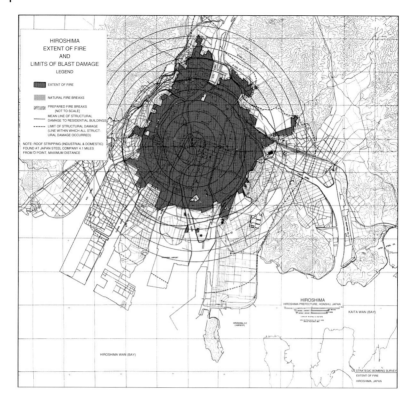

Figure 1.14 Impact of "Little Boy." Image is public domain.

On 10–11 May 1945 the Target Committee at Los Alamos recommended Kyoto, Hiroshima, Yokohama, and the arsenal at Kokura as possible targets. On 6 August 1945, a uranium-based weapon, "Little Boy," was dropped on the Japanese city of Hiroshima (Figure 1.14). Three days later, a plutonium-based weapon, "Fat Man," was dropped onto the city of Nagasaki. At least 100 000 Japanese were killed immediately by the heat, radiation, and blast effects of the nuclear weapons. President Truman's statement that the USA would continue with extensive use of nuclear weapons if Japan did not surrender immediately was in fact a bluff, as the USA had only one remaining completed uranium-gun type bomb.

On 1 January 1947 the Manhattan Program was turned over to the United States Atomic Energy Commission by the Atomic Energy Act of 1946. The Atomic Energy Act broke the partnership of the USA with the United Kingdom and Canada, which has been formed for the Manhattan program, and prevented the passage of any further information regarding nuclear weapons to them. It also ruled that nuclear weapon development and nuclear power management would be under civilian control.

1.4.3
Nuclear Arms Race

The Soviet Union was not invited join the Manhattan Program partnership and to share in the newly developed nuclear weapons. During World War II, however, several volunteer spies involved with the Manhattan Project passed information to the Soviet Union, and the Soviet nuclear physicist Igor Kurchatov, the appointed director of the Soviet nuclear program, was carefully watching the Allied weapons development. In the years immediately after World War II, the Soviets put their full industrial and manpower capabilities into the development of their own atomic weapons. The initial problem for the Soviets was primarily one of resources – they had not scouted out uranium resources in the Soviet Union and the USA had made deals to monopolize the largest known reserves in the Belgian Congo. In the early years mines in East Germany, Czechoslovakia, Bulgaria, and Poland were used as sources of uranium for the Soviet nuclear program.

Efforts in the nuclear program brought results, when the Soviet Union tested its first fission weapon on 29 August 1949 – years ahead of US predictions. The news of the first Soviet nuclear bomb was announced to the world first by the United States, which had detected the nuclear fallout it generated from its test site in Kazakhstan. The USA was shocked not only by the fast progress of the Soviet nuclear program but also by the fact that they had lost their monopoly on nuclear armament. Now both nations (USA and Soviet Union) faced a nuclear tit-for-tat situation. As a consequence, both started a competition for supremacy in nuclear armament, in which both built up a massive arsenal of nuclear weapons (Figure 1.15).

By the 1950s both the United States and Soviet Union had the power to obliterate the other side. Both sides developed a "second-strike" capability, that is, they could launch a devastating attack even after sustaining a full assault from the other side. This policy was part of what became known as Mutually Assured Destruction (MAD): both sides knew that any attack upon the other would be suicidal, and thus would refrain from attack.

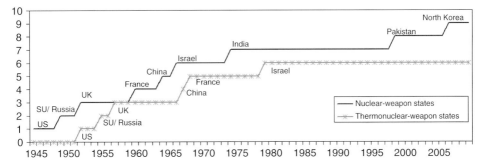

Figure 1.15 Timeline of states with a nuclear arsenal.

The third nation that developed nuclear weapon was the United Kingdom. The US Atomic Energy Act of 1946 prevented the passage of nuclear related information to the United Kingdom. Owing to its involvement in the Manhattan Program the UK had knowledge in some areas. The British nuclear program developed a modified version of the "Fat Man" plutonium based implosion type bomb. The British nuclear bomb was successfully tested in Operation Hurricane in Australia on 3 October 1952 (Table 1.5).

During the 1950s all three nuclear nations (USA, Soviet Union, and UK) worked on improving their weapon design as well as the development of the more powerful thermonuclear weapons or so-called "hydrogen bombs." Finally, the initial Anglo-American cooperation on nuclear weapons was restored by the 1958 US-UK Mutual Defense Agreement. As a result of this and the Polaris Sales Agreement, the United Kingdom has bought United States designs for submarine missiles and fitted its own warheads. This first wave of nuclear armament was characterized by competition between the two post World War II superpowers (USA and Soviet Union). The UK participated in this first wave as a spin-off of the Manhattan Project and due to its relationship with the USA.

The nature of this first nuclear wave, with the establishment of superpower nations, told other nations that building nuclear weapons, particular hydrogen bombs, is a form of increasing national self-expression. As a result several nations initiated their own nuclear programs and started a second wave of nuclear armament (Figure 1.15). In the 1950s the French Republic started a civil nuclear research program, which gained plutonium as a by-product. In 1965 a secret Committee for Military applications of Atomic Energy was formed. Under the presidency of Charles de Gaulle the final decision to build a nuclear bomb was taken in 1958. A

Table 1.5 Usage of nuclear weapons and first nuclear tests by known nuclear countries.

Country	Date of first nuclear weapon test	Date of first hydrogen weapon test	Date of usage of nuclear weapon
USA	16 July 1945 Trinity, first ever nuclear explosion	1 November 1952	6 August 1945 "Little Boy" on Hiroshima; first usage of nuclear weapon 9 August 1945 "Fat Man" on Nagasaki
Soviet Union/Russia	29 August 1949	22 November 1955	–
UK	3 October 1952	15 May 1957	–
France	13 February 1960	28 August 1968	–
China	16 October 1964	17 June 1967	–
India	18 May 1974	–	–
Pakistan	28 May 1998	–	–
North Korea	9 October 2006	–	–

successful test of a French nuclear bomb took place on 13 February 1960 and then on 28 August 1968 a hydrogen weapon was tested. Since then France has improved and maintained its own nuclear capability. China followed a path that is in some ways similar to the French nuclear program. In 1953 China started nuclear research under the umbrella of producing civilian nuclear energy. The Chinese program made rapid progress and, consequently, the first nuclear weapon was tested on 16 October 1964, followed by a hydrogen bomb on 17 June 1967 (Table 1.5).

Owing to intelligence reports Israel joined the group of nuclear weapon nations in 1966 and built thermonuclear weapons in the 1980s. Israel has never officially confirmed or denied that it possesses nuclear weapons, even though the existence of their Dimona nuclear facility was confirmed by the dissident Mordechai Vanunu in 1986. In addition, Israel never performed a full system test, which would have been detected by other nations. The last nation in the second wave of nuclear armament is India. It conducted an underground nuclear test, at Pokharan in the Rajasthan desert, code named the "Smiling Buddha." The government claims it was a peaceful test but it is in fact part of an accelerated weapons program.

The third wave of nuclear armament started at the end of the cold war and is characterized by the proliferation of nuclear weapons among lesser powers and for reasons other than the rivalry between the superpowers USA and the Soviet Union. Owing to competition between India and Pakistan and the success of the Indian nuclear program Pakistan started its own nuclear program. On 28 May 1998, after a series of nuclear tests in India, Pakistan joined the group of nuclear nations with an underground test of fission devices. The intense nuclear testing of nuclear weapons gave rise to concerns that they would use such weapons on each other.

In 2003 North Korea announced that it had performed several nuclear tests but they were never officially confirmed by experts. The first confirmed detonation of a nuclear weapon by the Democratic People's Republic of Korea took place on 9 October 2006.

> **Tsar Bomba – the Largest Nuclear Detonation**
>
> On 30 October 1961 the Soviet Union tested the AN602 hydrogen bomb (nicknamed Tsar Bomba). The Tsar Bomba is the largest, most powerful nuclear weapon ever detonated, and currently the most powerful explosive ever created by humanity. The original US estimate of the yield was 57 Mt, but since 1991 all Russian sources have stated its yield as 50 Mt. The fireball was 8 km in diameter, touched the ground, and reached nearly as high as the altitude of the release plane. The heat from the explosion could have caused third degree burns 100 km away from ground zero. The subsequent mushroom cloud was about 64 km high and 40 km wide. The explosion could be seen and felt almost 1000 km from ground zero in Finland, breaking windows there and in Sweden. Atmospheric focusing caused blast damage up to 1000 km away.

> The seismic shock created by the detonation was measurable even on its third passage around the Earth. Its Richter magnitude was about 5–5.25. The energy yield was around 7.1 on the Richter scale, but since the bomb was detonated in the air, rather than underground, most of the energy was not converted into seismic waves.

The dates of the first successful nuclear weapon test or the first test of a hydrogen weapon provides us with only one aspect of the nuclear arms race. Other aspects are the reasons for the nuclear weapon programs. As discussed above, the US Manhattan Program was established to investigate the general development of a new weapon technology and to obtain the nuclear weapon before Nazi Germany in World War II. The Soviet Union started their nuclear program to keep up with the other major post-World War II power, the USA. Great Britain and France have been powerful counties in Europe for centuries. Consequently, they saw a possibility to maintain this status post-World War II by building up nuclear capabilities. China's reasons were similar, as China wanted to rebuild its status as a ruling nation in Asia. In addition, those programs were part of the rising rivalry between the so-called western (USA) and communist or eastern (Soviet Union) blocks. The **Treaty on the Non-Proliferation of Nuclear Weapons** (NPT, nonproliferation treaty) of 1968 acknowledged the USA, Soviet Union, Great Britain, France, and China officially as nuclear states. The NPT tried to conserve this status as the five official nuclear states agreed not to transfer explosive devices or knowledge of their construction and the remaining member-states agreed not to acquire nuclear weapons. As evidence the official nuclear states are also permanent members of the United Nations Security Council (UNSC).

From this situation, less powerful nations draw the conclusion that the ownership of nuclear weapons makes a country more accepted and respected. As a consequence they saw nuclear armament as way of increasing the national self-expression. Israel started its nuclear weapon program to obtain a powerful option in the wars and competition with its Arab neighbors. At a time when nations with an industry capable of building and operating nuclear power plants gave up any plans for nuclear armament India and Pakistan continued their nuclear program. Since Pakistan has separated from India both nations compete with each other. Several times, disputes have come close to ending in war. Both nations believed that a nuclear arsenal would give them more independence from neighboring nations and more influence among "important nations." The last examples of such nuclear programs are North Korea, Iraq, and Iran. Iraq had to stop its nuclear program after the war with Kuwait. In 2008 North Korea demonstrated the status of their nuclear program by a nuclear test explosion. Officially, Iran declares that its nuclear program is a civil power generation program. However, there are concerns that Iran will build up a nuclear arsenal to counter Israel and to warn the USA.

A third aspect in analyzing the history of nuclear armament is the size of nuclear stockpiles (Figure 1.16). If we compare nuclear stockpiles we see that at each

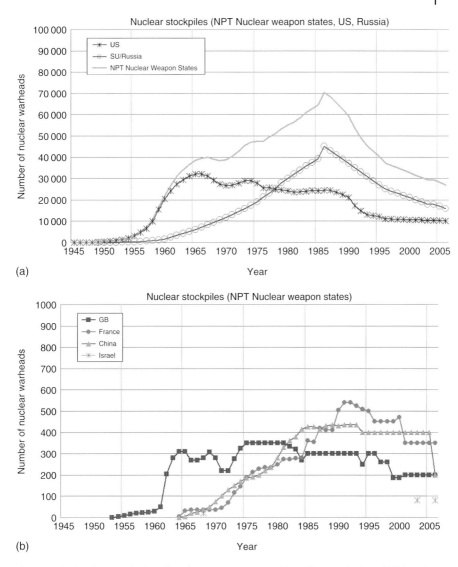

Figure 1.16 Nuclear stockpiles of nuclear weapon states: (a) nuclear stockpiles of USA and Soviet Union/Russia; (b) nuclear stockpiles of other nuclear states.

point in history more than 95% of nuclear warheads worldwide were owned by the USA and Soviet Union/Russia. During the first decade of the nuclear age the USA built up to 30 000 nuclear warheads. Since then the USA has improved its nuclear capability by modernization and has continuously decreased the number of warheads. Owing to problems accessing uranium, the Soviet Union needed two decades to reach the USA stockpile. The Soviet program continued for another decade to build more nuclear warheads. In 1986, the USA and Soviet Union owned more warheads than are needed to destroy the Earth several times over. This nuclear arsenal encompassed a mix of strategic as well as medium- and short-range tactical weapons. Owing to bilateral treaties on the reduction of nuclear arms both nations removed tactical nuclear weapons and reduced the number of warheads to approximately 10 000 each (9400 USA and 13 000 Russia) in 2009. The development of the nuclear stockpile of all other nuclear nations (Figure 1.16b) shows a different tendency. These nations processed a certain number of warheads within a decade and have maintained this number since then. Owing to the small number of warheads (less than 400) it is widely assumed that these arsenals consist of strategic warheads only.

If we compare the Manhattan Program with all other successful nuclear programs we see that the major logistical and technical challenges of nuclear weapon programs are:

- access to sources of uranium,
- production and purification of ^{235}U/^{239}Pu,
- weapon design.

As most parts of the first two challenges also occur in programs that develop nuclear power systems, any kind of nuclear program carries the danger of providing key knowledge towards a nuclear weapon program.

What is known as the *nuclear arms race* shows us that there are three constitutive stages of nuclear weapons (Figure 1.17 and Table 1.5):

- development (and testing) of a fission weapon,
- development (and testing) of fusion (thermonuclear) weapon,
- usage of nuclear weapon (only the USA has reached this stage).

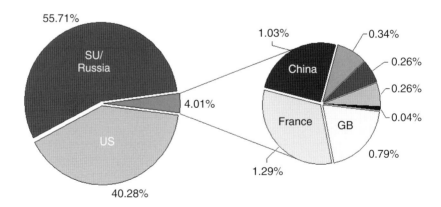

Figure 1.17 Nuclear stockpiles in 2009 [21].

1.4.4
Status of World Nuclear Forces

In each country, the exact number of nuclear assets as well as the current status is a closely held national secret. Despite this limitation, publicly available information and occasional leaks make it possible to estimate the size of the nuclear weapon stockpile (Table 1.6). Those estimates are regularly reported in the *Nuclear Notebook in the Bulletin of the Atomic Scientists* [21] and the nuclear appendix in the *SIPRI Yearbook* [22]. In 2009, approximately two decades after the cold war ended, the Stockholm International Peace Research Institute (SIPRI) reported that the world's combined stockpile of nuclear warheads remained at the high level of more than 23 300. SIPRI considered more than 8190 warheads operational, of which approximately 2200 US and Russian warheads are on high alert, for example, ready for use on short notice.

Owing to the Mutually Assured Destruction (MAD) policy approximately 96% of all nuclear warheads belong to the nuclear arsenal of Russia (55.71%) and the USA (40.28%). All other nuclear nations add up to only 4% (Figure 1.17).

In the current political situation most nations, including Russia and USA, want to avoid any kind of nuclear war. As a consequence even a country that owns a limited number of warheads (like North Korea, with less than ten warheads) poses a significant nuclear threat.

1.4.5
Radiological Warfare and Nuclear Terrorism

In addition to the military use of nuclear weapons, experts have been discussing the possibility that non-state terrorist groups could employ nuclear devices since the 1970s. In 1975 *The Economist* warned that a nuclear bomb can be built out of a few kilograms of plutonium [23]. Since by the mid-1980s power stations turned out

Table 1.6 Status of world nuclear forces[a] [21].

Country	Total nuclear inventory
Russia	13 000
United States	9400
France	300
China	240
United Kingdom	185
Israel	80
Pakistan	60
India	60
North Korea	<10

[a] All numbers are estimates.

many tons of plutonium that each year was transferred from one plant to another, as it proceeded through the fuel cycle, the dangers of robbery in transit became evident. In fact, though, there was a perception in Washington that the value of what is called *"special nuclear material"* – plutonium or highly enriched uranium – was so enormous that strict financial accountability of the private contractors who dealt with it would be enough to protect it from falling into the wrong hands. But it has since been revealed that the physical safeguarding of bomb-grade material against theft was almost scandalously neglected. The public focus on this issue changed after an NBC aired Special Bulletin, a television dramatization of a nuclear terrorist attack on the United States. As a result of the public discussion in 1986 a private panel of experts known as the *International Task Force on the Prevention of Terrorism* released a report urging all nuclear-armed states to beware the dangers of terrorism. The experts warned that the probability of nuclear terrorism is increasing and the consequences for urban and industrial societies could be catastrophic.

A report published in 2004 by the National Commission on Terrorist Attacks upon the United States showed how real the threat of nuclear terrorism became over the years [24]. In the report the Commission released information on an unsuccessful attempt by al-Qaeda to purchase uranium in 1994 and that al-Qaeda continues to pursue its strategic objective of obtaining a nuclear weapon. A May 2004 report [25] by Harvard University's Project on Managing the Atom found that a nuclear attack "would be among the most difficult types of attacks for terrorists to accomplish," but that with the necessary fissile materials, "a capable and well-organized terrorist group plausibly could make, deliver, and detonate at least a crude nuclear bomb capable of incinerating the heart of any major city in the world."

Nuclear terrorism employs nuclear weapons, which release energy in a huge nuclear explosion. As we have learned through the review of the history of nuclear warfare the design of such nuclear weapons belongs to a highly sophisticated level of engineering science.

In contrast a terrorist may use a radiological dispersal device (RDD) that simply scatters radioactive material. Evidently, the design of such a RDD is much easier than building a nuclear weapon. The main physical effect of radiological dispersal is the contamination of an area. Warfare involving deliberate radiation poisoning, without relying on nuclear fission or nuclear fusion, is summarized under the term radiological warfare. Consequently, any criminal and terrorist action using radiological dispersal can be called *radiological terrorism*. The fear of radiological terrorism arose in conjunction with the increasing use of radioactive isotopes in civil applications. For example, cesium-137, which is used in external beam radiation devices to treat cancers and equipment to monitor wells for oil, and cobalt-60, which is used in industrial radiography and cancer therapy, might be used as dispersal radioactive material.

The aftermath of several radiological accidents, such as those in Mexico City (1962), Algeria (1978), Morocco (1983), Ciudad Juarez in Mexico (1983), and Goiania in Brazil (1987) demonstrate the endangerment caused by disposed radioactive material. The accident in Goiania was one of the most serious radiological accidents. On 13 September 1987, a shielded, strongly radioactive cesium-137 source was removed from its protective housing in a teletherapy machine in an abandoned clinic in Goiania, Brazil, and subsequently ruptured. Consequently, 93 g of the radioactive cesium chloride salt was dispersed and many people incurred large doses of radiation, due to both external and internal exposure. Four of the casualties ultimately died and 28 people suffered radiation burns. Residences and public places were contaminated. Decontamination necessitated the demolition of seven residences and various other buildings, and the removal of the topsoil from large areas. In total about 3500 m^3 of radioactive waste were generated [26]. Nowadays the media often refer to RDD by the term *"dirty bomb."* This term focuses on a device in which powdered radioisotope surrounds chemical explosive. In fact many terrorist groups probably have the skill and materials to make the explosive part of the device; but it would be somewhat harder for them to obtain the radioactive material and convert it into powdered form. However, we should bear in mind that terrorists could also scatter radioactive material without an explosive.

1.4.6
Conclusions and Outlook

In this subsection we have learned that for the first time in our history we are able to eradicate ourselves. The unleashing of the nuclear power led to catastrophic consequences in Hiroshima and Nagasaki. During the cold war the nuclear arms race between United States and Soviet Union led to each having the power to obliterate the other side. Both sides had a "second-strike" capability, that is, they could launch a devastating attack even after sustaining a full assault from the other side.

Nowadays several other states are trying to enter the exclusive club of nations with nuclear weapons. Furthermore, the danger of nuclear terrorism cannot be denied. The combination of explosive devices with radiological material ("dirty bomb") may not have such apocalyptic consequences as the use of a nuclear bomb, but it could make large areas uninhabitable.

References

1. Mayor, A. (2003) *Greek Fire, Poison Arrows and Scorpion Bombs: Biological and Chemical Warfare in the Ancient World*, Overlook Duckworth. ISBN: 1585677348X.
2. Smart, J.K. (1997) in *Textbook of Military Medicine: Medical Aspects of Chemical und Biological Warfare* (eds R Zajtchuk and R.F. Bellamy), Office of the Surgeon General, US Department of the Army, Washington, DC, pp. 9–86.
3. James, S. (2005) *Carnuntum Jahrb.*, 189–206.
4. Martinetz, D. (1996) *Der Gaskrieg 1914 – 1918 – Entwicklung, Herstellung und Einsatz Chemischer Kampfstoffe*, Bernard & Graefe. ISBN: 978-3-7637-5952-1.
5. Trumpener, U. (1975) *J. Modern History*, 47, 460–480.

6. Harris, R. and Paxman, J. (1982) *A Higher Form of Killing. The Secret Story of Gas and Germ Warfare*, Chatto & Windus, London. ISBN-10: 0701125833.
7. Barker, A.J. (1968) *The Civilizing Mission: A History of The Italo-Ethiopian War of 1935–1936*, Dial Press, New York, pp. 241–244.
8. Deng, H. and Evans, P.M. (1997) *Nonproliferation Rev.*, Spring-Summer, **4**, 101–108.
9. Infield, G. (1988) *Disaster at Bari*, Bantam Books, New York, pp. 209, 230–231.
10. Shoham, D. (1998) *Nonproliferation Rev.*, Spring-Summer, **5**, 48–58.
11. Oldstone, M.A. (209) *Viruses, Plagues and History: Past, Present, and Future*, Oxford University Press, Oxford. ISBN: 0195327314.
12. Wheelis, M. (2002) *Emerg. Infect. Dis.*, **8** [serial online] September. Available at http://www.cdc.gov/ncidod/EID/vol8no9/01-0536.htm (accessed 12 January 2011).
13. Frischknecht, F. (2003) *EMBO Rep.*, **S4**, 47–52.
14. Harris, S. (1992) *Ann. N. Y. Acad. Sci.*, **666**, 21–52.
15. Regis, E. (1999) *The Biology of Doom – The History of America's Secret Germ Warfare Project*, Henry Holt, New York. ISBN: 978-0805057652.
16. Meselson, M., Guillemin, J., Hugh-Jones, M., Langmuir, A., Popova, I., Shelokov, A., and Yampolskaya, O. (1994) *Science*, **266**, 1202–1208.
17. Guillemin, J. (2005) *Biological Weapons: From the Invention of State-Sponsored Programs to Contemporary Bioterrorism*, Columbia University Press, New York. ISBN: 978-0231129428.
18. Atlas, R.A. (2001) *Crit. Rev. Microbiol.*, **27**, 355–379.
19. Leitenberg, M. (2001) *Crit. Rev. Microbiol.*, **27**, 267–320.
20. Meitner, L. and Frisch, O.R. (1929) *Nature*, **143** (3615), 239–240.
21. Norris, R.S. and Kristensen, H.M. (2009) *Bull. At. Sci.*, **65** (6), 86–98.
22. Gill, B., Cohen, R., Deng, F.M. et al. (2009) *SIPRI Yearbook: Armaments, Disarmament and International Security*, International Peace Research Institute, Stockholm. ISBN: 978-0-19-956606-8.
23. Smyth, H.D.W. (1945) *Atomic Energy for Military Purposes: The Official Report on the Development of the Atomic Bomb Under the Auspices of the United States Government*, Maple Press, York, PA. http://www.archive.org/details/atomicenergyform00smytrich (accessed 20 January 2011).
24. U.S. National Commission on Terrorist Attacks upon the United States (2004) Overview of the enemy, Staff Statement No. 15, June 2004.
25. Bunn, M. and Wier, A. (2004) Securing the Bomb: An Agenda for Action, Project on Managing the Atom, Harvard University, 109 pp.
26. International Atomic Energy Agency (1988) *The Radiological Accident in Goiânia*, IAEA, Vienna.
27. Khardori, N. (2006) *Bioterrorism Preparedness*, WILEY-VCH, Weinheim. ISBN: 3-527-31235-8.

2
International Treaties – Only a Matter for Diplomats?
Martin Schaarschmidt

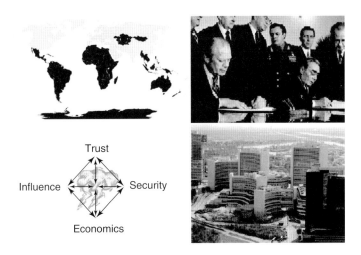

"*Armis bella, non venenis geri*" – "war shall be fought with weapons, not with poison." This sentence was spoken at the time of the Roman Empire. The temptation to use poisons or any dishonorable means of arms in armed hostilities to gain advantages can be found throughout history. Repulsion at the use these weapons and the horrors of the possible consequences has led humans to find rules to handle these weapons and to avoid the destruction of humankind. Therefore, the need to set up rules and to enforce these rules against resistance has increased with growing military knowledge and breakthroughs in science.

2.1
Introduction to the Minefield of Negotiations

In this section we discuss an aspect of chemical, biological, radiological, and nuclear (CBRN) weapons that differs from the scientific and historic treatment pursued in

CBRN Protection: Managing the Threat of Chemical, Biological, Radioactive and Nuclear Weapons,
First Edition. Edited by A. Richardt, B. Hülseweh, B. Niemeyer, and F. Sabath.
© 2013 Wiley-VCH Verlag GmbH & Co. KGaA. Published 2013 by Wiley-VCH Verlag GmbH & Co. KGaA.

other parts of this book. As with many subjects in society, the human urge is to set up rules, conventions, and ethical norms to formalize and regulate the matter at hand. With CBRN weapons and warfare these rules are usually in the domain of international law [1]. As diffuse and controversially discussed the application and enforcement between politicians and jurists may be [2], we will try to shed light on the internationally established system of treaties and their implementation.

However, covering every international treaty would easily go beyond the scope of this introduction. We therefore limit ourselves to a few important groundbreaking developments (Figure 2.1) and try to explain the basic ideas behind the system of treaties and the implications for international relations and also possible adverse effects caused by them.

> - **International treaties**: agreements in the domain of international law between (two or more) sovereign states and/or international organizations.

What is an international treaty? International treaties are agreements in the domain of international law between (two or more) sovereign states and/or international organizations. The act of their installment used to be part of customary international law but has been codified in 1980 by the Vienna Convention on the Law of Treaties [3]. Nevertheless the concept of international treaties has a long tradition of being considered a valid source of law in many jurisdictions. There is a wide terminology used to describe these agreements, amongst them convention, letters of intent, memorandum of understanding, treaty, protocol, covenant, and so on. The background and motives to regulate different aspects of CBRN related matters are multilayered (Figure 2.2). From the most basic wish to prevent the use of these weapons arises the further need to regulate also the possession, capabilities of production, and accessibility [4]. Other motives may be the avoidance of costly arms races or the possibility of gaining access to foreign territory through the

Figure 2.1 The way into the disarmament process – Gerald Ford and Leonid Brezhnev signing a joint communiqué on the SALT treaty (1974).

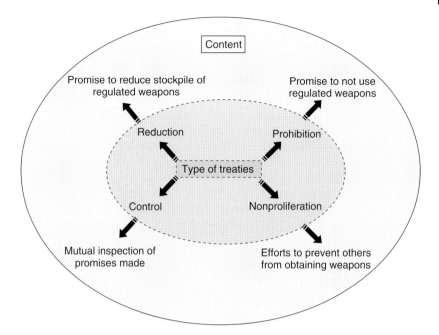

Figure 2.2 The four basic forms of weapons treaties and their main objectives.

means of verification inspections. It may also be the general wish to build trust and to advance communications between opposing power structures.

2.1.1
Arms Reduction and Prohibition of Use

Throughout history mankind has at times been disgusted by its own actions. After ever major military conflict attempts have been made to prevent further conflicts or at least to contain the damage and suffering involved [5]. Especially after the advent of weapons of mass destruction (WMD) great hazards have been foreseen. In the case of chemical weapons and the experiences of World War I the community of nations has set in motion efforts to constrain the use of such weapons. Insight into the potentially disastrous effects of biological and nuclear warfare has led to corresponding efforts – for the first time in human history – *before* widespread deployment in armed conflicts.

The foremost preventive action to prohibit the use of WMD is to prevent the availability of such weapons. Usually, the first steps in risk containment are bi- or multilateral treaties in which the participating parties promise each other to reduce arms according to the contract. Through these arms reduction treaties arises the need for both facilities for the safe disposal of biological and chemical weapons and also a suitable system of verification of compliance.

2.1.2
Arms Control and International Controlling Bodies

Owing to the nature of international law it has transpired that arms reduction treaties alone are not sufficient. Being contracts in the domain of public international law they are, realistically evaluated, mere declarations of intent, and suffer from the lack of a central body with appropriate means of control to enforce them.

Arms control with an international scope – unlike national arms control legislation such as the limitation of private access to firearms – usually governs the development, production, stockpiling, and use of weapons.

Contracting parties usually have a mutual interest in abiding to the treaties, which may be humanistic reasons or just to avoid costly arms races. To build confidence in adherence to the contract, often an international body with (often limited) means of verification and enforcement by sanctions in the case of violations is instated [6].

2.1.3
Nonproliferation

During all times, and especially during the cold war, the dangers of proliferation (i.e., the spread of weapons and technologies) have been acknowledged. It is feared that supplying key technologies to allies and the subsequent use in proxy wars will lead to an uncontrollable spread of such weapons [7]. It is therefore of vital interest to countries in possession of WMD to interdict proliferation. On a practical level this is done by nonproliferation treaties regulating trade with sensitive equipment and technologies. Another main aspect is the attempt to enforce limitations upon non-consenting parties by means of binding international law such as provided by organizations like the United Nations.

Challenges to proliferation control are the implications imposed on the civil economy as many technologies involved are of a dual use, like the civil harness of nuclear power.

In recent years the fear of proliferation has gained an additional dimension as not only states or state sanctioned organizations sought WMD but also independent groups, which are usually summed up as international terrorists.

2.2
Why It Is so Difficult to Implement International Regulations?

We might often wonder about the long negotiation time line before a treaty is ready to sign. However, manifold problems arise from the wish to implement international regulations. More than once have well-intended efforts proven futile. Sometimes cultural differences have to be overcome – perhaps between different parts of the world or only between politics and science. On occasions negotiations have failed due to the sheer impossibility of forging the abstract intentions into applicable regulations. Other treaties have been sacrificed on the altar of tactics in the battle between global power blocks to gain influence.

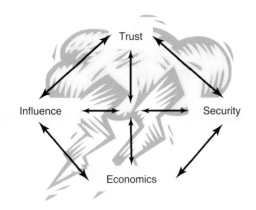

Figure 2.3 Areas of conflict in international arms control.

In most cases it boils down to a conflict of interests: trust versus security versus economics versus influence (Figure 2.3).

2.2.1
Trust – Devoid of Trust Every Effort Is Useless

The first prerequisite for any agreement is trust. In a (perhaps only perceived) point blank duel situation everyone involved is reluctant to make the first step. This fear of giving the opponent an advantage has to be overcome to start earnest attempts to resolve such a situation. The situation at the beginning of the 1960s in the middle of the nuclear arms race between the USA and Soviet Union is a good example [8]. Full-fledged nuclear armament started in the early 1950s with first long-range bomber airplanes and gained momentum with the advent of the first intercontinental ballistic missiles (ICBMs) in 1957. Even with the imminent risk of total mutual destruction it took another five years (and one major crisis, the "Cuban missile crisis") for the two superpowers to even talk to each other. Even when the "Memorandum of Understanding Regarding the Establishment of a Direct Communications Line" was signed in 1963 and the so-called "red telephone" was established, this link was opposed by some factions on both sides. Regardless of initial problems, the "red telephone" and its underlying philosophy was a huge success, pointing out the importance of communication and trust building measures. Many treaties after the end of the cold war were dedicated purely to trust building measurements, such as the Open Skies treaty of 1992.

2.2.2
Negotiation – Special Skills Are Required

Any international agreement travels a long way on going from an idea being proposed to a refined legal agreement being signed and ratified. Sometimes seemingly small differences on peripheral matters can cause delays. As an example we can analyze the development of the series of nuclear test ban treaties in the 1960s and 1970s. The question of disarmament was excluded from the consultations at a fairly early stage as it was obvious to the participants that no agreement would be reached. The exclusion of underground nuclear testing from the Partial Test Ban Treaty (PTBT) (1963), on the other hand, resulted from a disagreement over the possible technologies to employ in a verification regime. Resolving these technical differences and finally banning underground testing took until 1974 despite the desire of both parties to ban such tests since the beginning of the 1960s.

2.2.3
Dual Use – Good or Bad Technology?

The term *dual use* usually refers to substances, machinery, technology, and knowledge that have both civilian and military applications. The term is widely used in the scope of export control and export restrictions.

> - **Dual use**: machinery or technology, and so on, having both civilian and military applications.

Many regulated technologies and substances within the scope of CBRN are not only usable to produce WMDs. They also have legitimate applications in civil society (dual use). An obvious and much discussed example is the civil harness of nuclear power. An even wider field of this dual-use issue arises from chemical substances. Lost (sulfur mustards, mustard gas), for example, is not only one of the substances of most concern listed in Schedule 1 of the Chemical Weapons Convention (CWC) but is also used in chemotherapy to treat lung cancer. Some precursors in semiconductor growth are also explicitly regulated by the CWC. From such dual-use character arises the need for differentiated regulations that effectively interdict WMD while still allowing the beneficial civil use of substances and technologies [9].

The use and application of regulated substances and technologies, as pointed out above, and the resulting verification procedures burden national industries with the implementation of these regulations. This can lead to disadvantages in competition on the global market: Either the production costs rise due to additional legal requirements or operation in certain areas is prohibited as such. These considerations are made by a country when engaging in multinational consultations. The resulting treaties tread a thin line between effectively serving their original purpose and avoiding adverse side effects.

2.2.4
Verification – an Instrument for Trust Building

Verification and trust do not contradict each other: they are just two sides of the same concept. By allowing another party to verify my compliance on my territory I allow him to build trust in me. Allowing inspections poses the risk of not only exposing information concerning the inspection but also otherwise valuable intelligence. To appease these concerns, often the inspections are performed by accepted neutral organizations that are usually installed for just this purpose.

> - **Verification**: the act of proving or disproving the adherence of contracting parties of an agreement to the agreement.

Challenges for effective inspection and verification arise from the aforementioned dual use problem. Nuclear technology and material handling require highly sophisticated facilities. Therefore, in a cooperating host nation it is unproblematic in this sense as normally only a few locations have to be taken into account. However, monitoring chemical and biological weapons is more complicated. The ability to produce chemical and biological warfare agents is much more widely spread throughout modern industry. Many facilities are potentially able to produce them and, as pointed out above, there are legitimate reasons to produce precursors or even actual agents marked as biological or chemical weapons. This widespread availability of technologies as well as the possibility to produce small quantities with home-made equipment poses challenges not only to the international community but also to internal law enforcement. Most current treaties do not account for these recent changes. An example of a recent problem is the widespread use of microreactors in the chemical industry. With these installations it is possible to perform flexible small-scale batch production of nearly any chemical compound. The verification procedures of the CWC are insufficient to account for this new style of production.

2.2.5
Technological Advancement – Gain of Momentum

Up until the last decades of the twentieth century military armament was at the leading edge of technology and in many cases the driving force behind scientific progress. Towards the end of the cold war things changed dramatically. Through the rise in international trade and access to global markets, development resources could be bundled, resulting in ever shorter innovation cycles. Civilian technology advancement decoupled from military technology in the 1990s. The military with its long platform lifetimes was left behind by the state of the art. As a consequence many technologies arose that, while never intended for that purpose, could easily aid in warfare.

These changes influenced the structure and scope of international arms treaties. As an example of a treaty before the acceleration of development in the 1970s we can use the Strategic Arms Limitation Talks/Treaty (SALT)-1 treaty of 1972. In this treaty the maximum number of launch systems for nuclear devices (i.e., ICBMs) was limited. The subsequent development of multiple independent re-entry vehicles (MIRVs) circumvented this limitation and rendered SALT-1 meaningless. To counteract the new development two new treaties followed in 1979. One pursued the "classical" philosophy by limiting the number of delivery vehicles (SALT-2) and one employed a new approach: the Anti-Ballistic Missile Treaty (ABMT). The ABMT does not try to limit the actual number or technology of nuclear devices but, instead, limits the capabilities to defend against them. The ABMT limits missile defence systems to one for each power block (probably mainly introduced by the political elite to protect themselves). The different treaties will be explained in more detail later in this section. The idea behind this limitation is that without proper missile defence capabilities no side would see the need to further stockpile and improve their nuclear forces as the existing arsenal was sufficient to completely annihilate the opponent.

Reactions to the advancements in civil technology can be seen by comparing the Biological Weapons Convention (BWC) of 1972 and the CWC of 1993. The BWC gives an explicit list of regulated organisms and toxins (Chapter 4). In contrast the CWC does not even try to give a concise list of substances to regulate. It defines a chemical weapon not by its agent but rather defines legitimate and illegal uses of any substance. This accounts for the dual-use character of many substances [10]. However, in a fall back the CWC also gives a list of substances without any use other than as a chemical weapon (the "Schedule 1"). As science advances some substances on this list have civil applications nowadays. The most prominent example is the use of n-Lost in cancer treatment.

2.3
Historic Development of Treaties – the Link to the Incidents

In the previous section we learned about the historic use of chemical, biological, and nuclear substances in warfare and terrorist activities. Treaties are closely linked to these incidents to prevent further conflicts or at least to contain the damage and suffering involved. Therefore, we expand our journey through history with the development of treaties to ban the use of these deadly substances.

During medieval times where the rules of engagement were a mainly unwritten code, only sporadic attempts were made to write down these rules. One example is a regulation on poison bullets by the Strasbourg Agreement of 1675 between France and the Holy Roman Empire. At the end of the nineteenth century the first modern day attempts to regulate warfare and the use of certain weaponry were undertaken. Among the most notable were the first and second peace conferences in The Hague (Figure 2.4) (The Netherlands, 1899 and 1907 respectively) [11]. The first convention, entering into force in 1900, dealt in its second part ("On the Use of Projectiles the Object of Which is the Diffusion of Asphyxiating or Deleterious Gases") with chemical weapons for the first time. The use of such weapons was only prohibited in conflicts between member states of the contract, however. Owing to this limitation the widespread use in colonial conflicts by many countries could not be avoided.

Even though a control body in form of the Permanent Court of Arbitration was set up, the convention was not blessed with success. World War I (1914–1918) was the first conflict that saw massive use of chemical weapons, most notably (in public perception) mustard gas (Chapter 1).

Figure 2.4 The Hague Peace Conference (1899), Russian delegation.

After the experiences of World War I a few international attempt were made to prevent further large-scale armed conflicts. The World Disarmament Conference in Geneva in 1932–1934 was an attempt to control the emerging arms race at the dawn at World War II. It shared the fate of many attempts in the 1920s: the conference was doomed due to the lack of acceptance by many participating countries as well as the non-equal treatment of the German Reich and its allies from World War I.

Despite its failure to stop the arms race in Europe it argued against aerial bombardments as well as chemical and bacteriological warfare. Directly addressing WMD, the Protocol for the Prohibition of the Use in War of Asphyxiating, Poisonous or other Gases, and of Bacteriological Methods of Warfare (or Geneva Protocol) came into force in 1928. After the experiences of World War I it prohibited the use of all forms of chemical and bacteriological weaponry. However, the production and storage of such weapons was still allowed under the contract. The main criticism was that the non-use obligations only applied up to the point were prohibited agents were used by an opponent. The almost non-existent use of biological or chemical weapons in World War II was not so much due to conventions or treaties as to the deterrent potential of weapon stocks of the opposing armies [12].

2.4
Today's System of Treaties – a Global Network

After the brief history we will try to outline the system of treaties in effect today. Most of these international arrangements are under the auspices of the United Nations (UN) (http://www.un.org/disarmament/).

This section is organized as follows: First we give a rough timeline of the relevant agreements. Then the most important treaties are categorized by their main scope

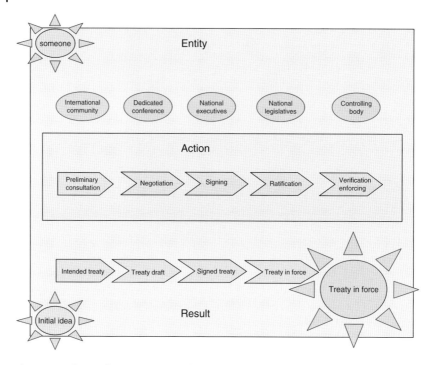

Figure 2.5 The cumbersome process that leads to an international agreement.

(biological, chemical, and nuclear weapons as well as deployment systems) and are presented in detail. Finally, we give an overview of the organizations and entities involved. The way an international treaty comes into existence is a delicate and often cumbersome process. Although there is no universal way, we will point out the main steps and milestones involved (Figure 2.5). Note that it is not unusual for international agreements to take several bends and turns during their development.

Once the initial idea for a treaty is born, preliminary (often unofficial) consultations between prospective parties start. When a certain level of agreement is reached and the participating parties come to the conclusion a formal agreement is feasible formal negotiations begin. Often these negotiations take place in a dedicated body, either existing (like the UN) or specifically spawned for that purpose. During these negotiations technical and legal questions are discussed at a very detailed level. The finalized draft treaty is then signed by the leading executives of the involved parties. This signing often takes place in a public ceremony and is perceived by the public as the final act instating a treaty. This is far from the truth. To be legally active, a treaty has to be ratified by the legislative bodies of the member states. More than once a treaty has failed at this final hurdle. To be fully functional most treaties install an international controlling body to supervise its installment and to control and verify the adherence of member states to the rules imposed by the treaty.

The progression of arms control treaties since 1900 shows key aspects of activities in:

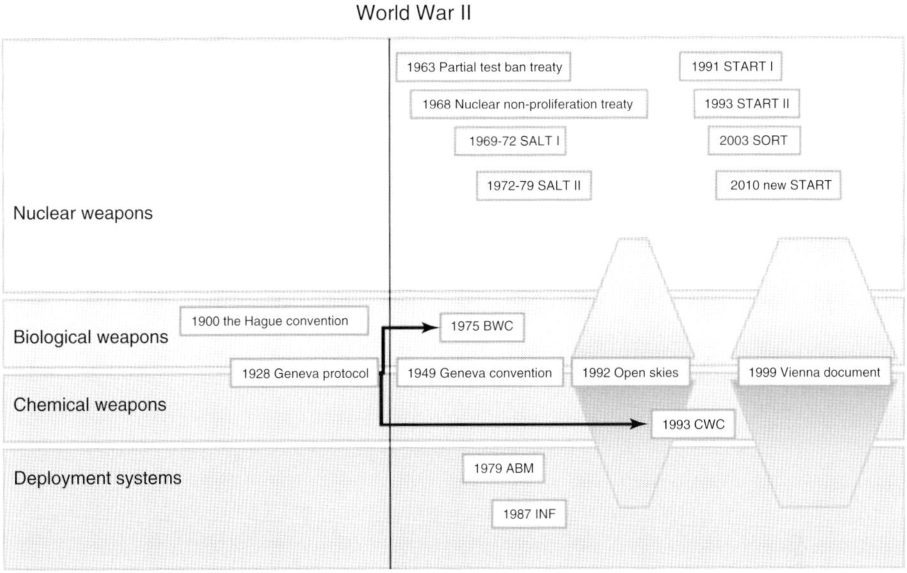

Figure 2.6 Timeline and classification of weapons treaties.

- deployment systems (Figure 2.6).
- biological weapons
- chemical weapons
- nuclear weapons.

Up to the end of the cold war in the late 1980s most treaties were weapon-class specific (atomic, biological, or chemical weapons). In the 1990s and 2000 treaties (like Open Skies or the Vienna Documents) became more focused on generic problems (like trust, proliferation, deployment systems) not properly addressed during the cold war decades.

Now as we understand more about the difficulties associated with the emergence of international treaties to ban or to hamper the use of CBRN substances, we will take a more detailed look on the most important treaties.

2.4.1
The Geneva Conventions – the Backbone for Further Treaties

The Geneva Conventions are by far the single most well-known international agreement. The term usually refers to The Fourth Geneva Convention relative to the Protection of Civilian Persons in Time of War and was finally negotiated in 1949 in the aftermath of World War II. It governs the rights and obligations of participants in modern warfare as well as humanitarian treatment of the victims of war and has been ratified by 194 countries. Its predecessors date back to 1864 (First Geneva Convention for the Amelioration of the Condition of the Wounded and Sick in Armed Forces in the Field), 1906 (Second Geneva Convention for the

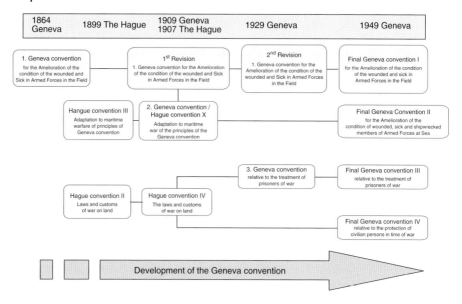

Figure 2.7 Development of the Geneva Conventions.

Amelioration of the Condition of Wounded, Sick, and Shipwrecked Members of Armed Forces at Sea), and 1929 (Third Geneva Convention relative to the Treatment of Prisoners of War) [13]. Figure 2.7 depicts the historical evolution.

In Article 51 and 54 the Geneva Convention forbids the indiscriminate attack of civilians and the destruction of food and water. In this scope it explicitly mentions the use of biological weapons and judges their use as a war crime.

2.4.2
Deployment System for Weapons – Control the Carrier Systems

The deployment system for WMDs has been identified as a vital component for the actual use of such weapons. The lack of such transportation systems renders otherwise accessible weapons useless. Constant attempts have been made to limit and control these deployment systems. In contrast to regulating technologies to produce WMDs the regulation of deployment systems has far less adverse effects on national industries as the dual-use potential of these deployment systems is limited. The main objects of regulations are ballistic missiles and airborne and submarine launch-systems (i.e., strategic bombers and nuclear submarines). The development of defence systems against ICBMs has been identified as a threat to the established system of mutually assured destruction between the two main power blocks and was consequently regulated to prevent a destabilization of the status quo. The treaties usually include means of verification of compliance.

In 1972 the ABMT was signed between the United States of America and the Soviet Union [14]. It limits the number of anti-ballistic missile (ABM) systems used

Figure 2.8 Signing of Intermediate-Range Nuclear Forces Treaty (INFT) in 1987. Source: White House Photographic Office.

to intercept ICBMs. It was feared that the advent of ABM-systems would endanger the fragile balance of power between the USA and USSR.

The ABMT marks a turning point in the regulation of nuclear weapons as it does not limit the weapons as such but, instead, the defence against them. It was therefore independent of possible future developments in delivery technology. The ABMT partially results from the failure of SALT-1 due to the advent of MIRVs.

In an unprecedented act the USA withdrew from the ABMT in 2002 – the first withdrawal from a major international arms treaty in recent history [15]. The subsequent efforts of the USA to establish a Missile Defence Shield has been a subject of potential conflict in international relations ever since.

In the Intermediate-Range Nuclear Forces Treaty (INFT) [14] of 1987 between the USA and USSR both signing parties agreed to eliminate completely their ground and air launched missiles with a range of 300–3400 miles (Figure 2.8). The treaty allowed both sides to make inspections of military installations on demand. As of 1991 the United States had destroyed 846 and the Soviet Union 1846 such weapon systems.

The Treaty on Open Skies was signed by 34 parties in 2002 [16]. It allowed unarmed aerial surveillance flights over the entire territory of the signing states. It was intended to enhance trust and confidence as well as openness and transparency of military deployments. The first attempts at such a treaty were made as early as 1955 but the final consultations between the main powers had to wait until the end of the cold war. The treaty was signed in 1992 in Helsinki.

The Vienna Document 1999 is a contract within the scope of the Organization for Security and Co-operation in Europe (OSCE) governing transparency and openness in Europe. It installs a system of information and verification for peaceful conflict management. It was negotiated between 1992 and 1999.

2.4.3
Biological and Chemical Weapons

Biological and chemical weapons are treated together here not to diminish their importance but to account for their similarity both in production-facilities and deployment systems and their differences to nuclear weapons. First, the two main conventions are presented. Then the problems of the implementations of both conventions as well as the adverse side effects are discussed.

The Convention on the Prohibition of the Development, Production, and Stockpiling of Bacteriological and Toxin Weapons and on their Destruction (BWC) [14] in 1972 was the first international treaty banning not only the use and proliferation but also the production of biological weapons. In this respect the treaty is a supplement to the Geneva Protocol of 1925. The convention went into force in 1975 when it was ratified by 22 countries. As of today it has been ratified by more than 160 governments.

In contrast to the CWC and corresponding nuclear treaties the BWC does not provide for formal verification of compliance by signing parties, which is its main drawback. In Article I the scope of prohibition of development, production, stockpiling, acquiring, and retaining is defined as microbial and other biological agents or toxins and their means of delivery. It is, however, permitted to retain small justifiable quantities for prophylactic, protective, and other peaceful purposes. The regulated substances are compiled in the so-called "dirty dozen" list (Table 4.2 in Chapter 4). Article II demands the destruction of existing stock and associated resources. Article III prohibits the proliferation of both biological agents and the knowledge to produce them. Articles IV–X are concerned with implementation of the convention. The only measure of verification is the possibility to consult the UN Security Council in the case of alleged breaches (Article VI).

2.4.4
Chemical Weapons Convention 1993 and Organization for the Prohibition of Chemical Weapons (OPCW)

The Convention on the Prohibition of the Development, Production, Stockpiling, and Use of Chemical Weapons and on their Destruction (CWC) [14] of 1993 prohibits the production, stockpiling, and use of chemical weapons. Currently 188 countries are member of the CWC. In the field of chemical weapons the CWC is a successor of the 1925 Geneva Protocol. It implements extensive verification measures and installs the independent Organization for the Prohibition of Chemical Weapons (OPCWs).

In the framework of CWC three classes of controlled substances are distinguished:

- **Chemicals with few or no uses outside of chemical weapons**: These may only be produced for research and medical or pharmaceutical use. Examples are precursors in semiconductor growth processes or mustard gas used in medical applications (chemotherapy). Production above 100 g per year has to be declared and possession above 1 ton is prohibited.
- **Chemicals with legitimate small-scale applications**: production has to be declared and export into non-CWC countries is prohibited.
- **Chemicals with large-scale uses apart from chemical weapons**: Production above 30 ton per year and site must be declared and are subject to inspection and export into non-CWC countries is restricted. To date thirteen countries have declared

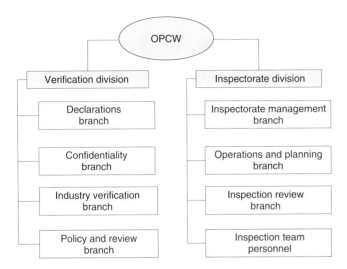

Figure 2.9 The Organization for the Prohibition of Chemical Weapons (OPCW) coordinates chemical weapons inspections.

production facilities for chemical weapons. Twelve are known by name, one is anonymous. All 65 declared facilities have been deactivated. A total of over 70 000 tons of chemical weapons have been declared, of which 40 000 tons have been destroyed (as of 2009).

The OPCWs was instated as part of the Chemical Weapons Treaty as an international control body in 1997 (Figure 2.9). It organizes inspection procedures to verify compliance with the treaty. Since 2000 the OPCW and UN officially coordinate their activities on the basis of a cooperation agreement.

2.4.5
Implications of the Chemical Weapons Convention (CWC) and the Biological Weapons Convention (BWC)

Both conventions call for the destruction of existing weapons but (in the case of the CWC) in nearly two decades just over half of the declared existing weapons have been destroyed. This cannot be attributed to unwillingness of the respective countries but solely to the simple lack of appropriate disposal facilities. As biological and chemical weapons are suitable for long-term storage no means of disposal were implemented during the build-up of production facilities of most countries. Only a few disposal plants exist (e.g., in the USA, Russia, and Germany). This limited capacity of environmentally safe disposal as well as the logistical problems of transportation of hazardous material to those facilities is responsible for the delayed destruction of biological and chemical weapons.

A second implication of the CWC and BWC is the restriction imposed on both the chemical and biological industry as well as on research [10]. Especially in chemistry no definite lines between legitimate civil products, dual use substances, and chemical weapons can be drawn. The CWC tries to introduce a classification system to account for this problem but the imposed restrictions and declaration procedures – undoubtedly necessary to effectively control chemical weapons – cause restrictions and a subsequent increase in production costs of certain products. As the positive effects greatly surpass the adverse effects most societies have decided to join the conventions.

As the CWC gives no definite list of regulated substances but rather introduces a classification regarding toxicity and intended use the need arises for new means of verification of compliance. The sheer amount of chemical substances available, with constant additions, makes keeping track a difficult task. In 2007 the European Union set into action a regulation called Registration, Evaluation, Authorization, and Restriction of Chemicals (REACH) [17]. The central principle of REACH is the restriction of marketability to substances with a full data set on record. This data set is to be provided by private industry and contains, among other, information regarding the toxicity and environmental harmfulness. The information gathered by REACH was not originally intended to be used in conjunction with the CWC but nevertheless is very valuable.

Declaration of regulated substances by a country is only the first step in compliance with a treaty. Independent means of verification and the possibility of securing proof of possible violations (forensics) are needed to establish a system of trust between participating countries. Especially in a biological and chemical context this is rather difficult.

2.5
Nuclear Weapons

Nuclear weapons differ from biological and chemical weapons in a few points that have affected the history of regulations and international agreements. Nuclear weapons production demands a fairly high level of scientific and engineering sophistication. The resources needed for production can usually only be provided by governmental agencies. This demands a proliferation control of technologies and equipment as well as precursors like enriched nuclear fuel.

Nuclear devices have a minimum explosion yield – a limited small-scale deployment is not possible. Together with the long lasting after effects, nuclear explosives cannot be used locally in time and space [18]. Nuclear explosions can be detected quite easily with seismic detectors and through analysis of airborne radioactive fission products. This makes secret tests of nuclear devices impossible and allows for very effective control of appliance to a test ban [19].

All these points led to a fairly early and concise regulation of nuclear weapons as well as the installations of weapon-free zones to limit the proliferation of nuclear weapons in proxy wars during the cold war.

Nuclear weapons treaties can be categorized as dealing with the following subjects:

- nonproliferation
- disarmament

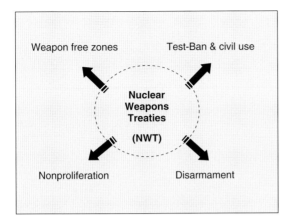

Figure 2.10 Categorization of nuclear weapons treaties.

- test-ban and civil use
- weapon-free zones (Figure 2.10).

2.5.1
Nonproliferation

Nonproliferation plays an important role in the area of nuclear weapons. On the one hand it is rather complicated to manufacture weapons-grade fissionable material. On the other hand, it is rather simple to construct a crude nuclear device once the nuclear fuel is obtained. A simple gun-type fission weapon (e.g., Reference [20]) is not very efficient but does not require great technological skills to build. The Treaty on the Non-Proliferation of Nuclear Weapons (NPT) (*http://www.un.org/disarmament/*) of 1968 has currently 189 member countries, among them all five official nuclear powers (USA, USSR/Russia, Great Britain, France, and China). The only remaining four non-member states (Israel, India, Pakistan, and North Korea) are known or believed to have nuclear weapons programs. The original limited duration of 25 years was extended indefinitely in 1995.

The treaty consists of three pillars:

1) **Nonproliferation**: the five official nuclear states agree not to transfer explosive devices or the knowledge of their construction. The remaining member states agree not to acquire nuclear weapons.
2) **Disarmament**: the NPT states the desire of signature states to halt production and to dispose of nuclear weapons. No hard limits or requirements are imposed and therefore all nuclear disarmament since has been due to other bi- or multilateral agreements.
3) **Right of peaceful use of nuclear technology**: The NPT governs the transfer of material and technologies between signature members of the treaty to

peacefully harness nuclear energy. The right of sovereign states to use nuclear energy is guaranteed; the trade and knowledge transfer, however, is regulated.

As international verification body the International Atomic Energy Agency (IAEA) [21] is installed by the NPT.

In particular, the right of peaceful use has caused international disputes in recent years. Countries that are believed to pursue military nuclear programs have in consequence been denied the right to use nuclear power by the community of states.

2.5.2
Disarmament

Nuclear disarmament talks during the cold war were a succession of consultations (SALT) and subsequent treaties during the late 1960s until the end of the cold war in the late 1980s between the USA and USSR. In the 1990s this series of talks continued, now between the USA and Russia.

2.5.2.1   Strategic Arms Limitation Talks/Treaty (SALT)

The first SALT bilateral talks between the two super powers of the cold war, USA and the Soviet Union, were held from 1969 to 1972 in Helsinki (Finland) and Vienna (Austria). The treaty limited the nuclear weapons stock by freezing the number of strategic ballistic missile launchers at existing levels. Though allowing for new submarine-launched ballistic missiles (SLBMs), these could only be built after reducing ICBMs accordingly. The formal result of the SALT was the Interim Agreement Between The United States of America and The Union of Soviet Socialist Republics on Certain Measures With Respect to the Limitation of Strategic Offensive Arms (SALT I) [22] of 1972.

After the SALT I a second round (SALT II) was initiated from 1972 to 1979. Through the advent of missiles with MIRV warheads in the late 1960s, the effectiveness of possible missile defence systems was severely reduced. To prevent a costly mass deployment of these defence systems in a new arms race, ABM systems were limited to two per country by the ABMT (http://www.un.org/disarmament/) of 1979, which was negotiated during the second SALT II . Both the USA and USSR did only deploy one such system each, one to protect Moscow and one to cover the Minuteman ICBMs site in North Dakota.

The introduction of MIRVs also rendered the metric applied at previous treaties (number of missile launchers) meaningless. The SALT II Treaty limited the number of delivery vehicles to 2250 on both sides. It also banned the development of new missile systems. The treaty in effect stopped development and deployment of Soviet SS-17, SS-18, and SS-19 heavy ICBMs and promoted the development of cruise missiles without first strike capabilities and alteration to the trident program in the USA. In a secret side protocol the USSR agreed to limit the production of its TU-22M bombers to 30 per year. The SALT II agreement was signed in 1979 by both heads of state but, due to the refusal of the US Congress to ratify the treaty and the Soviet invasion of Afghanistan, never went into force. Nevertheless the treaty was honored by both sides until the USA retreat from it in 1986.

2.5.2.2 Strategic Arms Reduction Treaty (START)

After the US retreat from SALT II and implementation of the Strategic Defense Initiative program (SDI) a new nuclear arms race unfolded in the 1980s, at the end of which both sides possessed tens of thousands of nuclear warheads. The Strategic Arms Reduction Treaty (START) [21] of 1991 established an upper limit of 6000 deployed nuclear warheads and 1600 deployment systems. It also limits conventional forces to 6000 fighter aircraft, 10 000 tanks, 20 000 artillery pieces, and 2000 attack helicopters. As START I was signed just five months before the final collapse of the Soviet Union in late 1991 it took ten years to fully implement its terms. As the largest arms control treaty in history it left the member states of the former Soviet Union (except Russia) nuclear disarmed and removed about 80% of strategic nuclear weapons from the USA and Russia.

The START II of 1993 banned the use of MIRVs on ICBMs and thereby severely limited first strike capabilities on both sides. START II was signed in 1993 (Figure 2.11) and ratified in 1996 and 2000.

It did not go into force, though, as the Russian ratification made this contingent on US Senate ratifying a September 1997 addendum to START II which included agreed statements on a future ABMT. In 2001 both the USA and Russia agreed to reduce their missile forces to nearly a third of the then forces. This reduction was completed in 2005. The 2002 withdrawal of the United States from the ABMT rendered the Russian ratification, and therefore the whole START II, obsolete.

START III negotiation aimed at a drastic reduction of the nuclear stockpile of both USA and Russia. The 2002 withdrawal of the United States from the ABMT ended all START III negotiations and led to the much weaker Strategic Offensive Reductions (SORT) agreement of 2003 [22]. As of 2010 the United States has 3696 and Russia has 4237 deployed strategic warheads. The total warhead count is circa 10 000 for the USA and around 16 000 for the Russian Federation. The New Strategic Arms Reduction Treaty (NEW START) [21] agreement signed in 2010

Figure 2.11 Signing Strategic Arms Reduction Treaty (START) II, the Presidents of the Russian Federation and the United States of America, 1993. Source: Executive Office of the President of the United States.

limits the deployed nuclear warheads to 1550 each and the deployment systems to 700 as well as establishing a new verification system.

2.5.2.3 Strategic Offensive Reductions (SORT) 2003

The Treaty between the United States of America and the Russian Federation on SORT (or "Moscow Treaty") [22] signed in 2002 and went into force in 2003 (on a ten-year term) reduces the nuclear arsenal of both countries to 2200 deployed nuclear warheads each. The SORT agreement substitutes the more strongly formulated, but failed, START II and the stalled START III agreements. Criticism of the SORT agreement includes the lack of verification provisions, and the fact that the reductions in warheads has to be performed at the end date of the agreement, at which point it loses force anyway. Nonetheless SORT led to a reduction of the US nuclear stockpile to half its 2001 level by 2007.

2.5.3
Test-Ban and Civil Use

Until recent years (and the development of high-performance computing capabilities) the actual test of nuclear devices was an essential part of the development cycle of such devices. It is therefore a logical step to limit and control nuclear tests. Furthermore, the adverse environmental effects of nuclear tests became apparent and public pressure grew to reduce these tests. The test-ban treaties have been rather successful as it is fairly easy to detect nuclear explosions and supervise adherence to the treaties (Figure 2.12) [19].

The Treaty banning Nuclear Weapon Tests in the Atmosphere, In Outer Space, and under Water (PTBT) [23] outlawed all nuclear tests except those performed underground. When it was signed by the USA, UK, and USSR in 1963 it was meant to slow down the nuclear arms race of the cold war and to limit the release of radioactive fission products into the atmosphere after the advent of nuclear

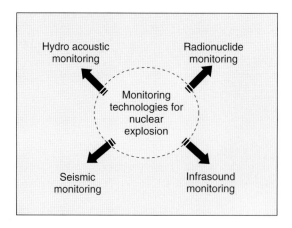

Figure 2.12 Monitoring technologies to detect nuclear explosions.

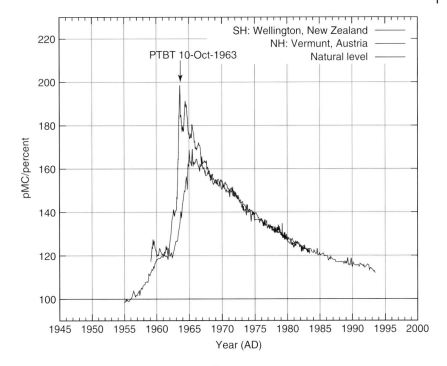

Figure 2.13 Atmospheric concentration of ^{14}C for the northern (upper curve) and southern (lower curve) hemisphere. It sharply rises with the onset of nuclear testing and exponentially decays after the ban of atmospheric nuclear explosions in 1963.

fusion devices (Figure 2.13). Initial consultations aimed at an agreement including nuclear disarmament but, due to a lack of agreement on actual terms, disarmament talks were finally excluded. Owing to different views on how a control mechanism should be implemented, the ban of underground nuclear tests was excluded from the negotiations in the early 1960s to prevent a deadlock.

The subject of underground nuclear testing excluded from the PTBT was the focus of the Treaty on the Limitation of Underground Nuclear Weapon Tests (Threshold Test Ban Treaty, TTBT) [23] signed in 1974 by the USA and the USSR. The TTBT limits underground explosions to a threshold yield of 150 kt. This threshold effectively limited the yield of new nuclear devices as testing was an integral part in developing nuclear explosives. This limitation served two purposes: limiting the first strike capabilities by limiting yield rates and allowing for an effective control mechanism as yield rates above 150 kt can be easily distinguished from natural phenomena like earthquakes. The treaty installed an extensive data exchange agreement including geological data of test sites and technical data on selected nuclear tests to allow for a decent calibration of seismic equipment in order to assess the yield of future nuclear tests. It was the first time the USSR and USA exchanged data on their nuclear programs. An additional protocol of understanding from 1976 allowed for small violations of yield threshold as the error margin in predicting the yield of an untested device can be fairly large.

During a short period during the 1960s and 1970s attempts were undertaken to harness the force of nuclear explosions for civil purposes like large-scale excavations and the exploitation of natural gas fields. To allow these uses and at the same time prevent military gain through them the Peaceful Nuclear Explosions Treaty (PNET) [21] was set into force in 1976. The PNET limits the individual yield as well as the total yield and number and repetition rates of these explosions to allow for reliable verification. Only six productive, non-test explosions ever happened before the programs for peaceful use of nuclear explosions were abandoned by both superpowers. The PNET became obsolete in 1996 with the ban of all nuclear explosions.

The Comprehensive Nuclear-Test-Ban Treaty (CTBT) [21] bans all nuclear explosions in all environments, for military or civilian purposes. The CTBT was adopted by the United Nations (UN) in 1996 and will, once in force, be binding to all member states. The failure to ratify the agreement by several countries, most notably China, the USA, India, and Pakistan, has so far postponed the CTBT.

The Comprehensive Nuclear-Test-Ban Treaty Organization (CTBTO) will be established once the CTBT enters into force. Provisional Commissions and Secretary were established in 1997 in Vienna. Its main purpose is coordination and verification of the CTBT once it is ratified through onsite inspections.

The CTBTO will operate the International Monitoring System (IMS) consisting of 180 seismic monitoring stations, 11 hydro-acoustic stations, 60 infra-sound stations as well as 80 radionuclide air sampling stations. Also part of the IMS is the Global Communications Infrastructure gathering data from all these stations and transmitting it to the CTBTO Data Centre in Vienna. The CTBTO is confident it can provide whole earth coverage with its IMS and to be able to detect nuclear explosions down to lowest yield explosions.

2.5.4
Nuclear-Weapon-Free Zones

To prevent the spread of nuclear weapons and their possible use in proxy wars as well as to ensure the safety and accessibility of their nuclear arsenal the states possessing nuclear capabilities have over the years agreed to several nuclear-weapons-free zones.

A nuclear-weapon-free zone is defined by the UN as:

> any zone recognized as such by the General Assembly of the United Nations, which any group of States, in the free exercises of their sovereignty, has established by virtue of a treaty or convention whereby:
>
> (a) The statute of total absence of nuclear weapons to which the zone shall be subject, including the procedure for the delimitation of the zone, is defined;
>
> (b) An international system of verification and control is established to guarantee compliance with the obligations deriving from that statute.

2.5 Nuclear Weapons | 61

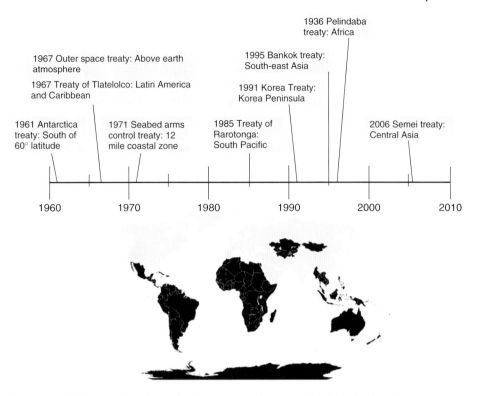

Figure 2.14 Treaties and members of nuclear-weapons-free zones (black) (not shown here: ocean floor and outer space).

These treaties leave the southern hemisphere as well as parts of central Asia free of nuclear weapons (Figure 2.14). Below is a short description of the individual treaties and their scope.

The Antarctic Treaty [23] of 1961 regulates the international relations toward earth's only continent without a native human population. Among other things the Antarctica Treaty forbids any military activity south of the 60°S latitude. Article 5 explicitly forbids any nuclear explosions and the disposal of nuclear waste in Antarctica.

The United Nations Treaty on Principles Governing the Activities of States in the Exploration and Use of Outer Space, including the Moon and Other Celestial Bodies (Outer Space Treaty) has formed the basis of international space law since 1967. To date 99 countries have joined the treaty, with another still in the ratification process. Among the regulations of the Outer Space Treaty is a ban on placing nuclear weapons or any other WMD in orbit of the Earth, installing them on the Moon or any other celestial body, or to otherwise station them in outer space.

The Treaty for the Prohibition of Nuclear Weapons in Latin America and the Caribbean (Treaty of Tlatelolco) [24]: This treaty prohibits the "testing, use, manufacture, production, or acquisition by any means whatsoever of any nuclear weapons" and the "receipt, storage, installation, deployment, and any form of possession of any nuclear weapons" in South America and the Caribbean. It was the first time a large populated area was declared a nuclear-weapons-free zone. Compliance to the Treaty is supervised by the OPANAL (el Organismo para la Proscripción de las Armas Nucleares en la América Latina y el Caribe) based in Mexico City.

The Treaty on the Prohibition of the Emplacement of Nuclear Weapons and other WMD on the Sea-Bed and the Ocean Floor and in the Subsoil Thereof (Seabed Arms Control Treaty) of 1971 bans the placement of nuclear devices and other WMD outside a 12-mile coastal zone. The treaty has been signed by 84 countries. Like the Antarctic and Outer Space treaties it succeeded in preventing an arms race on the ocean ground before it had a chance to start.

Since 1985 the South Pacific Nuclear Free Zone Treaty (Treaty of Rarotonga) bans the use, testing, and possession of nuclear weapons in the South Pacific [25]. It has been signed by most independent countries in the region as well as the major nuclear powers.

In late 1991/early 1992 North and South Korea signed the Agreement on Reconciliation, Non-aggression, Exchanges, and Cooperation and the Joint Declaration on the Denuclearization of the Korean Peninsula. Both treaties together (Korean Treaties) establish a nuclear-free zone banning both military and to some extent civilian use of nuclear technologies. Despite the establishment of a bilateral inspection regime the adherence of some signature parties to the treaty has been questioned in recent years.

The Conference on Disarmament is a multilateral disarmament negotiating forum. Founded in 1960 it is an international forum used mainly by the United Nations and its security council to consider specific disarmament matters. Once a certain level of agreement has been reached, the matter at hand is usually forwarded to a specifically spawned body for final negotiations and implementation.

The Southeast Asian Nuclear-Weapon-Free Zone Treaty (SEANWFZ, Bangkok Treaty) of 1995 between ten members of the Association of Southeast Asian Nations (ASEAN) implements a ban on development, manufacturing, and possession of nuclear weapons in South-East Asia.

The African Nuclear Weapon Free Zone Treaty (Treaty of Pelindaba) establishes a Nuclear-Weapon-Free Zone in Africa. It was signed in 1996 and came into effect in 2009. In addition to a ban on research, development, manufacture, stockpiling, acquisition, testing, possession, control, or stationing of nuclear explosive it also forbids the disposal of nuclear waste in Africa. Compliance is verified by the IAEA and the African Commission on Nuclear Energy.

The Central Asian Nuclear-Weapon-Free Zone (CANWFZ) treaty (The Treaty of Semei) between Kazakhstan, Kyrgyzstan, Tajikistan, Turkmenistan, and Uzbekistan forbids the possession of nuclear weapons. The Treaty of Semei was signed against the objections of the United States, United Kingdom, and France vocalized in the United Nations (UN) general assembly. These countries feared the restriction on the transit of nuclear weapons through the territory of the treaty.

2.6 Organizations

Most multilateral efforts in controlling CBRN weaponry (and arms-control as a whole) were and still are undertaken at the United Nations. Over the years many organizations were established (both within and outside the UN) that deal with different regional or technical aspects. Here we introduce the most influential and widely known organizations in the field of CBRN regulation that are not specific to a single treaty.

Replacing the League of Nations in 1945, the United Nations Organization (Figure 2.15a) is the most prominent international organization today. With currently 192 member nations it is supported by virtually every sovereign state in the world. The general assembly provides a platform for dialogue between the member states and other participating entities (e.g., the Holy See).

The United Nations (Figure 2.15a) represents manifold administrative bodies and agencies with the Security Council being the most prominent, serving as an institution to deal with imminent international crisis. Other well-known bodies and agencies are the World Health Organization (WHO) or the United Nations Children's Fund (UNICEF). Many efforts in arms reduction and control are negotiated and finalized under the auspices of the United Nations.

Figure 2.15 Organizations in the field of CBRN regulation that are not specific to a single treaty: (a) United Nations Headquarters, New York. (Image public domain, National Park Service.) (b) International Atomic Energy Agency (IAEA) Headquarters in Vienna. (Image public domain.) (c) OSCE Building in Vienna.

The IAEA [21] (Figure 2.15b), founded in 1957 by the IAEA Statute, is an international organization within the United Nations. Its main purpose is to promote the peaceful use of nuclear energy and to inhibit its use for military purposes. In response to the Chernobyl disaster in 1986 the scope of the IAEA was extended to cover nuclear safety efforts. The ruling entity of the IAEA is the board of Governors elected by the General Conference made up of all 150 member countries. In recent years the IAEA has attracted public attention through the work of its nuclear weapons inspection teams.

OSCE (Organization for Security and Co-operation in Europe) (Figure 2.15c) is an *ad hoc* organization under the United Nations. It was founded during the cold war in an attempt to loosen the grips of NATO and the Warsaw Pact on European countries and to serve as a forum for dialogue. It has played a vital role in many negotiations concerning disarmament and arms control in Europe as well as in the erosion of communist governments in Eastern Europe.

2.7
Conclusions and Where Does the Road Lead?

Over the last few decades, the world has changed. Long gone are the times of two dominant power blocks tightly controlling all aspects of the armament industry. On one hand, the situation today is marked by isolated military conflicts and solitary states opposing the community of states. On the other hand, non-state sponsored organizations try to gain international attention and influence by means of terrorist activities. These developments make further international regulations a challenging task as offenders are not easy to grasp with traditional treaties. The progress made in technology, communications networks, and globalized trade promotes a wide availability of knowledge and equipment for possible infringements.

But it is far from being a lost cause. There are encouraging examples like the recognized success of the CWC. It needs the enduring effort of the community of states as well as private organizations and industry to ensure a safe and peaceful future for our planet.

References

1. Shaw, M.N. (2008) *International Law*, 6th edn, Cambridge University Press. ISBN: 978-0521728140.
2. United Nations (1969) Vienna Convention on the Law of Treaties International Law Commission, United Nations Organization.
3. Aust, A. (2007) *Modern Treaty Law and Practice*, Cambridge University Press, Cambridge. ISBN: 978-0521678063.
4. Steinbruner, J.D. (2000) *Principles of Global Security*, Brookings Press. ISBN: 978-0815780953.
5. Burns, R.D. (2009) *The Evolution of Arms Control*, Praeger Security International. ISBN: 978-0313375743.
6. Dekker, G. (2001) *The Law of Arms Control, International Supervision and Enforcement*, Developments in International Law, vol. 41, Kluwer Law International. ISBN: 978-9041116246.
7. Clawson, R.W. (1986) *East-West Rivalry and the Third World: Security Issues and Regional Perspectives*, Scholarly Resources, Inc., Wilmington, DE, pp. 133–148.

8. Byrd, P. (2010) Cold War, *The Concise Oxford Dictionary of Politics* 3rd edn (ed. I. McLean and A. McMillan), Oxford University Press. ISBN: 978-0199205165.
9. Cressey, D. (2007) *Nature*, **448** (7155), 732–733.
10. Kelle, A., Nixdorff, K., and Dando, M. (2008) *A Paradigm Shift in the CBW Proliferation Problem: Devising Effective Restraint on the Evolving Biochemical Threat*, Deutsche Stiftung Friedensforschung.
11. Hudson M.O. (1931) *Am. J. Int. Law*, **25** (1), 114–117.
12. Tucker, J. (2007) *War of Nerves: Chemical Warfare from World War I to Al-Qaeda*, Anchor Books. ISBN: 978-1400032334.
13. Bartels, R. (2009) *Int. Rev. Red Cross*, **91** (873), 1–33.
14. Browen, W.Q. (2001) *Int. Aff.*, **77** (3), 485–507.
15. Bothe, M., Rosas, A., and Ronzitti, N. (eds) (1998) *The New Chemical Weapons Convention – Implementation and Prospects: Implementation and Prospectus*, Brill Academic Publishers. ISBN: 978-9041110992.
16. Dunay, P., Krasznai, M., and Spitzer, H. (2004) *Open Skies: A Cooperative Approach to Military Transparency and Confidence Building*, United Nations. ISBN-978-9290451648.
17. Knight, D.J. (2006) *EU Regulation of Chemicals: REACH*, Smithers Rapra Press. ISBN: 978-1859575161.
18. Glasstone, S. and Dolan P.J. (1977) *The Effects of Nuclear Weapons*, 3rd edn, U.S. Government Printing Office.
19. Moody, K.J. and Hutcheon, I.D. (2005) *Nuclear Forensic Analysis*, CRC Press. ISBN: 978-0849315138.
20. Wikipedia Gun-type fission weapon *http://en.wikipedia.org/w/index.php?title=Gun-type_fission_weapon&oldid=490051464* (accessed on 10.05.2012).
21. Hildreth, S.A. and Wolf, A.F. (2010) Ballistic Missile Defense and and Offensive Arms Reduction: A Review of the Historical Record, Congressional Research Service. Available at *http://www.fas.org/sgp/crs/nuke/R41251.pdf*.
22. SIPRI (2010) *SIPRI Yearbook 2010 – Armaments, Disarmament and International Security*, Oxford University Press Inc., New York. ISBN: 978-0-19-958112-2.
23. Dahlman, O., Mykkeltveit, S., and Haak, H. (2009) *Nuclear Test Ban: Converting Political Visions to Reality*, Springer. ISBN: 978-1402068836.
24. Robinson, P. (2008) *Dictionary of International Security*, Polity Press. ISBN-10: 0-7456-4028-1.
25. Pitt, D. and Thompson, G. (1988) *Nuclear Free Zones*, Routledge. ISBN: 978-0709940760.

Part II
CBRN Characteristics – Is There Something Inimitable?

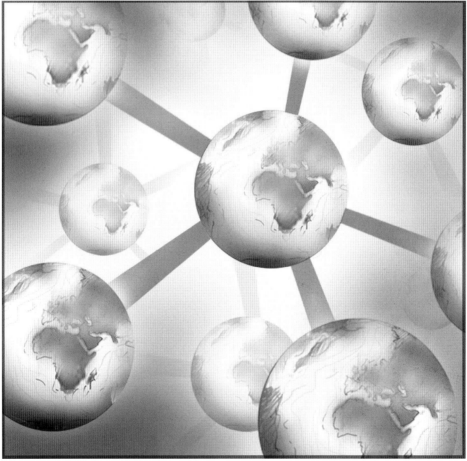

copyright by Jörg Pippirs, http://www.artesartwork.de

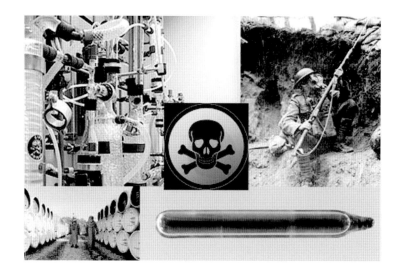

3
Chemical Agents – Small Molecules with Deadly Properties
Hans-Jürgen Altmann, Silke Oelze, and Bernd Niemeyer

Over the centuries, toxic substances have been used in assaults to poison enemies and to gain advantage without the fear of retaliation. With breakthrough in natural science the utilization of toxic substances in open battleground reached a first climax during World War I. Today, thousands of poisonous substances are known but only a few are considered to be suitable for chemical warfare. About 70 different chemicals have been stockpiled or even applied as chemical warfare agents during the twentieth century. Chosen for different tactic or strategic reasons, research into toxic compounds reached a peak during the cold war. It seems obvious that the same general principles are responsible for the mode of action of these chemicals in the body.

3.1
Are Special Properties Required for Chemical Warfare Agents?

Our understanding of chemical compounds that can be employed as chemical warfare agents is vital for the ability of military forces and first responders to cope with the possible threat due to the usage of these compounds by enemy forces or terrorists. Thus, this chapter gives the reader basic information about the classification, different types, their mechanisms of action, and hints to additional literature.

One of the most important questions we have to answer properly is: What are the chemical, physical, and biological properties of a chemical substance that lead to it being considered as a possible chemical warfare agent? What we can state is that chemicals have to fulfill several characteristics to make them considered as chemical warfare agents. Important properties are summarized and discussed. In Figure 3.1 important aspects for potential chemical warfare agents are summarized. In Table 3.1 we discuss these properties.

To give an impression of the toxicity of different chemical warfare agents we should be aware that only a small single droplet of VX (ethyl ({2-[bis(propan-2-yl)amino]ethyl}sulfanyl)(methyl)phosphinate) is enough for devastating effects on a single person. The necessary amount of different nerve agents and sulfur lost for a lethality via skin contact is visualized in Figure 3.2 The toxicity levels for various chemicals warfare agents for inhalation versus skin contact are given in Table 3.2.

CBRN Protection: Managing the Threat of Chemical, Biological, Radioactive and Nuclear Weapons,
First Edition. Edited by A. Richardt, B. Hülseweh, B. Niemeyer, and F. Sabath.
© 2013 Wiley-VCH Verlag GmbH & Co. KGaA. Published 2013 by Wiley-VCH Verlag GmbH & Co. KGaA.

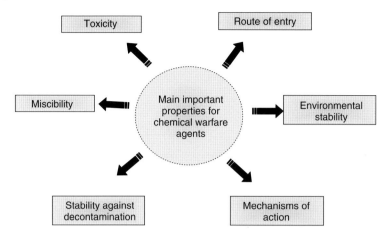

Figure 3.1 Main important issues of chemicals considered as chemical warfare agents.

Table 3.1 Main important properties for potential chemical warfare agents.

Property	Description
Toxicity	Substance must be able to act rapidly or delayed in small concentrations, producing damage or lethal effects on man and animals. Its inherent properties should be tailorable to maximum effects on personnel
Route of entry	For humans, a compound should have different routes of entry, such as skin penetration, assimilation on moist tissues, and absorption through the gastrointestinal tract
Mechanism of action	Effects via different organs and on various biochemical systems. For example, failure of organs and interference with special enzymes
Stability against decontamination	Matter should be resistant against common decontamination agents and procedures to complicate decontamination for the parties affected
Stability	Chemicals must be stable or capable of being stabilized by additives during the period between production and use. It should also be stable under the environmental conditions of application
Miscibility	Chemicals should be miscible with solvents and/or thickeners [e.g., poly(methyl methacrylate)s, chlorinated rubbers, waxes, etc.] as well as other compounds present in the ammunition

Substances used as chemical warfare agents normally are liquids or solids. However, some of the substances are fairly volatile and so the gas concentration after dissemination may become poisonous very rapidly due to high toxicity and high vapor pressures. Both solid substances and liquids can also be dispersed into the air as aerosols, which are small particles or droplets sprayed into the air but which behave like gases and can penetrate the body through the respiratory

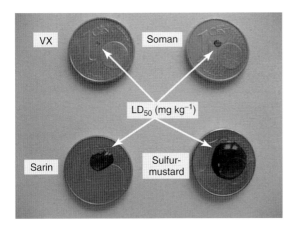

Figure 3.2 Comparison of the amount of indicated chemical warfare agent, applying skin contact effecting an LD_{50}-toxicity.

Table 3.2 Toxic levels for humans of several toxic compounds during inhalation and skin contact, projected from animal experiments in the early twentieth century (various sources). The values can differ due to the different experimental conditions.

	Inhalation, LCt_{50} (mg min^{-1} m^{-3})	Skin contact, LD_{50} (mg kg^{-1})
VX	50	0.14
Soman	70	0.7
Sarin	100	24
Sulfur mustard	1500	100
Phosgene	3000	–
HCN	11000	–

tract just like a gas. Therefore, the physical conditions, defining the compound's aggregate state at different seasons have to be considered in the discussion about the utilization of substances as possible chemical warfare agents. An illustration of the degree of vapourization of a solid/liquid agent in the four different seasons of the moderate climate zones is presented in the upper route of Figure 3.3. In the lower sketch the intensity of the darkness of the background colour reflects an improved gasification. Additionally the lengths of the radiating lines off the gridded rectangle, which stands for the hazardous substance also visualize the degree of a substance in the gas phase.

3.2
How can we Classify Chemical Warfare Agents?

Chemicals can be sorted in different ways. To classify chemical warfare agents we will use the most widespread criteria:

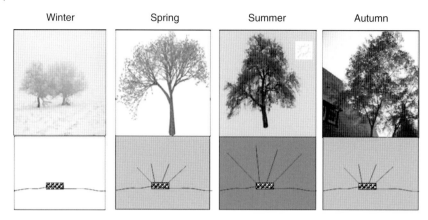

Figure 3.3 Aggregate states of chemical warfare agents as well as of other chemicals; the length of the radiating lines as well as the darkness of the gray colored background in the bottom sketches represent the intensity of gasification of substances transferred into the gas phase, depending on the season. The top line of pictures depicts the corresponding season (winter–spring–summer–autumn).

- A physicochemical behavior (natural scientific points),
- B route of entry into the body (protective measures),
- C organs to be affected (personal concern),
- D physiological effects on human (medical orientated),
- E identification according to the NATO Code (military background and disposal).

warfare agents according the above presented characterstics A to E. This classification automatically gives the interested reader hints for effective protective measures against the threat of chemical warfare agents.

3.2.1
A: Physicochemical Behavior

This criterion is based on physicochemical data of the relevant substances. The compounds can be divided into:

- volatile substances that mainly contaminate the air as they vaporize rapidly, especially when the temperature rises (short-term effect);
- persistent substances with low volatilities, which therefore mainly contaminate surfaces with a long-term effect.

> - **Volatility**: is the tendency of a solid or liquid substance to evaporate.

The volatility of a substance is directly related to its vapor pressure. In turn, the vapor pressure depends on the temperature. Increasing of the temperature

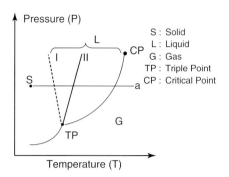

Figure 3.4 Schematic representation of a phase (pressure–temperature) diagram of a pure substance.

(e.g., as the sun shines or the change from winter to summer time or other geographical conditions) increases the vapor pressure and consequently heightens the volatility. In this way a compound that formerly was characterized as "low volatile" generates a higher vapor pressure, and consequently may contaminate and effect in another way. A long-term effective substance may change into medium or short-term effectiveness with rising temperature, due to different environmental conditions. Another parameter affecting volatility is the integration of additives. The (common) application of thickeners reduces the vapor pressure.

General information of the degree of volatility can be found from the phase diagram. In a temperature–pressure diagram (Figure 3.4), on the ordinate the pressure of a system containing a pure substance is indicated against the relevant temperature given at the abscissa.

The following areas or selected indicators are marked:

- **S Solid state of substances**: At sufficiently low temperature all compounds are solid (**S**), at medium and elevated pressure. On increasing the temperature we pass the melting curve ("solidus" line, indicated as I or II in Figure 3.4) into the liquid region **L** (follow line a leftward).
- **L Liquid phase**: Liquid phase L is present at characteristic temperatures and pressures for a given chemical. For most pure compounds the solidus line behaves like line II. Exceptions (line I) are detected for, for example, water, where we find a negatively inclined solidus line. This is connected with the phrase "anomaly of water." Most other substances show a descending of solid matter in its liquid phase.
- **G Gaseous phase**: Increasing the temperature still further (along line a) we reach the bubble point, which is situated at the bubble point line, reflecting the equilibrium of the gaseous and liquid phase. Going on further we enter the gaseous region **G**. This bubble point line starts from the triple point (TP) and ends at the critical point (CP). A direct phase transfer of solid substances into the gas phase can be realized by sublimation (line to left of and below the TP). Impressive examples are mothballs and also the diminishing of snow in winter times during longer periods of sunny days and temperatures constantly below 0 °C. Each phase transfer is related to an energy uptake (in the direction from left to right along line a) and vice versa.

- **TP**: Under this exceptional condition [e.g., for water $T_{TP} = 273.16$ K (0.01 °C) and $p_{TP} = 611.66$ Pa (around 6 mbar)] all three phases **S, L, G** are present at once. Under no other condition all three phases can be detected.
- **CP**: A compound behaves in a special way under the conditions above the critical point. For water these data are $T_{CP} = 647.1$ K and $p_{CP} = 22.1$ MPa (221 bar) [1–3].

3.2.2
B: Route of Entry into the Body

Naturally, the routes into the body are most important in terms of protection measures to be applied. Consequently, the technical development of protective material, activities to be taken during exposure, and, finally, the decontamination are often guided by them. The most important entry points of chemical warfare agents into the human body are the eyes, nose, mouth, and skin (Figure 3.5). They differ according to the amount and the mechanism of uptake of the biological, radiological, and nuclear warfare agents.

Entry via the eyes and mouth is preferred for gaseous and aerosol matter. Aerosols consist of small particulates such as small particles of liquid or solids. Additionally contaminated food, including beverages like water, infiltrates via the mouth into the gastrointestinal tract [4]. The entry of aerosols and gases through the nose is similar, as described in Section 4.4 – via breathing and inhalation. The skin can be affected by all of the mentioned aggregate states of chemical agents. Preferentially, poorly volatile chemicals, like thickened, sedentary chemical agents with low vapor pressure, such as S-Lost, bear on humans via the skin, the body's largest organ. Another mechanism of permeation is by direct contact of these compounds with unprotected parts of the body, enabling penetration and spreading of the hazardous effect.

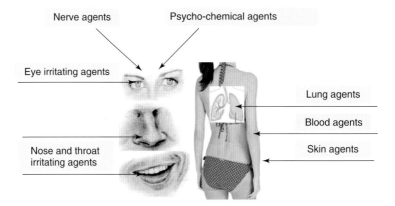

Figure 3.5 Routes of entry of chemical warfare agents into the human body and the organs affected.

3.2.3
C: Organs to be Affected

Several organs are affected by chemical warfare agents, which are intended to impact on the soldier in combat and civilian personal in terroristic attacks [5]. Besides these compounds a much greater number of chemicals exist that are applied in industry and which sometimes also exhibit a hazardous influence on health. They are summarized as toxic industrial chemicals (TIC). In general, a TIC often has a "lower impact" on human health than chemical warfare agents. Occasionally, TIC substances are simultaneously used as chemical warfare agents, such as phosgene ($COCl_2$), chlorine (Cl_2), hydrogen cyanide (HCN), and others. We focus here on chemical warfare agents, while considering TIC to be also dangerous in the same scenarios.

The two main differences between these compounds are the possible lethal and non-lethal effects, as seen from Figure 3.6, which illustrates major classes of chemical warfare agents.

The major level of discrimination is the various organs to be affected, and the effects of these substances, which can be ascertained in the third line of Figure 3.6. Direct contact with "vesicants" leads to blisters on the skin's surface [6]. "Blood agents" directly attack the ability of blood to transport oxygen (O_2) into, and carbon dioxide (CO_2) out of, cells [7]. "Lachrymators" are materials, that irritate the eyes to tears and finally to blindness [8]. Substances leading to vomiting are called "*sternutators*," which consequently affect the intestinal tract [9]. A lethal influence on the nervous system is conducted by "nerve agents" [e.g., tabun (GA), sarin (GB), soman (GD), cyclosarin (GF), and VX] [7]. They are the most effective substances

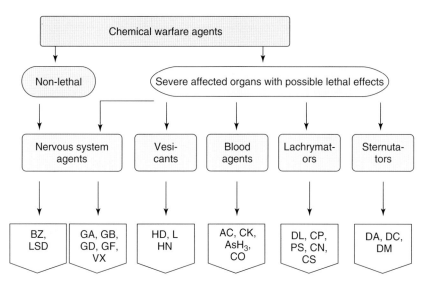

Figure 3.6 Division of chemical warfare agents into their effects on different human organs.

Table 3.3 NATO code for the most common, "pure" chemical warfare agents.

Common name(s)	NATO code
Mustard gas, yperite, sulfur mustard, S-lost, bis(2-chloroethyl) sulfide	HD
Sesquimustard, sesquiyperite	Q
Oxygen mustard, Oxol-lost, bis(2-chloroethylthioethyl) ether	T
Nitrogen mustard, bis(2-chloroethyl)ethylamine, NH-lost	HN1
Nitrogen mustard, bis(2-chloroethyl)methylamine, N-lost	HN2
Nitrogen mustard, tris(2-chloroethyl)amine, TBA	HN3
Phenyldichloroarsine	PD
Ethyldichloroarsine	ED
Lewisite	L
Hydrogen cyanide, prussic acid, HCN	AC
Cyanogen chloride	CK
Carbon monoxide	CO
ω-Chloroacetophenone, "tear gas"	CN
o-Chlorobenzylidene-malononitrile	CS
Adamsite	DM
Diphenylchloroarsine, Clark I	DA
Diphenylcyanoarsine, Clark II	DC
Tabun, O-ethyl-N,N-dimethylphosphoramide cyanide	GA
Sarin, O-isopropyl(methyl)phosphono fluoride	GB
Soman, O-pinacolyl(methyl)phosphonyl fluoride	GD
VX, ethyl ({2-[bis(propan-2-yl)amino]ethyl}sulfanyl)(methyl)phosphinate	VX
Chlorine	CL
Phosgene	PG
Diphosgene, trichloromethyl chloroformate	DP
Chloropicrine	PS
3-Quinuclidinyl benzilate	BZ

in relation to the concentration needed for effectuation on humans. "Non-lethal" compounds may exhibit psychogenic effects, which negatively influence a soldier's ability for combat. Adequate low concentrations even of "lethal" substances result in major physiological damage, although the patient may survive.

The last line of Figure 3.6 lists relevant compounds of the chemical classes according to the so-called "NATO code" (see Table 3.3). The different classes of substances as well as selected chemicals are described according this classification in Section 3.3.

3.2.4
D: Physiological Effects on Humans

Two main groups are defined, namely, those substances that are lethal and those that are incapacitating (non-lethal). This kind of classification is relevant for analysis and medication after exposure. A profound description of medical

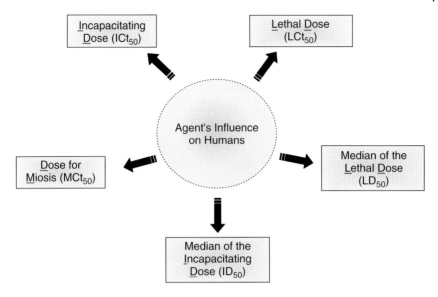

Figure 3.7 Data that indicate an agent's influence on humans.

diagnoses and medication of various chemical warfare agents is necessary for medical countermeasures [10]. The detrimental effects of chemical substances in general are characterized by the toxicity of the compound. Some characteristic data provide information on the agent's effect on humans (Figure 3.7).

> ICt_{50} gives the median *incapacitating dose* (*ct* product) of a chemical agent, agent vapor, or aerosol that affects 50% of applied persons.
> LCt_{50} represents the median *lethal dose* of the chemical warfare agent (as agent vapor or aerosol) for which 50% of contacted persons die.
> LD_{50} is the median of the *lethal dose* of chemical warfare agents.
> MCt_{50} stands for the median *dose for miosis*.
> ID_{50} is the median of the *incapacitating dose* for 50% of the affected population.

Vapor toxicity is expressed as LCt_{50}, ICt_{50}, and MCt_{50} and reported in terms of mg min m^{-3}, whereas the LD_{50} represents toxicity via percutaneous absorption, and it is reported in terms of milligrams. The terms LCt_{50}, ICt_{50}, and MCt_{50} are the lethal concentration × time, incapacitating concentration × time, and miosis concentration × time, respectively, for 50% of the population. LD_{50} represents the dose that causes lethality to 50% of the population. All values except MCt_{50} values are related to a "standard man" with a body weight of 75 kg and in sound condition. For the different chemical warfare agents these data will be reported separately.

3.2.5
E: Identification According to the NATO Code

Most chemical warfare agents are assigned a "NATO weapon designation" in addition to, or in place of, a common name (Table 3.3).

Another important point we have to address is the possibility of mixing pure chemical warfare agents with additives. The behavior and effects of "pure" substances are significantly altered by additives. This can lead to an increase in viscosity, which is adjusted to the desired grade. The resultant increased persistency and ability to stick to surfaces can complicate decontamination tremendously.

3.3
Properties of Chemical Warfare Agents

We will discuss the different agents of the above ordered groups. The structure of the presentation will be:

- general comments on the structural characteristic of the compounds in the group, chemical nomination on, and their physicochemical properties;
- toxic action of these substances, recognition of being affected, and the effect of these compounds in the human body.

3.3.1
Blister Agents (Vesicants)

This group consists of two different subgroups: the "alkylating agents" and the "arsenicals," which will be discussed separately. We start with the alkylating agents [4, 10].

The alkylating agents are divided into three different sub-classes, the sulfur, nitrogen, and oxygen Losts. The close chemical relation of these substances can be seen from the general structure of the nitrogen Losts, where the ligand R1 varies from an amino-ethyl, -methyl, and -ethyl-chloride group (Figure 3.8).

All of these agents consist of two or more chloro-alkyl groups, which provide, together with the more central positioned sulfur, nitrogen, and oxygen atom, respectively, chemically reactive leaving groups that exhibit, finally, their damaging impact (Figure 3.9).

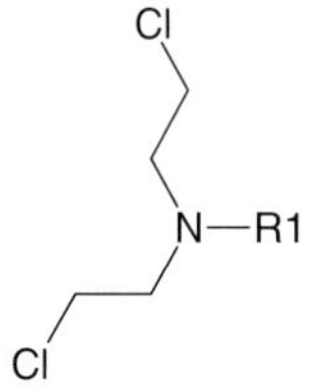

Figure 3.8 Generalized structure of nitrogen mustard, R1 = C_2H_5 (HN1), CH_3 (HN2), and C_2H_4Cl (HN3).

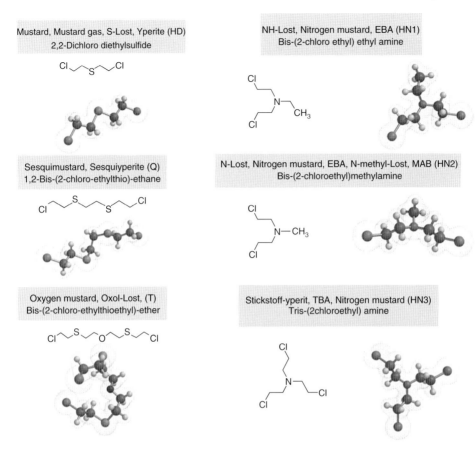

Figure 3.9 Nomination and molecular geometries of blister agents of the alkylating agents. The molecular geometries have been optimized based on molecular mechanics force field (MMFF) and DFT (density functional theory) calculations (hybrid density functional B3LYP/6-31G*, RMS gradient of 0.001), using the molecular modeling program Spartan '04 (Wavefunction, Inc. Irvine, CA). The molecules are represented in a ball-stick model enclosed by the calculated electron density (IsoVal = 0.03) [11].

The most well-known example of a blistering agent is 2,2-dichlorodiethyl sulfide, also-called *(sulfur) mustard*, or *mustard gas*, abbreviated with HD the NATO code for the technical product. It is a colorless and odorless liquid. Owing to decomposition products, the technical product smells like mustard or garlic. It is a strong vesicant. It is often thickened, which finally results in a pale yellow solid material. Additional information on structure, chemical reactions, and decontamination of mustard can be drawn from the literature [4, 5, 10]. A close relative of sulfur mustard is sesquimustard (Q). HN1, HN2, and HN3 are all called *nitrogen mustard*, although they differ in chemical structure (Figure 3.8) and, consequently, in their physicochemical as well as in their toxicological behavior.

Color is an important physicochemical property for the recognition of substances if suspicious substances are found and to be disposed. Mustards range from colorless to pale yellow (mustard gas) as well as to dark yellow for nitrogen mustard. They can be either odorless or smell like garlic or mustard (sulfur mustard) or fish (nitrogen mustard). Both types are persistent, even in soil and sea water (sulfur mustard in its thickened form). Nitrogen mustard vaporized into the air degrades to some degree within hours. As seen from the relative vapor density (in relation to air density) all vapors of blister agents are heavier than air and thus fill rooms from the bottom upwards.

Sulfur, oxygen, and HN3 mustard agents have low vapor pressures (Tables 3.4 and 3.5), based on their high boiling points. To increase the tendency of these substances to remain in one place, and in this way their long-term damaging effect, the boiling point is increased, resulting in a reduction of vaporization, by mixing with thickeners or with other chemical warfare agents providing this effect. Mixtures of H and Q are known as *HQ-mixtures*. Thus their intended impact is aimed more at the direct action of solid or liquid substances on the body than via the gaseous phase. We can see from Tables 3.4 and 3.5 that their solubility in water is generally low, and that they are liquid under standard conditions (20 °C and 1 bar air pressure).

Alkylating blister agents (e.g., mustard gas) cause little or no pain at the time tissue is exposed. This clinical symptom-free period of several hours may be not recognized by the patient and potentially spread the agent on the skin and elsewhere. As their name implies, these substances cause the formation of painful blisters and large bullae on the skin up to 24 h after exposure dependent on the penetrated dose. These blisters hurt, weep, coalesce, and always irritate, thus reducing the combatant's awareness.

Systemic effects after extensive exposure include bone marrow depression with a drop in the white blood cell count and gastrointestinal tract damage. Blister agents

Table 3.4 Physicochemical characterization of blister agents of the S- and O-mustards.

Property	HD (distilled mustard)	T (oxygen-mustard)	Q (sesqui-mustard)
Formula	$(C_2H_4Cl)_2S$	$(ClC_2H_4SC_2H_4)_2O$	$(ClC_2H_4S)_2C_2H_4$
Molecular weight (g mol^{-1})	159.1	263.3	219.2
Relative vapor density[a]	5.4	9.08 (calculated)	Not available
Liquid density (g cm^{-3})	1.27 (25 °C)	1.231 (20 °C)	1.229
Boiling point (°C)	217	174 (267 Pa) (decomp. above 170)	328.7
Melting point (°C)	14	10	–
Vapor pressure (Pa)	0.07 (20 °C)	2.9×10^{-5} (25 °C)	0.048 (25 °C)
Volatility (mg m^{-3})	610 (20 °C)	2.4 (25 °C)	0.4 (25 °C)
Solubility in water (%)	0.06–0.08 (20 °C)	Practically insoluble	0.03 (20 °C)

[a] Relative to air.

Table 3.5 Physicochemical characterization of blister agents of the nitrogen mustards.

Property	HN1 (nitrogen-mustard)	HN2 (nitrogen-mustard)	HN3 (nitrogen-mustard)
Formula	$C_6H_{13}Cl_2N$	$C_5H_{11}Cl_2N$	$C_6H_{12}Cl_3N$
Molecular weight (g mol^{-1})	170.1	156.1	204.5
Relative vapor density[a]	5.9	5.4	7.1
Liquid density (g cm^{-3})	1.09 (25 °C)	1.15 (20 °C)	1.24 (25 °C)
Boiling point (°C)	194 (1013 hPa), (decomp. above 120 °C)	75 (1333 Pa)	230–235 (1013 hPa) (decomp.; dimerization above 50 °C)
Melting point (°C)	−34	−65 to −60	–
Vapor pressure (Pa)	21.3 (20 °C)	38.7 (20 °C)	−4
Volatility (mg m^{-3})	1520 (20 °C)	2500–2600	70–76
Solubility in water (%)	Sparingly soluble	1.2 (20 °C)	0.008–0.016 (20 °C)

[a] Relative to air.

of the mustard group act as alkylating agents that covalently bind to DNA and consequently damage our basic biological information source, as the transcription mechanism of the DNA is disrupted. Their action on cell components results in inhibition of cellular division (mitosis) with decreased tissue respiration, finally leading to cell death. For this purpose some of the substances have found uses as chemotherapeutic agents in the treatment of cancer.

Owing to the immunosuppressive effects of mustards the risk of superinfection is significantly enhanced. Wound healing is dramatically impaired because of DNA damaged stem cells and altered protein synthesis. Thus, hospitalization of wounded patients lasts weeks to months depending on the burnt surface. Beside the skin, other areas of the body surface may also be affected. Eye, throat, lung tissue, and mucous membranes can be damaged. Pseudo-membrane formation in the bronchial tract may be life threatening due to airway obstruction. Alkylating blister agents (nitrogen and sulfur mustards) are carcinogenic and mutagenic and can alter membrane proteins, leading to altered signal transduction within the tissue as well as metabolic dysfunction as demonstrated in Figure 3.10 [6].

The median incapacitating dose (ICt$_{50}$) ranges from 50 mg min m^{-3} for oxygen mustard and sesquimustard to up to 1500 mg min m^{-3} for sulfur mustard (Table 3.6).

NH-mustard (HN1) is less persistent and has similar blistering potency than sulfur mustard. HN3 is more persistent and is equal to sulfur mustard in its blistering potential. Its properties and toxicity are also similar to the other mentioned agents (Table 3.7). Eye irritation and skin erythema occur sooner than with sulfur mustard upon exposure to nitrogen mustard, and eye lesions appear to be more severe. The substances are known carcinogens and can cause cancer even years after exposure.

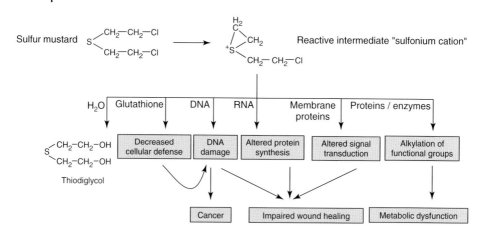

Figure 3.10 Effects of sulfur mustard on DNA, RNA, and proteins.

Table 3.6 Toxicity data of blister agents of the sulfur mustard and sesquimustard.

Variable	Sulfur mustard		Sesquimustard	
	Value (mg min m^{-3})	Application	Value (mg min m^{-3})	Application
LD$_{50}$	4000–7000 mg	Liquid on skin	200 mg	Liquid aerosol on skin
LCt$_{50}$	750–1500	Vapor	200	Vapor
ICt$_5$	50	Vapor	–	–
ICt$_{50}$	100–1500	Vapor	50	Vapor

Table 3.7 Toxicity data of blister agents: Nitrogen mustard.

Variable	Value (mg min m^{-3})	Substance	Application
ICt$_{50}$	200	HN1	Vapor
LCt$_{50}$	1500	HN1	Vapor
ICt$_{50}$	100	HN2	Vapor
LCt$_{50}$	3000	HN2	Vapor
ICt$_{50}$	200	HN3	Vapor
LCt$_{50}$	1500	HN3	Vapor
LD$_{50}$	7000–14 000 mg	HN3	Liquid on skin

The third important mustard variation is oxygen mustard, which is approximately three to four times more toxic to the skin than sulfur mustard and is used in combination with sulfur mustard as HT-mixtures. Oxygen mustard is an oily colorless liquid. Some sources state a garlic-like odor [8].

3.3.2
Arsenicals

Organic substances containing arsenic derivatives from arsenic(III) chloride, which itself is toxic to humans and very toxic to aqueous organisms with a long-term effect. This class of chemicals can be subdivided into arsenics and lewisites warfare agents – both are chloroarsines. We begin with the arsenics.

Phenyldichloroarsine (PD) and ethyldichloroarsine (ED) are the actual warfare agents of this subgroup. The older matter, which has been overcome at present, is methyldichloroarsine (MD). Figure 3.11 summarizes their names and structural properties.

The lewisite group consists of lewisite types I–III (called *L1 to L3* in the NATO code), which differ in the degree of substitution of chloro groups of arsenic(III)

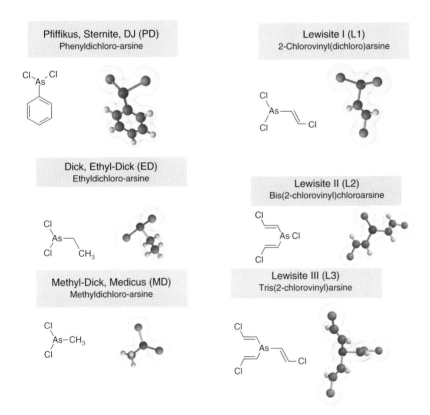

Figure 3.11 Nomination and molecular geometries of blister agents (arsenic and lewisite group). The molecular geometries have been optimized based on MMFF and DFT calculations (hybrid density functional B3LYP/6-31G*, RMS gradient of 0.001), using the molecular modeling program Spartan '04 (Wavefunction, Inc. Irvine, CA). The molecules are represented in a ball-stick model enclosed by the calculated electron density (IsoVal = 0.03) [11].

chloride by chlorovinyl substitutes (Figure 3.11). L1, an organic dichloroarsine, is the most common lewisite, which provides a high volatility and decomposes in aqueous media. Lewisites are colorless, odorless oily liquids when purified, but when produced with impurities or containing decomposition products they have a geranium-like smell.

As liquids, they can easily penetrate rubber, plastics, leather, and most fabrics and remain more dangerous as liquids than as vapor. Liquids with a high vapor pressure generally undergo constant evaporation and thus these substances are non-persistent. They react with water in such a manner that they lose their volatility and their blistering potency due to rapid hydrolysis (Table 3.8). Little is known about their stability in the natural environment. Phenyldichloroarsine, ethyldichloroarsine, and MD can be mixed with sulfur mustard similarly to lewisite–mustard mixtures. The arsenicals act as freezing point depressors for sulfur Lost and cause immediate action when contamination occurs. Mixtures of L and HD are known as *HL-mixtures*.

The toxic properties of arsenicals (Table 3.9) result in similar irritant effects to those from other blister agents. Contact with arsenics as well as lewisites results in effects on the skin (usually the first effect), eyes [with rapid damage from temporary to permanent blindness if immediate (<1 min) decontamination does not occur], and respiratory system, and also the stomach and intestines (main organs affected).

Lewisite acts first as a vesicant (since first contact is usually through dermal exposure). Within 5 min of dermal exposure a gray area of dead epithelium may develop, similar to that seen with corrosive burns. Skin penetration can result from contact with either liquid or vapor. Lewisite consecutively behaves as pulmonary irritant and finally as a systemic poison. It can cause violent sneezing and severe pain. It acts stepwise and leads to diarrhea, restlessness, hypothermia, weakness, acute lung injury, and respiratory distress syndrome, as well as hypotension, after severe inhalation, which finally may lead to shock and death. It differs in its clinical impact, which may even appear within seconds of exposure. Very small amounts (0.5 mL) can lead to systemic effects; as little as 2 mL can be

Table 3.8 Physicochemical characterization of the lewisite agent L1.

Property	L1 (lewisite)
Formula	$C_2H_2AsCl_3$
Molecular weight (g mol^{-1})	207.3
Relative vapor density[a]	7.1
Liquid density (g cm^{-3})	1.89 (20 °C)
Boiling point (°C)	190
Melting point (°C)	−18
Vapor pressure (Pa)	0.39 (20 °C)
Volatility (mg m^{-3})	4480 (20 °C)
Solubility in water (%)	0.05 decomposes rapidly

[a] Relative to air.

Table 3.9 Toxicological data of blister agents (arsenic and lewisite groups).

Variable	Value (mg min m^{-3})	Substance	Application
ICt_{50}	5–10	ED	Vapor
LCt_{50}	3000–5000	ED	Vapor
ICt_{50}	16–25	MD	Vapor
LCt_{50}	3000–5000	MD	Vapor
ICt_{50}	16–20	PD	Vapor
LCt_{50}	1300–2600	PD	Vapor
LD_{50}	7000 mg	L1	Liquid on skin
ICt_{50}	200–300	L1	Vapor
LCt_{50}	1200–1500	L1	Vapor
LD_{50}	1500 mg	L2	Liquid on skin
ICt_{50}	–	L2	–
LCt_{50}	1300	L2	Vapor
LD_{50}	–	L3	Lewisite III is expected to cause the typical symptoms of arsenic poisoning, such as damage to skin, nervous system, kidneys, and liver. Only a few data sets of toxicological data are available
ICt_{50}	–	L3	
LCt_{50}	–	L3	

fatal (Table 3.8, providing exemplarily data of Lewisite I). As few data on human exposure are available, most information on its clinical effects is based on animal studies.

Lewisites react with the thiol groups of proteins and thus interact with enzymes and finally interfere with the metabolism. Similar to Losts they also react with DNA by alkylation (Figure 3.10).

Recognition of being affected is through symptoms like itching and irritation, which may occur for about 24 h, with or without skin blistering. Dermal contact results in immediate stinging and burning sensations and erythema within 30 min. Formations of sharply defined and painful blisters, which may occur 2–13 h after exposure, with deep necrosis can be identified. In addition, a geranium-like odor may be detectable on the victim.

3.3.3
Blood Agents

These chemicals act in the blood, its compounds, and at the cell surface in order to reduce the body's oxygen uptake, O_2 utilization in cells, and the CO_2 release (Figure 3.12) [6].

They can be classified within three main groups:

- cyanides
- trihydrides
- carbonyls.

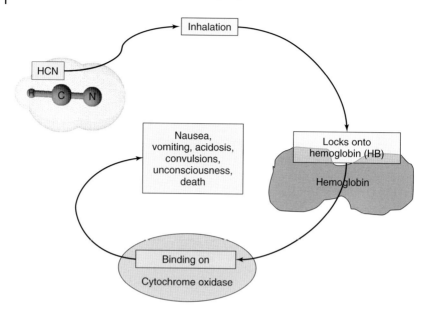

Figure 3.12 Effects of cyanide within the body. After the hydrogen cyanide gas (HCN) is inhaled, the molecule locks onto hemoglobin, the oxygen-carrying molecule in red blood cells, and is then distributed via the bloodstream to cells throughout the body. In the cells it binds to a metabolic enzyme called *cytochrome oxidase* and prevents cells from using oxygen to produce energy. In this way cyanide effectively chemically asphyxiates the body [12].

Some of these agents, which can be used as chemical warfare agents, are widely applied in the basic chemical industry. Hence these chemicals are widely available and thus they can be employed in asymmetric warfare.

Important substances of blood agents include hydrogen cyanide (AC), cyanogen chloride (CK), arsine (SA), and carbon monoxide (CO) (Figure 3.13).

These substances exhibit a high vapor pressure due to their low boiling points. Thus they are normally very volatile and therefore have a low persistency. Solid substances (e.g., cyanogen bromide) transfer directly from the solid state into the gas phase by sublimation, unter typical meterological conditions. Thus inhalation is the usual route of entry into the human body. Other entries of blood agents are ingestion, skin contact, or eye contact. Their solubility in water differs greatly, with the consequence that initial uptake via the mucosa differs accordingly.

Hydrogen cyanide (AC) is a colorless gas, which smells like "bitter almond" but it may not be easily recognized. Hydrogen cyanide gas in the air is explosive at concentrations over 5.6% (v/v), equivalent to 56 000 ppm (v/v); it is fully soluble in water (Table 3.10). Cyanogen chloride (CK) is a colorless, volatile liquid with strong irritant, choking, and corrosive effects. Today the substance is widely employed as a test substance for filter canisters. Cyanogen bromide (CB) is a crystalline, sublimating compound that is corrosive to copper and copper alloys, aluminum,

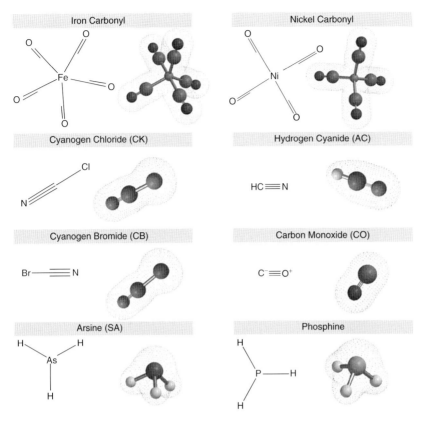

Figure 3.13 Nomination and molecular geometries of blood agents. The molecular geometries have been optimized based on MMFF and DFT calculations (hybrid density functional B3LYP/6-31G*, RMS gradient of 0.001), using the molecular modeling program Spartan '04 (Wavefunction, Inc. Irvine, CA). The molecules are represented in a ball-stick model enclosed by the calculated electron density (IsoVal = 0.03) [11].

and zinc. It has a pungent odor. Only pure, water-free cyanogen bromide is stable for several months when stored in tight containers. Carbon monoxide (CO) is colorless, odorless, and tasteless and does not irritate the respiration tract. It is slightly lighter than air and hardly soluble in water and it is flammable. It generates explosive mixtures in air. In the United States, the Occupational Safety and Health Administration (OSHA) limits long-term workplace exposure levels of carbon monoxide to 50 ppm [13]. The World Health Organization (WHO) also has guidelines for carbon monoxide [13]. In Germany the MAK-value (maximal concentration on workplaces) for carbon monoxide is set to 35 mg m^{-3}; both values reflect the hazardous effect of CO. The compound does not have the physical and chemical properties a chemical warfare agent usually has [7, 10]. Nevertheless it has detrimental effects of poisoning and it can be generated through side reactions, and thus potentially utilized for asymmetric warfare, and terroristic attacks.

Table 3.10 Physicochemical characterization of blood agents.

Property	AC (hydrogen cyanide)	CK (cyanogen chloride)	CB (cyanogen bromide)	CO (carbon monoxide)	SA (arsine)
Formula	HCN	ClCN	BrCN	CO	AsH_3
Molecular weight (g mol^{-1})	27	61	106	28	78
Vapor density (g cm^{-3})	0.99	2.1	3.4	1.145 (25 °C, 1 bar)	1.640 (−64 °C)
Liquid/solid density (g cm^{-3})	0.69[a]	1.18[a]	2.02[b]	0.814[a] (−195 °C)	1.34[a] (20 °C)
Boiling point (°C)	25.7	12.8	61–62	−191.5	−55 to −62.5
Melting point (°C)	−13.3	−6.9	49–52	−205	−117
Vapor pressure (Pa)	742 (25 °C)	1 000 (25 °C)	85–90 (20 °C)	Gas under standard conditions	1.47×10^6 (20 °C)
Volatility (mg m^{-3})	1 080 000 (25 °C)	2 600 000 (12.8 °C)	504 (20 °C)	–	31 000 (0 °C)
Solubility in water (%)	Completely soluble	6–7 (20 °C)	Limited, less than CK	0.0026 (20 °C)	0.07 (20 °C)

[a] Liquid density.
[b] Solid density.

Arsine is a very toxic compound, which acts mainly by inhalation. Ingestion is not considered as potential route of exposure. Owing to its physical and chemical properties pure arsine is difficult to handle.

Blood Agents – Interference with the Oxygen Transport

Blood agents interfere with red blood cells' ability to (i) transport oxygen (O_2) in the bloodstream to the cells and (ii) remove carbon dioxide (CO_2) from the cells into the lungs. Furthermore, they may block both the uptake of oxygen by cells into the body and oxygen utilization (Figure 3.12). Consequently, exposure to blood agents results in systemic effects and especially affects those organ systems that are most sensitive to low oxygen levels. These are:

- central nervous system,
- brain,
- heart and blood vessels,
- lungs.

> This leads ultimately to insufficient oxygen exploitation and the unnatural return transport of the oxygen in the venous blood. The symptoms of blood agent poisoning may vary a great deal from agent to agent.

Cyanide intoxication results in inhibition of the enzyme Cytochrome C Oxidase, which acts in the aerobic energy generation process of higher developed creatures. We can gauge the poisoning potential of AC by considering the very low lethal dosage on oral uptake of 50–60 mg per "standard man" of 75 kg weight; the lethal dose for intravenous application is, as expected, significantly lower. Table 3.11 shows the lethal concentration that leads to death within 30 Seconds. People affected by cyanides typically exhibit a light red color of the skin, which points to a high oxygen saturation of the venous blood stream, resulting from "inefficient" oxygen exploitation in the cells. Another effect is that the people concerned suffer from suffocation – thus these agents are also called *"asphyxiants."*

The toxic effect of carbon monoxide arises from its tight binding to hemoglobin, which is 325 times stronger than that of oxygen. Thus a carbon monoxide concentration of 0.1% in the breathing air deactivates around 50% of the red blood cells [14]. A carbon monoxide content of more than 2% leads to death within 1–2 min [15]. The half-life of carbon monoxide elimination is in the range of 2–6.5 h, depending on the amount of CO assimilated, the ventilation rate, and the oxygen concentration during ventilation. Carbon monoxide intoxication is indicated by cherry-red colored mucosa, which may be absent or not well developed in cases of light contamination. An irreversible effect of blood agents is the destruction of blood cells by arsine.

3.3.4
Tear Agents (Lachrymators)

Tear agents, or lacrimators (from the Latin *lacrima*, meaning "a tear"), commonly referred to as *tear gas* are a group of chemical agents with low toxicity. Their vapors can produce a sharp, irritating pain in the eyes resulting in an abundant flow of tears. There is no permanent damage to the eyes and the effects wear off quickly. For a short period clear visibility is not possible. A protective mask will provide a

Table 3.11 Toxicity data of the blood agent hydrogen cyanide (AC).

Variable	Value (mg)	Application
LD_{50}	50–60	Oral
LD_{50}	1	Intravenous
LD_{50}	7500	Liquid on skin
LD_{50}	100	Liquid on skin
LC_{50}	0.5–2%	Vapor on skin

complete safeguard. Some of the newer tear agents can cause runny noses, severe chest pains, nausea, and vomiting, too. These agents are also being employed against civil or prison riots by police and military to disperse crowds. Figure 3.14 shows the chemical structure and names of two well-known agents.

High purity chloroacetophenone (CN) is a white crystalline solid reminiscent of granulated sugar or salt. The technical product has a gray-green color. It has an odor similar to apple blossoms. Chlorobenzylidene-malononitrile (CS) is used in agents mixed with different additives and/or solvents to vary the persistency, way of dissemination, and rate of hydrolysis.

Tear agents exhibit high boiling points and melting points, resulting in very low vapor pressure and thus significantly low volatility (Table 3.12). Dispersion is realized as a solution in organic solvents, such as chloroform, for example, as a 30% solution of CN. They are disseminated as aerosols or in spray form by many police forces as an incapacitating agent to disperse civilian crowds, and to suppress persons who are violently aggressive. CS is up to ten times more effective than CN.

CN causes tears, irritation of the eyes, and a burning sensation on the skin. The primary effects of CN on the eyes are tears and photophobia (sensitivity to light). The median ICt_{50} concentration is described in literature as 80 mg · min · m^{-3}, the

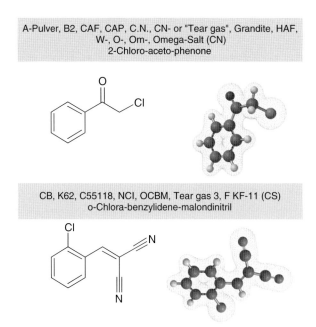

Figure 3.14 Nomination and molecular geometries of tear agents. The molecular geometries have been optimized based on MMFF and DFT calculations (hybrid density functional B3LYP/6-31G*, RMS gradient of 0.001), using the molecular modeling program Spartan '04 (Wavefunction, Inc. Irvine, CA). The molecules are represented in a ball-stick model enclosed by the calculated electron density (IsoVal = 0.03) [11].

Table 3.12 Physicochemical characterization of tear agents.

Property	CN	CS
Formula	C_8H_7ClO	$C_{10}H_5ClN_2$
Molecular weight (g mol^{-1})	155	189
Relative vapor density[a]	5.3	–
Liquid density (g cm^{-3})	1.32 (solid) (20 °C)	1.04 (20 °C)
Boiling point (°C)	249	310–315 (with decomposition)
Melting point (°C)	54	95–96
Vapor pressure (Pa)	0.0041 (20 °C)	0.00034 (20 °C)
Volatility (mg m^{-3})	34.3 (20 °C)	0.36 (20 °C)
Solubility in water (g l^{-1})	–	1–5 (16 °C)

[a] Relative to air.

Table 3.13 Toxicity data of tear agents.

Variable	Value (mg min m^{-3})	Substance	Application
ICt_{50}	80–100	CN	Aerosol
LC_{50}	7000–25 000	CN	Aerosol
TC_{50}	0.3–0.4 mg m^{-3}	CN	Aerosol/eye and respiratory tract
ICt_{50}	20	CS	Aerosol
LCt_{50}	10 000–61 000	CS	Pyrotechnically generated aerosol
TC_{50}	0.023 mg m^{-3}	CS	Aerosol

lethal dose of 50% of the probe animals (LCt_{50} mg · min · m^{-3}) is reported to be 3700 for rats, 73 500 for mice, and 35 000 for guinea pigs. The incapacitation dose of these chemicals is low, while at the same time the lethal dose is significantly high for both compounds (Table 3.13). The newly introduced TC_{50} value is also given in Table 3.13. It reflects the concentration required to produce a detectable irritation within 1 min for 50% of the population studied. This extremely low value proves the effectiveness of the tear agents, if employed.

CS may create severe pulmonary damage. It can also severely damage heart and liver and there is a possibility that exposure could cause lethal effects. When CS is metabolized, cyanide can be detected in human tissue. According to the United States Army Center for Health Promotion and Preventive Medicine, CS emits "very toxic fumes" when heated to decomposition and can be an immediate danger to life and health at specified concentrations. Persons exposed to CS even in low concentrations should contact medical personal immediately due to reported severe traumatic injury and possible mutagenic effects [16].

Although described as a non-lethal weapon for crowd control, many studies have raised doubts about this classification. Such an evaluation easily derives from the potential impacts cited from the literature above. The employment of tear agents is a widespread police doctrine. Domestic utilization of CS by police forces is legal in many countries.

3.3.5
Vomiting Agents (Sternutators)

The most widely known vomiting agents are organoarsenic substances (Figure 3.15). The main and intended effect on humans is the "mask breaking" action. The victims feel sickness and are vomiting. At this point the victims are forced to unmask, thereby exposing themselves unprotected to the environment. Consequently, these substances were intended to be employed in combination with more hazardous compounds, like nerve agents, so that the unprotected persons might be attacked.

DA appears as colorless crystals, DC as a white solid, and DM as light yellowish-to-greenish crystals dependent on purity. DA and DM are odorless, while DC reportedly smells similar to garlic or bitter almonds. All three agents are insoluble in water, and soluble in organic solvents, such as acetone, dichloromethane, ethanol or diethyl ether. They have extremely low vapor pressures (Table 3.14). Consequently, they are typically disseminated as aerosols, by spraying or adding them [up to 50% (in solid state)] to pyrotechnic mixtures (like nitrocellulose,

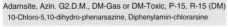

Figure 3.15 Nomination and molecular geometries of vomiting agents. The molecular geometries have been optimized based on MMFF and DFT calculations (hybrid density functional B3LYP/6-31G*, RMS gradient of 0.001), using the molecular modeling program Spartan '04 (Wavefunction, Inc. Irvine, CA). The molecules are represented in a ball-stick model enclosed by the calculated electron density (IsoVal = 0.03) [11].

Table 3.14 Physicochemical characterization of vomiting agents.

Property	DM	DA	DC
Formula	$C_{12}H_9AsClN$	$C_{12}H_{10}AsCl$	$C_{12}H_{10}AsCN$
Molecular weight (g mol^{-1})	277.6	264.6	255.2
Density (g cm^{-3})	1.65	1.42	1.45
Boiling point (°C)	410 (decomp.) sublimes readily	333 (decomp.)	377
Melting point (°C)	195	38–45	31.5–35
Vapor pressure (Pa)	2.67×10^{-11} (20 °C)	0.067 (20 °C)	0.027 (20 °C)
Volatility (mg m^{-3})	0.02	0.3–0.68 (20 °C)	0.1–0.15 (20 °C)
Solubility in water (%)	Practically insoluble	Practically insoluble	Practically insoluble

Table 3.15 Toxicity data of vomiting agents.

Variable	Value (mg min m^{-3})	Substance	Application
LD	620 mg m^{-3}	DM	Aerosol (30 min inhalation)
ICt$_{50}$	2–150	DM	Aerosol
LCt$_{50}$	11 000–30 000	DM	Aerosol
ICt$_{50}$	10–30	DC	Aerosol
LCt$_{50}$	10 000	DC	Aerosol
LC	60 mg m^{-3}	DA	Vapor (30 min inhalation)
ICt$_{50}$	10–30	DA	Vapor
LCt$_{50}$	15 000	DA	Vapor

or nitropenta). Thus, the primary route of absorption is through the respiratory system. Exposure can also occur via ingestion, dermal absorption, or eye impact.

DM is now regarded as obsolete. But for some aggressors adamsite could still be in use, as stockpiles may be available.

The toxic effects of vomiting agents are similar to those caused by typical "riot control agents" (e.g. CS). The action may vary, as seen for DM, which is slower in onset and longer in duration, often lasting several hours. Five to ten minutes after agent pick up, irritation of the eyes, lungs, and mucous membranes develop, followed by headache, nausea, and persistent vomiting. As shown in Table 3.15 the incapacitation dose of these chemicals is as low or even lower than that of the tear agents, while at the same time the lethal dose is significantly high, similar to that of CN and CS (refer to Table 3.13). Vomiting agents are even more effective than tear agents while providing similar low fatal potential.

3.3.6
Nerve Agents

Nerve agents belong to the class of phosphorous-oxygens, containing various organic compounds (Figure 3.16). A more detailed classification allocates them into the phosphono-group (Figure 3.17). Tabun is a phosphono-cyanide, while soman and sarin are phosphono fluorides, and VX is a member of the phosphono-thiolate group, according to the ligands of the general chemical structure of these substances (Figure 3.16).

In terms of physicochemical behavior nerve agents show high boiling points, ranging from 150 to 300 °C, and low melting points down to less than −50 °C. Thus, although they are also called nerve "gases" these chemicals are liquid at

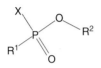

Figure 3.16 General chemical structure of nerve agents; R1 and R2 reflect the hydrocarbon ligands, while X represents the heteroatom function.

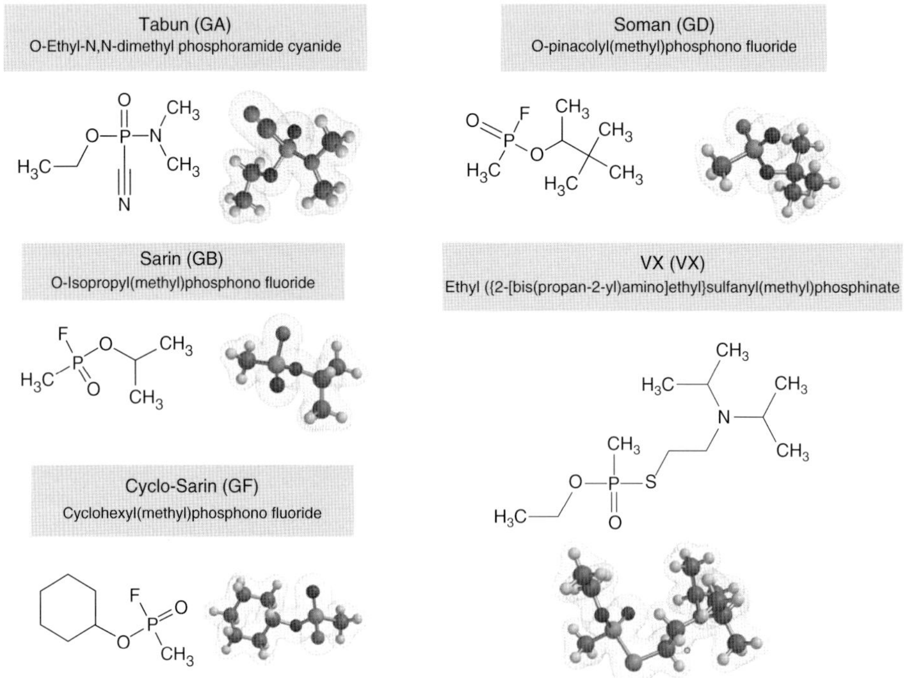

Tabun (GA)
O-Ethyl-N,N-dimethyl phosphoramide cyanide

Soman (GD)
O-pinacolyl(methyl)phosphono fluoride

Sarin (GB)
O-Isopropyl(methyl)phosphono fluoride

VX (VX)
Ethyl ({2-[bis(propan-2-yl)amino]ethyl}sulfanyl)(methyl)phosphinate

Cyclo-Sarin (GF)
Cyclohexyl(methyl)phosphono fluoride

Figure 3.17 Nomination and molecular geometries of nerve agents. The molecular geometries have been optimized based on MMFF and DFT calculations (hybrid density functional B3LYP/6-31G*, RMS gradient of 0.001), using the molecular modeling program Spartan '04 (Wavefunction, Inc. Irvine, CA). The molecules are represented in a ball-stick model enclosed by the calculated electron density (IsoVal = 0.03) [11].

Table 3.16 Physicochemical characterization of nerve agents.

Property	Tabun (GA)	Sarin (GB)	Soman (GD)	Cyclohexyl-sarin (GF)	VX
Formula	$C_5H_{11}N_2O_2P$	$C_4H_{10}FO_2P$	$C_7H_{16}FO_2P$	$C_7H_{14}FO_2P$	$C_{11}H_{26}NO_2PS$
Molecular weight (g mol^{-1})	162	140	182	180	267
Density (g cm^{-3})	1.073	1.089	1.022	1.120	1.008
Boiling point (°C)	247	147	167	298	300
Melting point (°C)	−50	−56	−42	<−30	−39
Vapor pressure (Pa)	9 (24 °C)	387 (25 °C)	53.3 (25 °C)	6.67 (25 °C)	0.09 (25 °C)
Volatility (mg m^{-3})	600 (25 °C)	22 000 (25 °C)	3 900 (25 °C)	0.44 (25 °C)	0.01 (25 °C)
Solubility in water (%)	10 (25 °C)	∞	2 (25 °C)	~2	3 (25 °C)

Table 3.17 Estimated toxicity data of nerve agents.

Variable	Value (mg min m^{-3})	Substance	Application
LD$_{50}$	4000 mg	GA	Contact on human skin
LD$_{50}$	1700 mg	GB	
LD$_{50}$	300 mg	GD	
LD$_{50}$	10 mg	VX	
LCt$_{50}$	200	GA	Inhalation
LCt$_{50}$	100	GB	
LCt$_{50}$	100	GD	
LCt$_{50}$	50	VX	

room temperature. They provide a wide variety in volatility as well as solubility in water, which might be of interest for the method of decontamination. The broad range of vapor pressure classifies the different chemicals into more sedentary (e.g., VX) and more volatile agents, such as sarin (GB) (Table 3.16).

The toxicity of nerve agents is the highest of all chemical warfare agents [10]. Clinical effects of nerve agents depend on:

- the route of entry,
- concentration of the agents,
- time of exposure,
- kind of agent.

Tabun (GA) has a much less effect than VX, which provides the most hazardous effect. Table 3.17 summarizes the estimated toxic impact of nerve agents on humans. Action from contact on skin is listed in the top half of the table, while the impact from inhalation is shown in the lower half. Exposure to low concentrations of nerve agent vapor produces immediate ocular symptoms, rhinorrhea (filling of the nasal cavities with mucus fluid), and, in some patients, dyspnea (shortness of breath). These ocular effects are secondary to the localized uptake, for example, of sarin (GB) vapor across the outermost layers of the eye, causing tears, pupillary sphincter contraction (miosis), and ocular pain. As the exposure increases, dyspnea and gastrointestinal symptoms ensue.

Exposure to high concentrations of nerve agent vapor causes immediate loss of consciousness, followed shortly by convulsions, flaccid paralysis, and respiratory failure. These generalized effects are caused by the rapid absorption of nerve agent vapor across the respiratory tract, producing maximal inhibition of acetylcholine esterase within seconds to minutes of exposure [7, 10].

The biomolecular mechanism of action of nerve agents is the disruption of information between nerve cells (Figure 3.18, right column). It is caused by blocking the acetylcholine esterase, an enzyme that degrades the neurotransmitter acetylcholine, after being emitted by the synapses for information transfer between

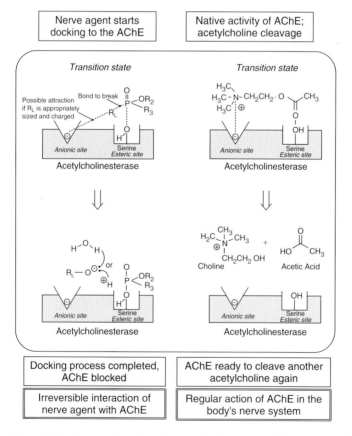

Figure 3.18 Natural effect of native acetylcholinesterase (AChE) (right-hand column) and blocking of AChE due to nerve agents (left-hand column).

cells. Clinical effects from nerve agent exposure are therefore caused by excess acetylcholine due to the blocked ability of the enzyme acetylcholine esterase to cleave acetylcholine at muscarinic, nicotinic, and central nervous system synapses into choline and acetic acid. Nerve agents act by first binding and then irreversibly inactivating acetylcholine esterase (Figure 3.18, left column).

Excessive acetylcholine at these cholinergic receptors may account for the spectrum of clinical effects observed in nerve agent exposure [17], namely, miosis and glandular hypersecretion, sweating and initial defasciculation followed by weakness and flaccid paralysis. At central nerve system cholinergic receptors, nerve agents produce irritability, giddiness, fatigue, lethargy, amnesia, ataxia, seizures, coma, and respiratory depression.

3.4
Choking and Irritant Agents

Choking agents inflict injury mainly on the respiratory tract, that is, the nose, throat, and especially the lungs. Attacked persons suffocate of suffer from reduced oxygen into their bloodstream. Victims typically inhale these agents, which cause the alveoli to secrete a fluid flow into the lungs, essentially "drowning" these persons internally. Table 3.18 gives examples and the nomenclature of choking agents, as well as a summary of their physicochemical properties. As some of these substances are mass produced in the chemical industry their threat is mainly through asymmetric attacks. At present, they have less military significance than blister and nerve agents. Thus they will be described only exemplarily.

Choking agents are gaseous or liquids with high volatility. They provide specific optical recognition, and odors that are often reminiscent of suffocation. Their gas density is higher than air – they sink into and fill depressions, like trenches or deeper bunker areas. They are more or less soluble in water. For instance 2.3 L chlorine may be dissolved in 1 L water. Phosgene reacts in water, and thus rapidly hydrolyzes, even at low temperatures; the hydrolysis of 1 g of phosgene even at 0 °C takes less than 20 seconds. This feature can be used to prevent exposure.

The effect on humans is mainly suffocation, choking, and poisonous impact. The strong irritation is mainly based on the on the high oxidation power of the related substances, mainly, on CL, with long lasting damage at the body. These substances burn tissues in the eyes and respiratory tract and thus also lead to lachrymation, (e.g., DP). These attributes may serve for "personal detection" of choking agents. The poisoning effect of CG and related compounds, such as DP and PS, is based on the decay of the molecule into CO (carbon monoxide). Such impacts of CO parallel to those seen with the blood blocking effect (see blood agents), namely, prevention of oxygen transport to, and removal of CO_2 from, cells. All these chemicals act rapidly (burn of tissue), although poisoning may be delayed for more than 3 h after exposure with minimal contact (especially for DP). PS is more toxic than chlorine but less toxic than phosgene (CG) (Table 3.19).

Table 3.18 Nomenclature and physicochemical characterization of choking agents and irritant agents.

	Agent			
	Phosgene (CG)	Chlorine (CL)	Diphosgene (DP)	Chloro-picrine (PS)
Chemical name	Carbon oxide dichloride, carbonyl chloride	Chlorine	Trichloromethyl chloroformate	Trichloro-nitromethane
Chemical formula	$OCCl_2$	Cl_2	$Cl_3\text{-}CO\text{-}COCl$ ($C_2Cl_4O_2$)	Cl_3CNO_2
Aggregate state and optical recognition (20 °C)	Colorless gas	Greenish yellow gas	Yellow to dark brown or black liquid	Colorless to light green/brown oily liquid
Odor	Green corn or recently mown hay (low concentration); unpleasant and strong (high concentration)	Pungent	Green corn or new mown hay	Intense and penetrating
Persistence at 4–16 °C	1 h	2–3 d	1–4 h	–
Persistence at 21–32 °C	30 min	18–36 h	30 min to 3 h	–
Time of action	Rapid	Rapid, but may be delayed	Rapid	Very rapid
Symptoms of eye and skin exposure	Burns to the skin (according to CL, as CG dissociates into CL and CO. Chronic skin contact may result in dermatitis. Acute eye contact results in conjunctivitis (inflammation of the inner eyelid surface, and outer eye layer, respectively), lachrymatory; chronic eye contact may result in conjunctivitis	Corrosive to skin and eyes, causing pain, blurred vision, and severe deep burns	Mid-level exposure causes eye tearing and irritation of skin. Direct eye exposure to liquid can cause corneal abrasions, ulcers (purulent abscess), or perforation	Chemical burns. Eyes will have pain, redness, and tearing. Prolonged eye exposure causes blindness

Table 3.19 Toxicity data of choking and irritant agents.

Variable	Value (mg min m^{-3})	Substance	Application
LD_{50}	500 ppm min^{-1}	CG	Inhalation (vapor/standard man)
LC_{50}	3200 ppm m^{-3}	CG	Vapor/standard man (75 kg)
ICt_{50}	1600	CG	Vapor/standard man (75 kg)
LCt_{50}	3200	CG	Vapor/standard man (75 kg)
LD_{50}	2900 mg m^{-3}	CL	Lethal after a short exposure time
ICt_{50}	1800	CL	Vapor/standard man (75 kg)
LCt_{50}	19 000	CL	Vapor/standard man (75 kg)
LD_{50}	Same data as CG	DP	Vapor/standard man (75 kg)
ICt_{50}	1600	DP	Vapor/standard man (75 kg)
LCt_{50}	3000	DP	Vapor/standard man (75 kg)
LD	2300 mg m^{-3}	PS	Vapor/standard man (75 kg); 1 min inhalation
ICt_{50}	50–100	PS	Vapor/standard man (75 kg)
LCt_{50}	2000–20 000	PS	Vapor/standard man (75 kg)

3.5
Incapacitating Agents

A wide range of possible nonlethal chemicals with a highly incapacitating impact with psychological behavioral action has been developed. This class includes fentanyls, cannabioles (THC), phenothiazines, LSD, chemically related substances, and ketamine. As they are of lesser importance, only examples of such agents are described here.

3-Quinuclidinyl benzilate, (BZ) is a crystalline, odorless substance, which is slightly soluble in water, soluble in diluted acids, and soluble stable in most solvents, for example, alcohol, acetone, trichloroethylene, and dimethylformamide. It is insoluble in aqueous alkaline solutions. It has a half-life of three to four weeks in moist air and is extremely persistent in soil, water, and on most surfaces. BZ is an anticholinergic agent that affects both the peripheral and central nervous systems. As a potent psychomimetic, only small doses are necessary to produce incapacitation, but with delayed symptoms developing several hours after contact. BZ is a glycolate anticholinergic compound related to scopolamine, hyoscyamine, atropine, and other deliriants acting as a competitive inhibitor of acetylcholine at postsynaptic receptors, and decreases the effective concentration of acetylcholine seen by receptors at these sites. Impacts seen include panoramic illusions, hallucinations, and confusion [7, 10].

3.6
Dissemination Systems of Chemical Warfare Agents

The application of chemical warfare agents requires a spreading system. Thus some aspects of this topic are discussed here, although the subject is of minor interest

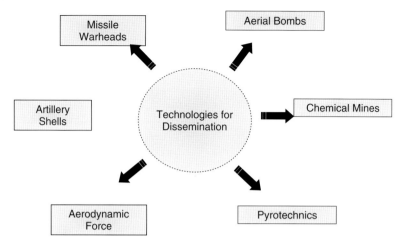

Figure 3.19 Overview of the main technologies for dissemination of chemical warfare agents.

in terms of the main topic of this chapter. One of the most important factors in the effectiveness of chemical weapons, and thus relating to the threat to soldiers, is the efficiency of dissemination. Various technologies have been developed and can be used to disseminate chemical warfare agents. Here we address only some of the main technologies [18]. The possible arsenal of chemical warfare munitions includes various technologies (Figure 3.19).

For the different types of chemical warfare agents and according to environmental conditions on the battlefield the different technologies are more or less applicable. In distributing the contaminant, losses of substances are considered to be to the advantage of the attacked soldiers. Additional contaminated areas may be involved, which is not directly obvious. A high impact on the troops occurs when the intended particle size of the solid or liquid agents is realized by the explosion. In addition, contamination densities over the area should be homogeneous. Explosive dissemination produces an uncontrollable droplet size, usually having fine and coarse particles. For example, the "central burster technology" in artillery shells provides contamination densities from a few milligrams per square meter to several hundred grams per square meter around the point of detonation. This has to be considered for protection as well as for decontamination. Another aspect is agent loss by incineration in the initial blast, which is normal; additionally, agent may be forced onto and in the ground by the blast, which has to be considered for decontamination later. Air burst munitions overcome these aspects and thus provide a higher threat and impact on the attacked soldiers and civilians. Besides dissemination by explosion, dispersal can be carried out by spraying from tanks or aircrafts via non-explosive dispersion. Additional factors, such as wind speed and direction, changes of the wind, the delivering vehicle, and so on, influence the operation and the attacked troops significantly.

3.7
Conclusions and Outlook

In this chapter we have brought out the special properties of chemical warfare agents. We have discussed the mechanisms of action of the main chemical warfare agents and gained an impression of the possible devastating effects of these compounds. Based on this information we should be able to follow the possible chemical, biological, radiological, and nuclear (CBRN) countermeasures discussed later.

References

1. Klotz, M.I. and Rosenberg, R.M. (2008) *Chemical Thermodynamics: Basic Concepts and Methods*, John Wiley & Sons, Inc., Hoboken, NJ, USA. ISBN: 978-0471780151.
2. Weingärtner, H. *et al.* (2000) Water to zirconium and zirconium compounds, *Water*, Ullmann's Encyclopedia of Industrial Chemistry, vol. 39, Wiley-VCH Verlag GmbH, Weinheim.
3. Dohrn, R., Peper, S., and Fonseca, J.M.S. (2010) *Fluid Phase Equilib.*, **288**, 1–54.
4. Hoenig, S.L. (2008) *Compendium of Chemical Warfare Agents*, Springer-Verlag GmbH, Heidelberg, Germany. ISBN: 978-0387346267.
5. James, A., Romano, J.R., Kukey, B.J., and Salem, H. (2007) *Chemical Warfare Agents: Chemistry, Pharmacology, Toxicology, and Therapeutics*, CRC Press, Boca Raton, FL, USA. ISBN: 978-1420046618.
6. Kehe, K. and Szinicz, L. (2005) *Toxicology*, **214** (3), 198–209.
7. Chauhan, S. *et al.* (2008) *Environ. Toxicol. Pharmacol.*, **26** (2), 113–122.
8. Gallant, B.J. (2006) Toxicology and medical monitoring, *Hazardous Waste Operations and Emergency Response Manual*, John Wiley & Sons, Inc., Hoboken, NJ, USA. ISBN: 978-0471684008.
9. Wecht, C.H. (2004) *Forensic Aspects of Chemical and Biological Terrorism*, Lawyers & Judges Publishing, Tucson, AR, USA. ISBN: 978-1930056671.
10. Marrs, T.C., Maynard, R.L., and Sidell, F. (eds) (2007) *Chemical Warfare Agents: Toxicology and Treatment*, John Wiley & Sons, Ltd., Chichester, ISBN: 978-0470013595.
11. Kong, J. *et al.* (2000) *J. Comput. Chem.*, **21**, 1532–1548.
12. Holstege, C.P., Neer, T.M., and Saathoff, G.B. (2010) *Criminal Poisoning: Clinical and Forensic Perspectives*, Jones & Bartlett Publishers, Burlington, MA, USA. ISBN: 978-0763744632.
13. Han, X. and Naeher, L.P. (2005) *Environ. Int.*, **32** (1), 106–120.
14. Collman, J.P., Brauman, J.I., Halbert, T.R., and Suslick, K.S. (1976) *Proc. Natl. Acad. Sci. USA*, **73** (10), 3333–3337.
15. Fauci, A.S. *et al.* (2008) *Harrison's Principle of Internal Medicine*, McGraw-Hill Professional, Columbus, OH, USA. ISBN: 978-0071466332.
16. Hu, H. *et al.* (1989) *J. Am. Med. Assoc.*, **262** (5), 660–663.
17. Lee, E.C. (2003) *J. Am. Med. Assoc.*, **290** (5), 659–662.
18. Ledgard, J. (2006) Dissemination Techniques and Munition, *A Laboratory History of Chemical Warfare Agents*, Jared Ledgrad, USA, section VII, ch. 12, 246 ff. ISBN-10: 0-615-13645-1.

4
Characteristics of Biological Warfare Agents – Diversity of Biology

Birgit Hülseweh

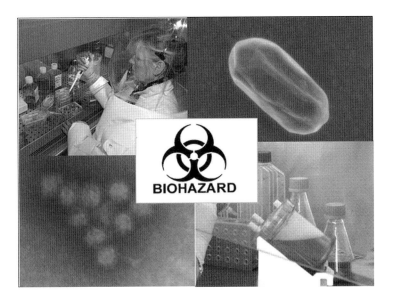

Microbiological agents are universally distributed in nature and make up the predominant part of life on earth. They occur in soil and water as well as in plants and animals, but also in the most hostile habitats like hot springs and deep ocean vents. The great majority of microbial agents are harmless and many are beneficial. Some of their properties have been used for fermentation in food processes for centuries, while others are applied in biotechnological processes. However, a few hundred of these agents are pathogenic or toxic, which means they are capable of causing disease in man, plants, or animals. Some of them can to be considered as biological warfare agents.

4.1
What Is Special?

Biological agents (B-agents) generally considered as suitable for biological warfare include *viruses, bacteria, toxins, and fungi*, but only the use of a biological agent in combination with a system for the deliberate dissemination turns the B-agent into a biological weapon (B-weapon) or biological warfare agent (BW-agent).

Not all pathogenic microorganisms and their toxic products are considered as potential biological warfare agents. First, certain properties and characteristics make them attractive for abuse (Figure 4.1).

In this section we hope to give an understanding of the uniqueness of biological agents. On the one hand, this is necessary for proper safety- and counter-measures in case of a biological incident. On the other hand, risk- and quality management efforts require a basic biological knowledge (Chapter 2 and Part Five).

Compared to chemical agents or nuclear materials, biological agents are unique in their diversity. For technical and medical B-protection, their classification is important with reference to detection, identification, prophylaxis, and treatment.

Below some of the general features of the four relevant biological agent categories are defined consecutively in more detail [1–5].

4.2
Types of Biological Agents

B-agents like bacteria, viruses, toxins, and fungi differ significantly in size (Figure 4.2). While most agents can be identified by microscopic techniques,

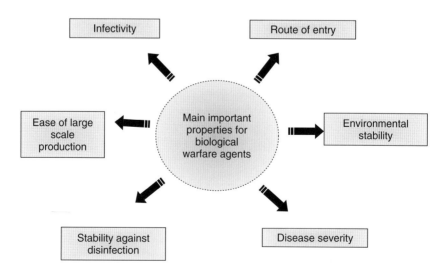

Figure 4.1 Main important properties of biological agents desired for biological warfare.

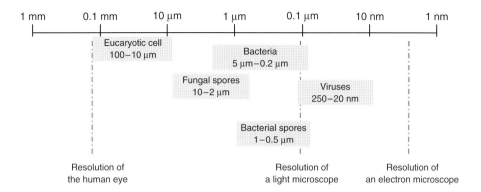

Figure 4.2 Comparison of proportions and size of biological agents.

biological toxins are at the molecular level and well below the resolution of an electron microscope (EM).

4.2.1
Bacteria

Bacteria are single-cell microscopic organisms that possess cell walls and cell membranes where the cell wall maintains the shape. Two different types of cell wall in bacteria exist, called *Gram-positive* and *Gram-negative* (Figure 4.3). The names originate from the reaction of cells to the Gram-staining method, a basic test long-employed for the classification of bacterial species. The difference in Gram reaction of these two groups of bacteria is thought to be due to a difference in the structure of their cell walls and explains the distinct susceptibility to antimicrobial agents as well as the different resistance to physical disruption, drying, and sodium azide.

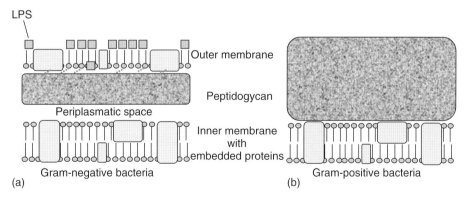

Figure 4.3 Comparison of the cell-wall structure of Gram-negative (a) and Gram-positive (b) bacteria.

As illustrated in Figure 4.3 the Gram-positive cell wall consists of thick multi-layered peptidoglycan and does not possess a lipid outer membrane. In contrast, Gram-negative cell walls have only one or a few layers of peptidoglycan but also possess an outer membrane consisting of various lipid complexes that confine the perisplasmatic space to the outer milieu.

The difference in cell-wall structure does not refer to the electrical charge of the bacteria. Both Gram-negative and Gram-positive bacteria are negatively charged.

Usually, bacteria have a deoxyribonucleic acid (DNA) genome, transcribe their necessary genes for reproduction and metabolism to ribonucleic acid (RNA), and translate them to proteins. Bacteria are generally much smaller than eukaryotic cells (Figure 4.2) but despite their small size (0.2–5 µm) they are very complex in physiology and cell biology.

Most bacteria replicate by cell division and grow under simple, well-defined conditions. They are either spherical or rod-shaped and occur nearly everywhere in nature. Classical examples of bacteria with potential as BWAs (biological warfare agents) are the species *Bacillus anthracis* (Figure 4.4), *Yersinia pestis*, and *Francisella tularensis*, and also *Coxiella burnetii* and *Brucella melitensis*. Moreover, some Gram-positive bacteria, like bacilli or clostridia, produce endospores (Figure 4.5),

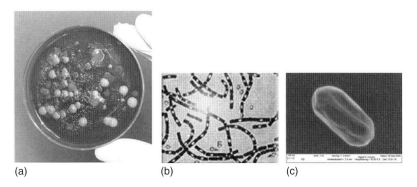

Figure 4.4 Example of a biological agent; (a) *Bacillus anthracis* colonies on blood agar; (b) sporulating *Bacillus* cells; (c) electron microscopic picture of a single *Bacillus* spore.

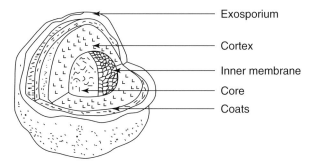

Figure 4.5 Structure of an endospore. Reproduced and modified according to Reference [6].

a kind of highly resistant and permanent mold, others are selective parasites, produce gelatinous slimes or capsules – all features and bacterial characteristics that influence the efficiency of disinfection (Section 14.5).

In the electron microscopic picture (Figure 4.4c) spores show a thin outer spore coat, thick spore coat, thick spore cortex, and an inner spore membrane surrounding the chromosomal DNA located in the core region. Furthermore, the spore's chromosomal DNA is protected by small acid-soluble spore proteins (SASPs). Spores are generally formed in response to nutritional deprivation and enable, for example, bacilli to survive in an environment without metabolism. In addition, spores are extremely resistant to chemical and physical stress (Section 14.5).

4.2.2
Viruses

Viruses are non-cellular infectious agents consisting of a nucleic acid containing the genetic information, a strand of DNA *or* RNA, surrounded by a protein capsule (capsid). At the simplest level the capsid protects the viral genome from physical and chemical damage as well as from enzymatic damage. Moreover, many viruses have viral envelopes covering the protein capsid. These envelopes typically consist of phospholipids and proteins and derive from the host cell membranes; some envelopes also include viral glycoproteins. In general the outer surface of the virus is responsible for the recognition of the host cell, which normally takes place by binding of a specific virus-attachment protein to a cellular receptor (Figure 4.6).

Viruses are much smaller than bacteria, typically submicroscopic in size and are only visible by EM. Since viruses cannot reproduce independently they are not considered as living. They replicate using the host's metabolic processes and cellular components. In sharp contrast to bacteria, virus structures vary greatly.

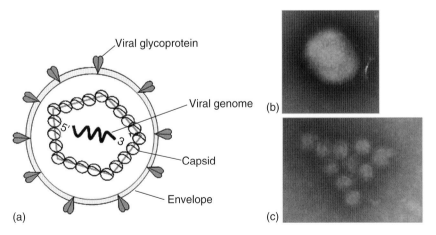

Figure 4.6 Schematic presentation of an enveloped virus (a) and electron microscopic pictures of an orthopoxvirus (b) and alphavirus (c).

Table 4.1 Classification of selected viral families with potential for biological warfare.

Virus family	Envelope	Symmetry/structure	Size (nm)	Biological warfare-relevant viruses
Orthopox	Enveloped viruses		200 × 300–450	Variola major virus, Variola minor virus, Monkeypox virus
Toga			60–70	Venezuelan equine encephalitis virus (VEEV), Eastern equine encephalitis virus (EEEV), Western equine encephalitis virus (WEEV)
Flavi			40–50	Dengue virus, West Nile virus, Yellow fever virus (YFV)
Filo			80 × 790–14000	Ebola virus, Marburg virus
Bunya			90–120	Krim Kongo virus, Sin Nombre virus
Arena			50–300	Lassa virus
Paramyxo			150–300	Nipah virus, Hendra virus
Picorna	Non-enveloped		28–30	Poliovirus, Foot and mouth disease virus

They come in many shapes that refer to the structure of the viral capsid and taxonomy and classification are in part based on these distinctions: polyhedral, helical, or binal. In addition, classification is based on size, nucleic acid, replication method, and the presence or absence of a viral envelope (Table 4.1).

Examples of viruses with potential as biological warfare agent are variola major and variola minor virus, Venezuelan equine encephalitis virus (VEEV), yellow fever virus (YFV), Ebola virus, and hantavirus.

4.2.3
Toxins

Toxins of biological origin are not viable and do not reproduce. Therefore, these B-agents are neither infectious nor contagious. Biological toxins are either simple

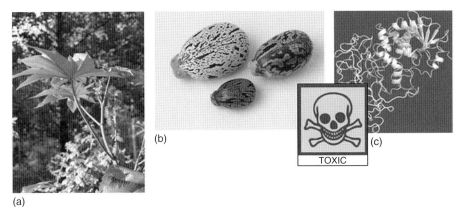

Figure 4.7 Castor bean plant *(Ricinus communis)* (a), with its toxic seeds (b), and the three-dimensional structure of the heterodimeric toxin (c).

metabolic products from microorganisms or are poisons extracted from plants, marine organisms, snakes, or insects and share many characteristics like the absence of smell, taste, and color with chemical warfare agents. The structure of biological toxins varies from relatively simple to very complex polypeptides or proteins. Toxic substances can be classified according to the physiological effects they have on the human body – many cause stomach pains, diarrhea, and vomiting, and also muscle weakness up to respiratory paralysis.

Classical examples of toxins with potential as BWAs include some of the botulinum toxins produced by the bacteria *Clostridium botulinum*, ricin derived from castor beans (Figure 4.7), various staphylococcal enterotoxins from *Staphylococcus aureus*, and saxitoxin produced by blue green algae. Biotoxins act on the human body as either cyto- or neurotoxin. While the former causes cellular destruction, the latter affects the central nervous system.

4.2.4
Fungi

Fungi can be generally divided into microscopic small yeast and molds. While the former are unicellular organisms the latter are filamentous and almost entirely multicellular and produce asexual endospores. Fungi are heterotrophic organisms, which means they derive their energy from another organism, independent of whether the organism is alive or dead. They are typical pathogens that can be weaponized for use against crops to cause diseases and loss of food plants, like rice blast or cereal rust (Figure 4.8).

An infected leaf has diamond-shaped or elliptical or spindle-shaped spots with gray or white centers and brown margins. The spots may merge and lead to a complete drying of the infected leaf. The infected panicle turns white and dies before being filled with grain.

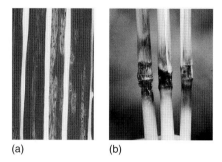

Figure 4.8 Blast of rice caused by *Pyricularia grisea*. Infection is characterized by the appearance of lesions on the aerial parts of the rice plant: (a) typical lesions of leaves; (b) typical infection on plant nodes.

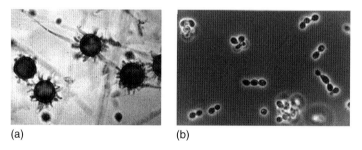

Figure 4.9 *Histoplasma capsulatum* takes on a mycelial (filamentous) form (a) at 25 °C and grows as yeast (b) at body temperature in mammals [7].

Fungal pathogens considered of significance as biological agents against humans are *Coccidioides immitis* and *Histoplasma capsulatum* (Figure 4.9). Both fungi are dimorphic and people are infected by inhalation of spores. Infection is characterized by flu-like symptoms and the cognate disease primarily affects the lungs.

4.3
Risk Classification of Biological and Biological Warfare Agents

Several attempts have been made to classify pathogens according to the infection risk they present to the health of laboratory workers and the public should these agents escape from the laboratory. The result is a national and European categorization of biological agents into four subgroups. Class 1 agents present no risk for workers and the community whereas class 4 agents present the highest individual and community risk.

The biological lists are revised on a regular basis for taxonomy and biological risk class and the classification and related employment regulations allow staff to handle samples, organisms, and toxins in a safe manner.

Risk Classification of Biological Agents

- **Class 1 biological agents** are unlikely to cause human disease.
- **Class 2 biological agents** can cause human disease and might be hazardous to workers. They are unlikely to be spread to the community. Effective prophylaxis or treatment is available.
- **Class 3 biological agents** can cause severe human disease and present a serious hazard to workers. They have a moderate risk of spreading to the community; there is usually effective prophylaxis or treatment available.
- **Class 4 biological agents** causes severe human disease and present a serious hazard to workers. They present a high risk of spreading to the community. There is usually no effective prophylaxis or treatment available (Figure 4.10).

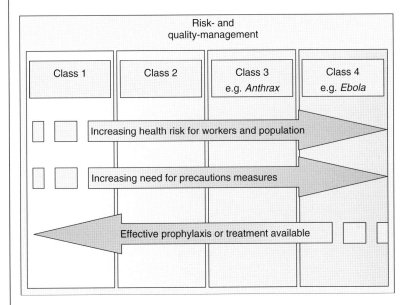

Figure 4.10 Classification of biological agents according to international conventions like the European directive 2000/54/EC.

4.3.1
Risk Classification of Potential Biological Warfare Agents

Criteria for the weighting and evaluation of potential biological threat agents are essential properties shown in Figure 4.1, but also their military and public health impact influences the evaluation.

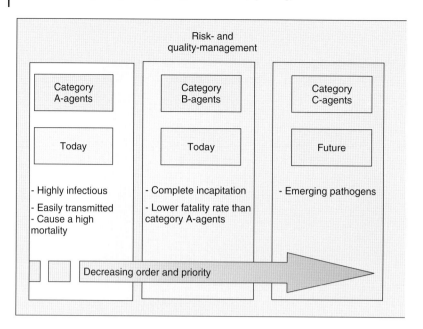

Figure 4.11 Categories of biological agents considered for biological warfare.

However, to date, no consensus list of potential biological warfare agents exists between nations. Most scientific, military, and security communities rely on the classification and public listing maintained by the Atlanta-based Centers for Disease Control (CDC). Its classification distinguishes three categories of B-agents, A, B, and C, in decreasing order and priority (Figure 4.11). At present, only six B-agents meet the criteria for category A.

Categories of Biological Agents, Considered for Potential Biological Warfare

- **Category A agents** are highly infectious, easily transmitted, and cause a high mortality.
- **Category B agents** – the second highest priority agents – completely incapacitate the victims but with a lower fatality rate than category A agents.
- **Category C agents** include emerging pathogens that might be engineered for mass dissemination in the future because of their availability, ease of production and dissemination, and potential for high morbidity and mortality, and major health impact.

Other national lists for orientation are the German list of "the dirty dozen" [8] and the Australian warning list of security-sensitive biological agents, a core list for export control that was recently updated [9] (Table 4.2).

Nevertheless, all these lists take the following factors and parameters into consideration:

Table 4.2 Comparison of different biological agent (B-agent) warning lists.

Biological-agents list		
By the CDC	By the Australian group	"The dirty dozen"

By the CDC	By the Australian group	"The dirty dozen"
Category A agents *Bacillus anthracis,* *Clostridium botulinum* toxins, *Yersinia pestis,* *Francisella tularensis,* Smallpox (variola major), Viral hemorrhagic fever viruses **Category B agents** *Brucella* species, *Salmonella* species, *Escherichia coli* O157:H7, *Shigella* species, *Burkholderia mallei,* *Burkholderia pseudomallei,* *Chlamydia psittaci,* *Coxiella burnetii,* *Vibrio cholerae,* *Cryptosporidium parvum,* *Rickettsia prowazekii,* Ricin toxin from *Ricinus communis,* Staphylococcal enterotoxin B, Epsilon toxin of *Clostridium perfringens,* Viral encephalitis viruses (e.g., VEEV, EEEV, and WEEV) **Category C agents** Nipah virus, Hantavirus	**Bacteria** *Bacillus anthracis, Brucella melitensis,* *Chlamydia psittaci,* *Francisella tularensis,* *Burkholderia mallei,* *Salmonella typhi,* *Yersinia pestis,* *Brucella abortus,* *Brucella suis,* *Clostridium botulinum,* *Shigella dysenteriae,* *Burkholderia pseudomallei,* *Vibrio cholerae,* *Clostridium perfringens,* Epsilon toxin producing types, *Escherichia coli,* serotype O157, and others **Viruses** Chikungunya virus, Congo-Crimean hemorrhagic fever virus, Dengue fever virus, Venezuelan and Eastern equine encephalitis virus, Ebola virus, Hantaan virus, Junin virus, Lassa fever virus, Lymphocytic choriomeningitis virus, and others **Toxins and subunits thereof** Botulinum toxins, *Clostridium perfringens* toxins, *Staphylococcus aureus* toxins, Conotoxin, Saxitoxin, Tetrodotoxin, Shiga-like proteins, Ricin, Shiga toxin, Verotoxin **Rickettsiae** *Coxiella burnetii,* *Rickettsia quintana,* *Rickettsia prowazekii,* *Rickettsia rickettsii*	*Bacillus anthracis,* *Yersinia pestis,* *Francisella tularensis,* *Brucella* species, *Burkholderia mallei,* *Burkholderia pseudomallei,* *Coxiella burnetii,* Smallpox (variola major), Venezuelan equine encephalitis virus (VEEV) *Clostridium botulinum* toxins, Ricin toxin from *Ricinus communis,* Staphylococcal enterotoxin B

- pathogenicity of the organism,
- mode of transmission and host range,
- availability of effective preventive measures (e.g., vaccines or medical treatment),
- availability of effective treatment (e.g., antibiotics).

Based on the properties mentioned above some of the B-agents have a greater risk of being used as a bioweapon than others. *Bacillus anthracis*, for example, along with variola virus (smallpox virus) and *Yersinia pestis* are more likely candidates than hemorrhagic fever viruses. In addition, research programs of the United States of America before 1969 and the former Soviet Union indicate that pathogens are considered as effective biological weapons.

Table 4.3 provides a starting point for information about some of the most virulent and prevalent biological warfare agents. It gives the reader a more concrete idea about some of the diseases caused by potential biological warfare agents, the countermeasures, and prophylaxis.

In public discussions the question often arises, "What is the most effective toxin or pathogen?" However, it is impossible to answer this question properly since the effectiveness of a toxic or pathogenic agent depends on diverse factors (Figure 4.1). In general severe health effects depend on the dose of a biological agent as well as on the time of exposure (Tables 4.4 and 4.5). While the ID_{50} (infectious dose 50%) level describes the dose of an infectious organism or pathogen that is required to produce infection in 50% of the subjects, the LD_{50} (lethal dose 50%) serves as the preferred scale of toxicity.

Because the selected biological toxins (Table 4.4) are extremely hazardous, even in minute quantities, they require strict safeguards against inhalation, absorption through skin or mucous membranes, and ingestion. Specific inactivation and disposal requirements have to be in place for them. A pathogen's infectious dose can vary from one to hundreds of thousands of units or particles and, as mentioned before, some of the B-agents considered here will have only incapacitating and weakening effects on man.

In general, all published data on infectious or lethal doses of B-agents should be regarded with great caution since most of them are extrapolated from either incident rates of epidemiological studies or animal experiments. At best they are worst case estimates since quantitative values for most human doses are not feasible. In addition, the route of exposure, anatomical differences of the respiratory tract, and agent potency can cause variations in the ID_{50} and LD_{50} levels of pathogens and toxins of several orders of magnitude.

4.4
Routes of Entry

Exposure to B-agents may occur in many ways, namely, by inhalation, injection, and ingestion (Figure 4.12). However, the most severe infections from B-agents come from aerosol and droplet infections caused by the natural process of breathing.

Table 4.3 Potential Biological Warfare Agents, the caused diseases and potential medical countermeasures.

Agent	Disease	Symptoms	Transmission	Treatment
Bacteria				
Bacillus anthracis	Anthrax	Fever, headache, malaise, and cough or macule, papule, ulcer, and black eschar	By inhalation, direct contact, and ingestion	Antibiotics available, effectiveness depends on type of disease. US FDA licensed vaccine available
Francisella tularensis	Tularemia	Fever, headache, malaise, and cough	By inhalation, direct contact, ingestion, and vector	Antibiotics available; no effective vaccine
Yersinia pestis	Plague	Fever, headache, malaise, cough, and bloody sputum	By inhalation and vector	Antibiotics available
Burkholderia mallei	Glanders	Fever, diarrhea, headache, muscle pain, and many others	By inhalation and direct contact	Antibiotics available
Brucella species	Brucellosis	Fever, headache, malaise, and cough	By inhalation, direct contact, and ingestion	Antibiotics available
Coxiella burnetii	Q fever	Influenza-like symptoms	By inhalation and direct contact	Antibiotics available
Viruses				
Ebola virus and Marburg virus	Hemorrhagic fever	Fever, chills, headache, malaise, myalgia, nausea, vomiting	By unknown vector and by direct contact	No effective antiviral treatment available, no effective vaccine available
Arenaviruses (Lassa v. Machupo virus)	Hemorrhagic fever	Fever, headache, malaise, fatigue, weight loss, sweats	By inhalation, direct contact, ingestion	Antiviral therapy with restricted effectiveness
Variola major virus	Smallpox	Fever, headache, and backache	By direct contact	Vaccine available

(continued overleaf)

Table 4.3 (continued)

Agent	Disease	Symptoms	Transmission	Treatment
VEEV	Encephalitis	Fever, neurological signs, malaise, vomiting	By vector (mosquito) or inhalation	No other treatment than supportive care; vaccine restricted to the USA
Tick-borne viruses	Hemorrhagic fever or encephalitis	Fever, malaise, headache, muscle aches, vomiting, nausea	By vector (tick)	Vaccine available
Flaviviruses (yellow fever and dengue viruses)	Yellow fever or dengue fever	Fever, headache, muscle pain, nausea, vomiting, bleeding, liver, and bladder problems	By vector (mosquito)	No effective antiviral treatment available for dengue fever, vaccine available for yellow fever
Nipah virus	Encephalomyelitis	Fever, headache, disorientation	By direct contact	Antiviral therapy with restricted effectiveness
Toxins				
Clostridial toxins	Intoxication	Weakness, paralysis, dilated pupils, dry mouth double vision	By ingestion	Antitoxin available; no treatment available
Staphylococcus enterotoxin B (SEB)	Intoxication	Nausea, vomiting, cramps, diarrhea, severe weakness	By ingestion	A few antibiotics may be effective, supportive care
Ricin	Intoxication	Nausea, vomiting, cramps, diarrhea, difficult breathing	By ingestion	No treatment available

Table 4.4 Selected toxins and their oral LD_{50}.

Agent	LD_{50} ($\mu g\,kg^{-1}$)	Time of onset (h)	Source
Ricin	3–15	15–20	Plant, castor bean
Abrin	10	7–36	Plant, rosary pea
Staphylococcus enterotoxin B	10–20	1–12	Bacterium
Botulinum toxin A	1–3	12–72	Bacterium
Saxitoxin	5–12	1–2	Algae

Data compiled from open source data and References [10–13].

Table 4.5 Lethal or incapacitating dose of selected viral and bacterial agents.

Agent	Aerosol ID_{50}	Incubation period (days)
Variola major	Unknown	7–17
Venezuelan equine encephalitis virus	1–10 PFU[a]	1–5
Ebola virus	Unknown	6–7
Bacillus antracis	8000–55 000 spores	1–7
Yersinia pestis	100–500 CFU[b]	1–6
Francisella tularensis	10–50 CFU[b]	1–10
Coxiella burnetii	1–30	4–21
Brucella melitensis	10–100 CFU[b]	5–21

[a] PFU, plaque forming unit.
[b] CFU, colony forming unit.
Data compiled from open source data and References [10–18].

Thereby, a continuing influx of the biological agent is inhaled and a direct pathway to the systemic circulation is provided. In this way low numbers of highly infectious microorganisms are already sufficient to cause disease by inhalation. As illustrated in Table 4.5 some threat agents like *Francisella tularensis* or VEEV require only 1–10 organisms for disease-causing infection. Aerosolized B-agents in the size range 1–10 μm are of most concern because these particles are efficiently retained in the human respiratory tract and deposited in the lung (Figure 4.12). Biological particles above this size range either settle out on the ground or are filtered out in the upper respiratory tract and disgorged. Particles below this size range are mainly transported by diffusion and mostly exhaled.

Another portal of entry for B-agents is the oral route by ingestion of contaminated food, water, or even medicine. The B-agent can be either deliberately sprinkled on or mixed in food or alimentary may be exposed to a B-agent during an aerosol attack. Classical symptoms of food poisoning include stomach ache, diarrhea, and vomiting.

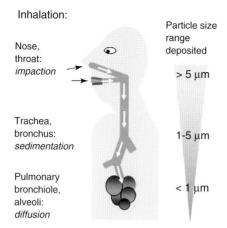

Figure 4.12 Deposition of biological agents in the human respiratory tract.

A third way of entry that can be utilized for biological agents is the skin although, normally, particular biological agents cannot pass the intact skin. However, penetration of the skin becomes possible if the skin is either injured or may be accomplished by the bite of an arthropod vector (carrier) or an intentional injection. The most prominent death by injection occurred in 1968 in England when Georgi Markov, a Bulgarian journalist, was killed by ricin. By using a spring-loaded umbrella a small metallic ball was injected into the back of his leg while he was standing on a street corner.

4.5
Origin, Spreading, and Availability

There are several key factors and features that make a pathogen or toxin suitable for use as a bioweapon. First, the availability or ease of production of B-agents in sufficient quantity is decisive (Figure 4.13). For example, the estimated amount of agent needed to cover a 100 km^2 area and cause 50% lethality is 8 metric tons for even a "highly toxic" toxin such as ricin versus only 1 kg of anthrax that would be needed to achieve the same coverage [19, 20].

As mentioned at the beginning of this chapter, pathogenic or toxic B-agents are universally distributed and, especially, some of the bacteria can be easily cultivated because they have only low nutritional requirements.

The castor bean plant (*Ricinus communis*), for example, originally a native tropical plant, is cultivated and grown in many European gardens and parks. The seeds of the plant are commercially available and the process necessary to harvest the toxin is technically straightforward. Viral and bacterial strains as well as toxins are legally available from culture repositories as well as from clinics, universities, governmental, and industrial laboratories where bioweapon activities can be easily

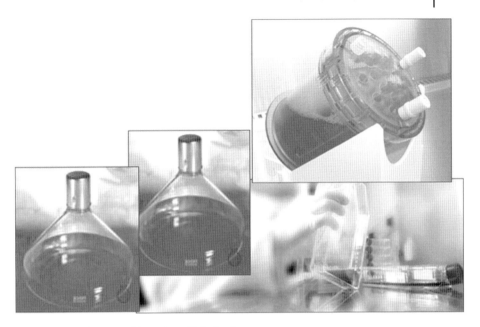

Figure 4.13 Cultivation of bacteria to high density.

hidden within legitimate research. Moreover, information about how to obtain and prepare bioweapons is increasingly available on the Internet and in open scientific literature.

Furthermore, a vast number of pathogens could be obtained from infected humans or animals in endemic areas or during outbreaks. Outbreaks of infectious disease and intoxications occur naturally around the world – whether of Ebola virus in Africa or of food and mouse disease in Europe. Further examples are the emergence or reemergence of diseases like tuberculosis (TB) and the microevolution of Chikungunya viruses (CHIKVs) causing the Indian Ocean outbreaks from 2005 to 2007 [21, 22].

Most bacterial and viral pathogens recognized as potential bioweapons occur primarily in countries of the third world. They appear in the form of classical tropical diseases with nonspecific symptoms, including muscle and joint pains as well as a strong headache and fever. Many emerging or reemerging infectious diseases start out as zoonotic disease or zoonoses, which means that the disease and infection are transmitted between vertebrates and man, often mediated by arthropod vectors. Humans have always lived with bacteria-like anthrax and glanders in their environment and *Bacillus anthracis* as well as *Burkholderia mallei* are typical soil dwellers. Therefore, one of the most common natural sources of human infection is the close community and cohabitation of man and animal. Transmission of infectious disease can occur by contact with excrement, by nasal and oral secretions, or by injuries from infected animals.

4.5.1
Methods of Delivery

A biological warfare attack is likely to be covert and the biological agents may be delivered as wet aerosol or dry dust. Biological agents can be dispersed by aerosolization or by an explosive device like a bomb or bomblet; in addition, spraying delivery systems for agriculture or rural development might be misused (Figure 4.14). However, biological agents do not necessarily need a highly sophisticated delivery system to cause public fear, as was demonstrated by "anthrax letters" in 2001.

Without question the production of B-agents with an appropriate particle size of 1–5 µm is preferred (Section 4.3). Their environmental stability plays an enormous role and limits the application of microorganisms, viruses, and toxins as bioweapons.

To maintain virulence, production, storage, and dissemination is critical. Exposure to ultraviolet (UV) radiation of daylight carries an increased risk of inactivation, as does the influence of high or low relative humidity or the chemical composition of the surrounding air. Moreover, production of biological warfare agents may include their modification so that the agent has non-natural characteristics.

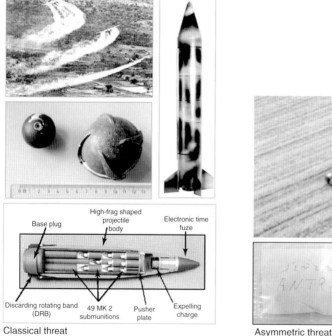

Figure 4.14 Potential delivery systems for B-agents.

4.6
The Biological Event – Borderline to Pandemics, Endemics, and Epidemics

A small outbreak of illness might be an early warning of a serious biological attack and basic epidemiological and medical skills are essential to differentiate a spontaneous endemic outbreak from an intentional attack. Once the case definition and attack rate have been determined, the outbreak can be characterized in context of time, place, and person. This basic epidemiological approach does not differ significantly from a standard epidemiological investigation. Defining an outbreak as endemic, epidemic, or pandemic can be subjective, depending in part on what is "expected." For a correct definition and discrimination of these incidents, as well as for exemplified events, the reader is referred to Part One. Since endemics describe diseases that are restricted but common to a local and particular geographic area or population, epidemics refer to diseases that either involve many more people than usual in a particular community or spread into regions in which they naturally do not occur. A biological attack is normally not immediately obvious but many BW-agents, except toxins, are highly contagious so that a disease can spread from single exposed persons to a broader population. However, the period between exposure and the onset of clinical signs can last days to weeks (Table 4.3), although sometimes sick or dead animals could provide an early warning to humans. A further indication that a biological attack has occurred may be a growing number of patients/soldiers suffering from a similar disease. In addition, an unusual geographic or seasonal distribution of disease might be suspicious.

4.7
The Bane of Biotechnology – Genetically Engineered Pathogens

The modern era of biotechnology and genetic engineering began in the late 1970s and early 1980s. Since then these technologies have made significant technical and scientific advances in regard to gene synthesis, mutagenesis, *in vitro* cell culture, and *in vitro* production [23–27]. In parallel the risk of misuse has increased and today, besides natural pathogens, genetically engineered pathogenic wild-type strains and also completely genetically engineered organisms are conceivable (Figures 4.15 and 4.16).

In the following we give the reader an initial idea of the enormous dual-use potential of these new technologies. Although there are still technical barriers for bioterrorists it is undeniable that the described techniques can pose severe safety and security risks.

In general, genetically modified B-weapons are not new and the Soviet Civilian Biological Warfare Agency, Biopreparat, had experimented already from 1973 with various harmful and antidote-resistant organisms, including a combination of smallpox with Venezuelan equine encephalitis, known as "*Veepox*" [25]. Russia also developed "Obolensk" anthrax – a *Bacillus anthracis* strain resistant to both vaccines and antibiotics [25]. Whether the new form of anthrax is able to elude

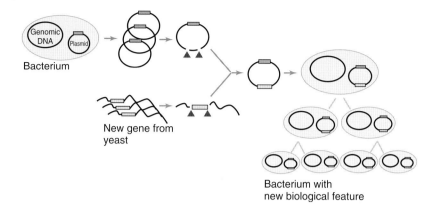

Figure 4.15 Principle of genetic engineering shown by bacteria, the first organisms modified in the laboratory due to their simple genetics.

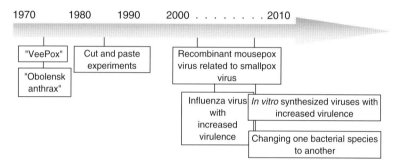

Figure 4.16 Changing biological features of B-agents – examples from the past up to the present day.

the vaccine that American troops receive is unanswered. Early experiments in the 1980s were restricted to identifying and to "cutting and pasting" single genes from one cell to another, but since then many new frontiers have opened up. These include the possibility of manipulating genes and of creating novel treatment- and detection-resistant biological warfare agents as well as those with enhanced hardiness [25–31].

These days chimera B-agents are close to reality. Multiple virulence and toxic properties can be placed within a single organism and genes for toxic proteins can be fused and expressed together functionally.

In 2003, for example, scientists succeeded in combining defined gene fragments of the extremely virulent influenza virus strain from 1918 with a standard laboratory influenza strain of low virulence [27]. Mice infected with a certain chimera died earlier and in higher number. This clearly indicated that genetic engineering allows the creation of microorganisms with an increased virulence. In addition, an Australian research team reported in 2001 a modified mousepox virus that does not infect humans but is closely related to the smallpox virus [30].

Today viruses can be modified as vectors to alter their pathogenicity in animals and man and they can act as carriers for genes or toxins. Conceivably, plant pathogens can be modified as biological warfare agents against crops. Moreover, our expanding and profound knowledge of microbial genomes and progress in chemical gene synthesis offers these days the possibility to produce single genes as well as complete genomes. Diverse databases provide information about genes involved in pathogenic mechanisms, virulence, immune response, and antibiotic resistance. One of the first viruses fully synthesized *in vitro* was the poliovirus and the 1918 Spanish influenza pandemic virus, both having a RNA genome of only a few kilobytes [27]. In addition, the power of gene synthesis was spotlighted by J. Craig Venter. He reported the construction of the first complete bacterial genome and also demonstrated that replacing a genome could give a microorganism a new identity, which represents a first step toward creating artificial life [26].

However, the more complex a pathogens' genome is the more complications arise during gene synthesis. In addition, the difficulty of producing or synthesizing a pathogen genome and creating a viable pathogen does not depend only on its genome size.

4.8
Conclusions and Outlook

We hope that the contents of this chapter will contribute to a deeper understanding of the diversity of biology. We have taken a closer look at the unique characteristics of B-agents with potential as B-weapons. For some of them we explained the symptoms of disease, their effective dose, possibilities as well as their routes of entry. Moreover, we discussed their risk classification, possibilities of dissemination, and availability. We have pointed out the future threat of genetically engineered pathogens. Hopefully, the contents will help support us in future to take appropriate countermeasures in case of a biological attack.

References

1. Cossart, P., Boquet, P., Normark, S., and Rappuoli, R. (2000) *Cellular Microbiology*, ASM Press, Washington, DC.
2. Watson, D.J., Baker, T.A., Bell, S.P., Gann, A., and Losick, R. (2008) *Molecular Biology of the Gene*, Benjamin Cummings.
3. Engelkirk, P.G. and Burton, G.R.W. (eds) (2007) *Burton's Microbiology for the Health Sciences*, Lippincott Williams and Wilkins.
4. Madigan, M., Martinko, J., and Dunlap, P. (eds) (2009) *Brock Biology of Microorganisms*, Benjamin Cummings.
5. Holt, J.G. (ed.) (1984) *Bergey's Manual of Systematic Bacteriology*, Williams and Wilkins, Baltimore.
6. Foster, S.J. and Johnstone, K. (1990) *Mol. Microbiol.*, **4** (1), 137–141.
7. Volk, T. (2000) Tom Volk's Fungus of the Month for January 2000. Available at *http://botit.botany.wisc.edu/toms_fungi/jan2000.html* (accessed 1 May 2011).
8. Sohns, T. (1999) *ASA Newsl.*, **5**, 99.
9. Security sensitive biological agents (2011). *http://www.health.gov.au/SSBA* (accessed May 2011).

10. Committee for an Assessment of Naval Forces' Defence Capabilities Against Chemical and Biological Warfare Threats, National Research Council (2004) *Naval Forces' Defense Capabilities Against Chemical and Biological Warfare Threats*, The National Academies Press.
11. USAMRIID (2005) USAMRIID's Medical Management of Biological Casualties Handbook (Bluebook) 6th edn, US Army Medical Research Institute of Infectious Disease. Available at *http://www.usamriid.army.mil/education/bluebookpdf/USAMRIID%20BlueBook%206th%20Edition%20-%20Sep%202006.pdf* (accessed 23 February 2012).
12. Health Canada (2001) Material Safety Data Sheets, *http://www.hc-sc.gc.ca/pphb-dgspsp/msds-ftss/index.html* (accessed 20 August 2009).
13. Gill, D.M. (1982) *Microbiol. Rev.*, **46**, 86–94.
14. Langford, R.E. (ed.) (2004) *Introduction to Weapons of Mass Destruction: Radiological, Chemical and Biological*, John Wiley & Sons, Inc., Hoboken.
15. Johnson, B. (2003) *Appl. Biosaf.*, **8** (4), 160–165.
16. Franz, D.R., Jahrling, P.B., Friedlander, A.M., McClain, D.J., Hoover, D.L., Byrne, W.R., Pavlin, J.A., Christopher, G.W., and Eitzen, E.M. (1997) *J. Am. Med. Assoc.*, **278** (5), 399–411.
17. Franz, D.R. (1997) in *Textbook of Military Medicine: Medical Aspects of Chemical and Biological Warfare* (eds F.R. Sidell, E.T. Takafuji, and D.R. Franz), Office of The Surgeon General, Department of the Army, Falls Church, ch. 30, pp. 603–619.
18. Leitenberg, M. (2001) *Crit. Rev. Microbiol.*, **27**, 267.
19. Spertzel, R.O., Wannemacher, R.W., Patrick, W.C., Linden, C.D., and Franz, D.R. (1992) Technical ramifications of inclusion of toxins in the chemical weapons convention (CWC). Technical report no. MR-43-92-1, U.S. Army Medical Research Institute of Infectious Diseases, Fort Detrick.
20. World Health Organization Group of Consultants (1970) Health Aspects of Chemical and Biological Weapons, World Health Organization, Geneva.
21. Vazeille, M., Moutailler, S., Coudrier, D., Rousseaux, C., Khun, H., Huerre, M., Thiria, J., Dehecq, J.S., Fontenille, D., Schuffenecker, I., Despres, P., and Failloux A.B. (2007) *Public Libr. Sci.*, **14**, 2 (11), 1168.
22. Renault, P., Solet, J.L., Sissoko, D., Balleydier, E., Larrieu, S., Filleul, L., Lassalle, C., Thiria, J., Rachou, E., de Valk, H., Ilef, D., Ledrans, M., Quatresous, I., Quenel, P., and Pierre, V. (2007) *Am. J. Trop. Med. Hyg.*, **77** (4), 727–731.
23. Alberts, B., Bray, D., Lewis, J., Raff, M., Roberts, K., and Watson, J.D., Recombinant DNA Technology, chapter 7 (1994) *Molecular Biology of the Cell*, Part II: Molecular Genetics, 3rd edn. Garland Science, New York
24. Pennisi, E. (2007) *Science*, **316**, 1827.
25. Alibek, K. and Handelman, S. (1999) *Biohazard: the Chilling True Story of the Largest Covert Biological Weapons Program in the World – Told from Inside by the Man Who Ran It*, Random House Inc., New York.
26. Lartigue, C., Glass, J.I., Alperovich, N., Pieper, R., Parmar, P.P., Hutchison, C.A. III, Smith, H.O., and Venter, J.C. (2007) *Science*, **317** (5838), 62832–62832.
27. Conenello, G.M., Zamarin, D., Perrone, L.A., Tumpey, T., and Palese, P. (2007) *PLoS Pathog.*, **3** (10), 1414–1421.
28. Block, S.M. (1999) in *The New Terror: Facing the Threat of Biological and Chemical Weapons* (eds S.D. Drell, A.D. Sofaer, and G.D. Wilson), Hoover Institution, Stanford, pp. 39–75.
29. Nowak, R. (2001) *New Sci.*, **169**, 4–5.
30. Jackson, R.J., Ramsay, A.J., Christensen, C.D., Beaton, S., Hall, D.F., and Ramshaw, I.A. (2001) *J. Virol.*, **75**, 1205–1210.
31. Wheelis, M. and Dando, M.R. (2000) *Disarmament Forum*, **4**, 43–50.

5
Characteristics of Nuclear and Radiological Weapons

Ronald Rambousky and Frank Sabath

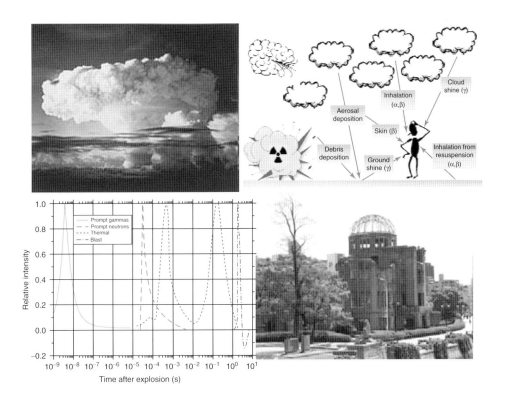

Since the detonation of nuclear weapons over Japan in 1945, nuclear weapons have been the most feared kind of weapons of mass destructions, due to their capability to cause enormous destruction. In fact, nuclear weapons are relatively small, light, and inexpensive compared to the conventional explosives needed to destroy large area targets. A single nuclear warhead can destroy an entire city. Even with a small stockpile of such devices a state or subnational group will boost its importance. Most of today's nuclear stockpile is owned by two nations, Russia and the USA.

Owing to the enormous capability and political importance of such weapons, six further countries maintain a small number of nuclear warheads and several nations are planning to obtain nuclear weapons as a result of their own nuclear programs.

This chapter focuses on the description of the characteristics of nuclear and radiological weapons, like the fundamental mechanism of nuclear explosion, the different forms in which the energy is released, basic design concepts of nuclear weapons, and how a nuclear explosion or radiological material may affect living organisms and materials.

5.1
Introduction to Nuclear Explosions

A nuclear explosion occurs as a result of the rapid release of energy from an intentionally high-speed nuclear reaction. Today, in nuclear warheads nuclear fission, nuclear fusion, or a multistage combination of both is employed as the driving reaction. Like weapons based on conventional explosives, the effects of nuclear weapons are dominated by blast and thermal radiation. The two aspects that make nuclear weapons special are (i) the huge amount of energy produced and (ii) the (initial and residual) nuclear radiation.

Before we discuss the effects of a nuclear explosion in detail, we shall take a brief look into the physics of the nuclear reactions used and the basic design principles of nuclear weapons in order to understand their different impacts and effects.

5.1.1
Nuclear Fission

When a free neutron enters the nucleus of a fissionable atom, it can initiate the splitting of the nucleus into two or more smaller parts (Figure 5.1). This process is called the *fission reaction*; the lighter nuclei that result from the fission reaction are

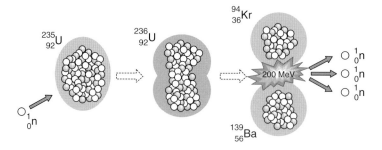

Figure 5.1 Principle of nuclear fission reaction.

called *fission products*. The fission reaction is accomplished by the release of a large amount of energy.

The fission of an uranium or plutonium nucleus releases two or three neutrons that can initiate further fission. Thus, in principle, a single neutron could start a chain of nuclear fissions, the so-called nuclear chain reaction. Although there are many different ways in which the nuclei of uranium and plutonium can split into their fission products, the total amount of energy liberated per fission does not vary greatly from 200 MeV. This fission energy is approximately distributed as shown in Figure 5.2 and Table 5.1.

5.1.1.1 Critical Mass for a Fission Chain

Not all neutrons that are produced in a fission reaction are available to cause further fission. Some of the fission neutrons are lost by escape, whereas other are lost in various non-fission reactions. To sustain a fission chain reaction, for each neutron absorbed in a fission reaction at least one fission neutron must be available to cause further fission. To achieve a nuclear explosion it is essential to establish conditions under which the loss of neutrons is minimized, so that on average more than one neutron is produced by each absorbed neutron.

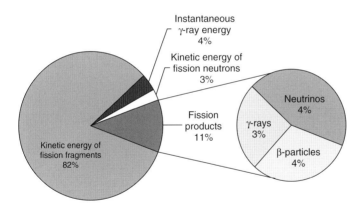

Figure 5.2 Energy of nuclear fission.

Table 5.1 Distribution of fission energy [8].

	Energy (MeV)	% of total
Kinetic energy of fission fragments	165 ± 5	82
Instantaneous γ-ray energy	7 ± 1	4
Kinetic energy of fission neutrons	5 ± 0.5	3
Beta particles form fission products	7 ± 1	4
Gamma rays from fission products	6 ± 1	3
Neutrinos from fission products	10	4
Total energy per fission	200	100

Generally, the escape of neutrons occurs at the boundary of the fission material. Therefore, the rate of loss is determined by its surface area. As the fission process takes place over the whole fission material, the rate of produced neutrons is a function of the mass. If the mass of a given fission material is increased, the ratio of the surface area to the mass and therefore the rate of escaped neutrons to neutrons produced by fission is decreased. Consequently, by further increasing of the mass a point will be reached at which the ratio of escaped neutrons equals the ratio of newly produced neutrons and the chain reaction becomes self-sustaining. The literature refers to this state as a critical mass of the fission material. The same result will be achieved if the surface area of a given (constant) mass is decreased by compressing it to a smaller volume (higher density).

Another source of neutron loss is impurities, which can absorb neutrons in non-fission reactions. Among other aspects (e.g. above 50 km) the critical mass is a function of the fissionable isotope, its density, and its purity (Table 5.2).

5.1.2
Nuclear Fusion

Nuclear fusion is a process during which light atoms fuse to form heavier ones (Figure 5.3). Theoretical considerations showed, that the fusion of two nuclei with a mass lower than iron generally releases energy, while the fusion of nuclei heavier than iron absorbs energy. Consequently, fusion occurs for lighter elements only.

In 1939 Hans A. Bethe suggested that much of the energy output of the sun and other stars results from reactions in which two hydrogen nuclei form one helium nucleus while releasing a large amount of energy. Later research proved that fusion

Table 5.2 Critical mass of fission material at normal density [17].

Isotope	Purity (%)	Critical mass (kg)
^{235}U	93.5	49
^{235}U	20	400
^{239}Pu	93	8

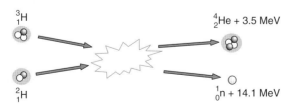

Figure 5.3 Principle of deuterium (D)-tritium (T) fusion.

Table 5.3 Energy released for selected fusion reactions.

Fusion reaction	Released energy per fusion (MeV)
$^2_1H + ^3_2He \rightarrow ^4_2He + ^1_1p$	18.3
$^2_1H + ^3_1H \rightarrow ^4_2He + ^1_0n$	17.6
$^1_0n + ^6_3Li \rightarrow ^4_2He + ^3_1H$	4.8
$^2_1H + ^2_1H \rightarrow ^3_1H + ^1_1p$	4.0
$^2_1H + ^2_1H \rightarrow ^3_2He + ^1_0n$	3.3

reactions can indeed occur only at many millions of degrees kelvin, when the electrostatic forces of repulsion that result from the presence of positive electric charges in both nuclei can be overcome so that the nuclear forces of attraction can perform a fusion. Creating these conditions on earth is very difficult and in fact it has not been accomplished in a systematic, stable, and controlled way for any isotope, including hydrogen.

In thermonuclear weapons the energy that is released by an uncontrolled nuclear chain reaction (nuclear explosion of an atomic bomb) is employed to heat and compress a fusion fuel (isotopes of hydrogen and lithium) to the point of ignition. At this point, the released energy is able to maintain the fusion reaction (Table 5.3).

5.1.3
Weapon Design

Basically, there are three design types for nuclear explosive modules [7, 18, 19]:

1) pure fission weapons
2) fusion-boosted fission weapons
3) thermonuclear weapons.

5.1.3.1 Pure Fission Weapon

Gun-type weapons are fission-based nuclear weapons whose design assembles their fissile material into a supercritical mass by shooting one piece of sub-critical material into another. Typically a hollow projectile is shot onto a spike that fills the hole in its center (Figure 5.4). Because the material is shot through an artillery barrel like projectile, the design is the called "gun" method. Alternative arrangements may include firing two pieces into each other simultaneously, though whether this approach has been used in actual weapons designs is unknown.

The main advantage of the gun-type weapon design is the simple and robust set up. Practically, plutonium cannot be used as fissile material as the speed of assembly is too low. In addition, the required amount of uranium is relatively large, and the efficiency is relatively low (with the tendency to decrease with increasing yield).

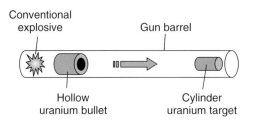

Figure 5.4 Gun-type design.

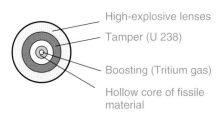

Figure 5.5 Principle of implosion design.

As a consequence, present nuclear weapon programs employ the implosion-type weapon only. Implosion-type weapon achieve supercritical condition and a nuclear initiation by squeezing fissile material (^{235}U, ^{239}Pu, or a combination of the two) and increasing its density. The principle design consists of a subcritical sphere of plutonium surrounded by high explosive (Figure 5.5). The required symmetry of the inward traveling shock wave is formed by a high explosive lens system, a combination of faster and slower explosives. The key to the great efficiency of the implosion-type design is the inward momentum of a massive tungsten or uranium tamper (which surrounds the core but does not undergo fission in the case of uranium). Once the chain reaction has started in the core of fissile material, the momentum of the implosion has to be reversed before expansion can stop the fission. Owing to its mass the tamper holds everything together for additional couple of 100 ns and therefore increases the efficiency.

5.1.3.2 Fusion-Boosted Fission Weapon

A fusion-boosted fission weapon is a type of fission weapon that applies a small amount of fusion material to increase the amount of free neutrons and thus the rate (and yield) of the fission reaction. The high pressure and temperature at the center of an exploding fission weapon is employed to heat and compress heavy isotopes of hydrogen and start a fusion reaction. As each neutron generated in the fusion reaction starts a new fission chain reaction, the fission is accelerated and the amount of unused fissile material is reduced. Fusion-boosting of a fission weapon can more than double the released energy.

5.1.3.3 Thermonuclear Weapons

The thermonuclear weapon design appropriates the energy released by a fission weapon (first stage) to create the conditions needed to ignite the fusion reaction in a second stage. The term *thermonuclear* refers to the high temperatures required to

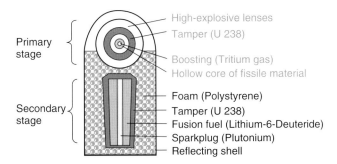

Figure 5.6 Schematic diagram of the Teller–Ulam design.

initiate fusion but it ignores the equally important factor of pressure (which was considered unknown at the time the term became current). The more correct term for this weapon design is fission-boosted fusion weapons.

The essential features of this design (Figure 5.6), which was originally developed by Edward Teller and Stanislaw Ulam, are [19]:

1) The separation into two stages, with a primary fission based stage that triggers a more powerful secondary fusion based stage.
2) Compression of the secondary stage by X-rays generated by the fission reaction of the primary stage. This radiation implosion utilizes the temperature difference between the hot surrounding radiation channel (polystyrene foam) and the relatively cool interior of the secondary stage. The temperature difference is maintained by a massive head barrier that also serves as an implosion tamper.
3) The secondary stage is heated by a second fission explosion (sparkplug) inside the secondary stage.

As it is also the most efficient design concept for small nuclear weapons, today virtually all nuclear weapons deployed by the five major nuclear-armed nations use the Teller–Ulam design.

5.1.4
Effects of a Nuclear Explosion

The energy released from a detonated nuclear weapon can be divided into the four basic categories (Fig. 5.7) [7]:

1) initial nuclear radiation,
2) thermal radiation,
3) blast and shock,
4) residual nuclear radiation.

Table 5.4 provides the general distribution of energy of these categories. Depending on the design of the weapon and the environment in which it is detonated the particular energy distributed to these categories might differ from the numbers shown.

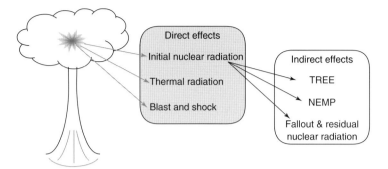

Figure 5.7 Effects of a nuclear detonation.

Table 5.4 General distribution of energy-to-effect categories [20].

Direct effects	Energy distribution		Indirect effects
	Fission weapon (%)	Fusion weapon (%)	
Initial nuclear radiation	5	50	Transient radiation effects on electronics (TREE). Nuclear electromagnetic pulse (NEMP)
Thermal radiation	35	20	–
Blast and shock	50	29	–
	10	1	Residual nuclear radiation

Most effects of nuclear weapons depend on the type of burst or explosion. Therefore, it is helpful to give a brief overview of the different types of bursts relevant in supposable nuclear scenarios. The most important types for destruction purposes are the *surface burst* and the *low air burst*. In a surface burst the nuclear warhead explodes on or slightly above the surface of the earth so that the fireball touches the land or the water surface. Therefore, a surface burst is accompanied by a large amount of radioactive fallout in the vicinity and downwind to ground zero. An *air burst* is a nuclear weapon explosion in the air at an altitude below 30 km but at sufficient height that the fireball does not contact the surface of the earth. Because blast and thermal effects on the surface can be maximized with an air burst, tactically this type is the most likely to be used against ground forces.

A *high altitude burst* is a nuclear explosion outside the earth's atmosphere at an altitude greater than 30 km. The prominent effect of a high altitude burst is the electromagnetic pulse (EMP) produced over a large area of the earth's surface, while blast, thermal effects, initial radiation, and fallout are of no concern. A *subsurface burst* is a nuclear explosion below ground or water. A subsurface burst will cause a large crater and strong local fallout when the burst penetrates the surface. These effects have to be taken into account when talking about nuclear bunker breaking weapons. A subsurface sea or water burst mainly causes water shockwaves and a large amount of radioactive splash water.

5.2
Direct Effects

> Effects that are immediately caused by the energy released in a nuclear explosion (nuclear radiation) or its interaction with the material in the immediate vicinity are denoted as **direct effects** of a nuclear explosion (Table 5.5).

5.2.1
Thermal Radiation

In conventional bomb explosions, temperatures of several thousand kelvin can be generated by chemical processes, that is, by interactions of the electrons in the shells of the atoms or molecules. Because of the vast energy release of nuclear fission or fusion processes the temperatures reached by nuclear explosions can exceed 100 million kelvin. This high temperature is the reason for the characteristic highlighted fireball accompanying nuclear explosions. Depending on the design of the nuclear warhead about one-third of the total energy is released as thermal

Table 5.5 Direct effects of nuclear explosions under certain conditions [8].

	Effect radius (km)			
	Weapon yield (HoB in m)			
	1 kt (HoB 200 m)	20 kt (HoB 540 m)	1 Mt (HoB 2000 m)	20 Mt (HoB 5400 m)
Blast effects				
Total destruction (140 kPa)	0.2	0.6	2.4	6.4
Extensive destruction (34 kPa)	0.6	1.7	6.2	17
Moderate destruction (6.9 kPa)	1.7	4.7	17	47
Thermal radiation effects				
Conflagration	0.5	2.0	10	30
Third degree burns	0.6	2.5	12	38
Second degree burns	0.8	3.2	15	44
First degree burns	1.1	4.2	19	53
Initial radiation effects				
Lethal dose (10 Gy)	0.8	1.4	2.3	4.7
Acute dose (1 Gy)	1.2	1.8	2.9	5.4

radiation, including electromagnetic infrared-, visible-, ultraviolet-, and low-energy X-ray radiation. Because electromagnetic radiation propagates at the speed of light, the lightning flash and thermal radiation affect locations away from ground zero a few seconds before the blast. Thermal effects of a nuclear explosion are maximal with a low air burst, depending of course on the total yield of the nuclear device. Because of the thermal radiation all combustible objects are inflamed in the effective area before the blast wave arrives. These ignited fires will then be again partially extinguished by the arriving blast, although the wind following the shock wave can, on the other hand, cause existing fires to spread out over large areas. This can lead to extensive firestorms.

Thermal radiation from nuclear explosions has various effects on humans. The enormous light density of the fireball can lead to temporary or permanent blindness even at great distances from ground zero. Because of the temperature (millions of kelvin) in the emerging fireball, which is 10 000–20 000 times greater than the surface temperature of the sun (about 5800 K), the fireball appears much brighter than the sun. The released thermal radiation causes severe skin burns that decrease with increasing distance to ground zero (Table 5.6). In the hypocenter of the explosion thermal radiation is so high that nearly all material is vaporized. The distance for different thermal radiation effects of course depend on the total weapon yield and the height of burst (HoB). In addition, other environmental conditions like air density, air humidity, or cloud cover will also influence the dependence of thermal radiation effects on distance. For example, high humidity and a large amount of dust in the air will decrease thermal radiation while snow, ice, bright sand, or cloud cover above the actual explosion height will enhance thermal radiation within a certain distance. To get an idea of the expected thermal radiation effect radii, for a 1 Mt air burst at optimized HoB in a cloudless sky and standard visibility range third-degree skin burns will happen at a distance of about 12 km, second-degree skin burns at about 15 km, and first-degree skin burns at about 19 km. A high chance of inflammation of ambient objects will occur within 10 km. Table 5.6 shows the energy fluence of the thermal radiation of different nuclear weapon yields for skin burns from first to third degree. At first glance it seems confusing that the energy fluence for a certain skin burn effect is not constant for different weapon yields. The reason for a higher energy fluence at higher yields

Table 5.6 Energy fluence of thermal radiation ($J\ cm^{-2}$) and resulting degree of skin burn; the distance in km up to which the skin burns occur is given in parentheses.

Skin burn	Weapon yield		
	20 kt	1 Mt	20 Mt
Third degree	34 (2.5)	40 (12)	50 (38)
Second degree	21 (3.2)	25 (15)	36 (44)
First degree	11 (4.2)	13 (19)	21 (53)

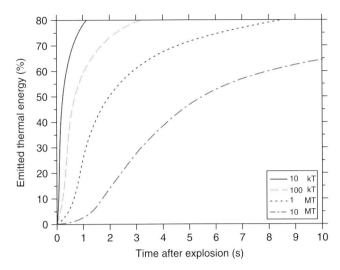

Figure 5.8 Time dependence of relative thermal energy emitted for various weapon yields in an air burst with standard visibility. (Adopted from [8]).

(and therefore also at a greater distance) for the same skin burn effect is that at higher yields the duration of irradiation is longer than for lower yields (Figure 5.8). Hence the energy fluence rate of the thermal radiation is higher for smaller yields for the same fluence, leading to the known effect that the same dose of radiation is biologically more effective when applied in a shorter time. Additionally, the skin burn symptoms appear at a greater distance for higher yields and the spectrum of the thermal radiation also changes with distance, which is caused by frequency dependant scattering processes in the atmosphere.

We now look at the physics of thermal radiation in nuclear explosions in more detail using air bursts as an example. After the explosion of a nuclear device, temperature equilibrium is rapidly achieved in the residual material. Within the first microseconds about 70%–80% of the energy is emitted as primary thermal radiation [8]. Most of this thermal radiation consists of soft X-rays. Almost all the rest of the energy is delivered from the kinetic energy of the weapon debris at this time. The soft X-ray radiation is absorbed within a layer of air near the explosion. This absorbed energy is then immediately reemitted from the fireball as secondary thermal radiation with a longer wavelength in the ultraviolet, visible, and infrared regime. As a result, energy distribution and temperature are nearly uniform throughout the volume of the hot gas and the fireball is called an *isothermal sphere*. While growing in size, the temperature of the fireball cools to about 300 000 K and a shock wave develops at the fireball front. This is called the *hydrodynamic separation*. For a 20 kt burst this happens about 0.1 ms after explosion when the fireball has a radius of about 15 m (Figure 5.9). At about 1 μs after explosion an inner shock wave, formed by the expanding weapon debris, expands outwards in the isothermal sphere. Shortly after the hydrodynamic separation, the inner shock wave overtakes the outer shock wave of the fireball front. The propagating shock wave in front

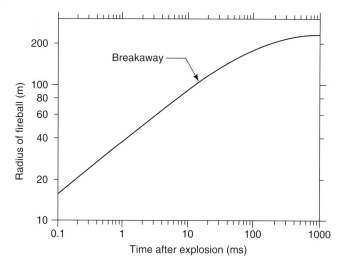

Figure 5.9 Radius of the luminous fireball as a function of time for a 20-kt air burst.

of the fireball causes an enormous compression of the surrounding air, leading to a luminous shell advancing the actual fireball. This is called the *hydrodynamic phase* of the nuclear explosion. The temperature of the outer luminous shell is higher than 8500 K and therefore the air is effectively opaque in this region at this point in time. Consequently, the hotter isothermal sphere is not visible through the outer luminous shell. For some time the fireball continues to grow in size at a rate determined by the propagation velocity of the shock front. Of course the temperature in the shock front decreases during this process. The moment when the temperature in the shock front has decreased so that the air becomes transparent and the inner hotter isothermal sphere can be seen again is called *breakaway*. This happens for a 20 kt air burst at about 15 ms after explosion when the fireball has a radius of about 100 m (see arrow in Figure 5.9). After the breakaway, the visible fireball continues to increase at a lower rate than before and will reach its maximum size after about a second for a 20 kt air burst.

The described processes are fairly complicated in detail. Roughly, though, there are two surface temperature pulses generated and therefore the emission rate of the thermal radiation also consists of two pulses (Figure 5.10). The first pulse is of very short duration, of about 11 ms for a 20 kt and 100 ms for a 1 Mt explosion and delivers only about 1% of the total thermal radiation energy. Because of the physics and temperature of the fireball at this time scale the radiation consists of a large amount of ultraviolet light. This is why the first thermal pulse is significant mainly for the permanent or temporary eye blindness effects.

The second thermal radiation pulse will last for a much longer time and carries about 99% of the total thermal radiation energy of the nuclear explosion. For a 1 Mt explosion the second pulse lasts for about 10 s for an air burst. Because the temperature of the fireball is much lower at this time most of the thermal radiation consists of infrared and visible light. It is this radiation that mainly causes the skin

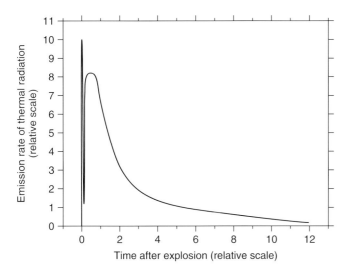

Figure 5.10 Exemplary emission of thermal radiation for an air burst.

burns and which is responsible for the large-scale ignition of flammable objects and the firestorms in the affected area.

5.2.2
Blast and Shock

Much of the destruction caused by a nuclear explosion in an ambient media, for example, air, surface, or sub-surface type, is due to the blast and shock effects. The shock wave of an air or low altitude burst contains 50–60% of the nuclear explosion's energy, depending on the design, size, and yield of the weapon. Owing to the physical mechanism that generates the shock wave, the blast fraction is higher for low yield weapons. The strength of the generated shock wave decreases at high burst altitudes because there is less air mass to absorb radiation energy and convert it into blast wave. Practically, the blast effect vanishes for burst altitudes above 30 km.

In a nuclear detonation at low altitude molecules of the surrounding material, for example, air, in the immediate vicinity are heated to high temperatures by absorption of gamma radiation. The extremely high temperatures and pressure of the material in the resulting fireball causes a radially outward moving, thin and highly dense shell called the *hydrodynamic front*. The hydrodynamic front acts like a piston that pushes against and compresses the surrounding medium to generate a spherically expanding shock wave. The main characteristic of this shock wave is that the pressure at the moving front rises very sharply and falls off toward the interior region of the explosion. In the early stages, pressures at the shock front can be three times as large as the already very high pressures in the interior of the fireball.

The blast effects are associated with two important quantities of the blast wave:

- **Static overpressure**, that is, the sharp increase in pressure exerted by the shock wave. The overpressure at any given point is directly proportional to the density of the air in the wave.
- **Dynamic pressures**, that is, drag exerted by the blast winds required to form the blast wave. These winds push, tumble, and tear objects.

Most of the material damage caused by a nuclear air burst is caused by a combination of both the high static overpressures and the blast winds. The long compression of the blast wave weakens structures, which are then torn apart by the blast winds.

From a practical point of view it is of interest to examine the variation over time of static overpressure (overpressure) and dynamic pressures (wind) at a fixed observation point (Figure 5.11). As it takes some time for the shock wave to reach the observation point, there will be no change in the ambient pressure for a short time interval after the detonation. As the speed of the shock wave increases with the overpressure at the shock front, a shock wave of an explosion of higher yield will arrive at a given observation point earlier than one from a lower yield. The initial velocity of the shock wave might be several times the speed of sound in air. As the blast wave moves outward the pressure at the front decreases and consequently its

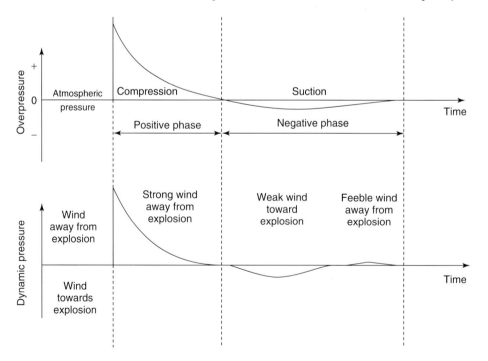

Figure 5.11 Time variation of overpressure and dynamic pressure at a fixed location (Adopted from [8]).

speed falls off. At long ranges the speed of the blast wave decreases to the ambient speed of sound.

When the shock front arrives at the observation point it acts like a moving wall. The overpressure as well as the dynamic pressure increase rapidly to their peak value and fall off behind the shock front (Figure 5.11). The first positive (or compression) phase is characterized by high static overpressure conjunct with strong winds in the direction away from the nuclear explosion. The strong winds can be greater than the strongest hurricane and may reach several hundred kilometers per hour. The end of the positive phase is marked by a drop of the overpressure to zero and a calm phase. The duration of the positive phase increases with the yield of the explosion and the distance of the observation point from it.

After the overpressure has reached zero, it will continue to decrease due to fluid dynamics. As a consequence, in this negative phase (or suction phase) the overpressure in the shock wave is less than the ambient atmospheric pressure. As depicted in Figure 5.11 the static underpressure is very small compared to the peak value in the positive phase. The underpressure draws air in from outside and the dynamic pressure decreases and the wind starts to blow in the opposite direction (i.e., towards the point of nuclear explosion). As the overpressure minimum is reached the wind reverses direction and blows again in the direction toward the point of the nuclear explosion. The negative phase usually lasts longer than the positive phase and can last for several seconds. When the negative phase is ended, the pressure returns to the ambient atmospheric value and the air is calm again the blast wave has passed the observation point.

The effects of the shock wave and blast winds on urban structures are depicted in a sequence of pictures shown in Figure 5.12. The pictures were taken at the Nevada test site during the Annie shot of Operation Upshot-Knothole nuclear test series on 17 March 1953. The wood-frame house shown was located 1.1 km from ground zero. The nuclear detonation with a yield of 16 kt took place at a high of 90 m caused a peak overpressure of 35 kPa and a surface wind speed of 250 km h^{-1}.

Table 5.7 lists typical effects of a shock wave of a given peak value in urban areas. The data clearly indicate that shock waves with a peak value of 35 kPa or higher result in severe destruction of buildings and lethal injuries. If such a shock wave hits a human body it causes pressure waves that travel through the tissue and mostly damages tissue junctions of different density (e.g. bones and muscles) or interfaces between tissue and air. Therefore eardrums, lungs, and the abdominal cavity are expected to be damaged by the shock wave.

The curves in Figure 5.13 show isobars of the overpressure as a function of the distance D from ground zero and the height of burst (HOB) at observation points near the ground surface for a 1-kt air burst. The corresponding data for explosions with other yields can be obtained by employing the scaling relation:

$$\frac{D_{kT}}{D} = \frac{HoB_{kT}}{HoB} = \sqrt[3]{W_{kT}} \qquad (5.1)$$

where, for a given overpressure:

Figure 5.12 Effect of the thermal radiation and blast wave on a wood-frame house caused by a 16-kt detonation at a height of 90 m and a distance of 1.1 km.

D denotes the distance from ground zero for a 1-kt air burst,
HoB is the height of burst for 1-kt air burst,
D_{kT} denotes the distance from ground zero for a W_{kT} kt air burst,
HoB_{kT} is the height of burst for a W_{kT} kt air burst.

5.2.3
Initial Nuclear Radiation

The initial nuclear radiation, in short initial radiation, is generally defined as the ionizing radiation emitted from the fireball and the radioactive cloud within the first minute after explosion. This somewhat arbitrary time period of 1 min was originally derived from the following nuclear scenario. For a 20-kt explosion the effective range of gamma radiation is about 3 km. The radioactive cloud of a 20-kt air burst at optimized HoB takes about 1 min to rise to a height of 3 km. Consequently, for the mentioned weapon yield there will be no significant radiation effects from the fission products of the radioactive cloud on the surface after about 1 min.

Table 5.7 Typical effects of blast waves in urban areas [8].

Overpressure (kPa)	Effects in urban areas
1.4	Typical window panes break
7	Window smashes, injuries by splinter possible
21	Dwelling houses (light construction method) severely damaged or destroyed, numerous serious casualties, isolated deaths
35	Destruction of most unreinforced buildings, numerous deaths
70	Destruction or heavy damage of armored concrete constructions, death of most inhabitants
140	Destruction or heavy damage also of heavy concrete structures, hardly any survivors (hypocenter of Hiroshima: approximately 210 kPa)
350	Entire destruction of all surface buildings (hypocenter of Nagasaki: approximately 420 kPa)
2000	Complete plantation of the scenery (hypocenter of the "Tsar bomb")

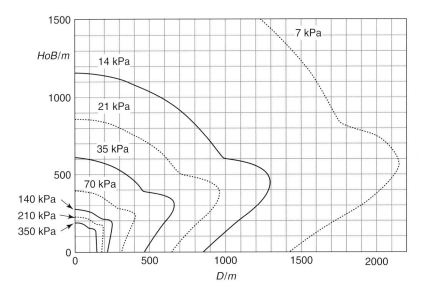

Figure 5.13 Overpressure of a 1-kt air burst as a function of burst height *(HoB)* and distance *(D)* to ground zero [8].

> **Ionizing radiation** consists of subatomic particles or electromagnetic radiation that possess enough energy to detach electrons from atoms or molecules (ionization process) on interacting with matter.

According to this definition initial radiation includes not only neutrons and gamma radiation produced nearly instantaneously at the beginning of the nuclear explosion but also the gamma radiation produced by the fission products and other radioactive particles in the rising cloud. The explosion of a nuclear weapon is associated with the emission of a mixture of ionizing radiation, consisting of neutrons, high energy photons, and alpha and beta particles. Because of the restricted range of alpha and beta radiation (Chapter 9) only neutron and gamma radiation are relevant for initial nuclear radiation.

Concerning pure fission, nearly all of the neutrons and part of the gamma radiation are produced simultaneously with the nuclear explosion caused by the fission process in the nuclear material, which consists of highly enriched uranium or plutonium. Therefore, this neutron and gamma radiation is called *prompt ionizing radiation*. Some of the neutrons produced in the fission process, which are fast neutrons, are immediately captured by the surrounding material, others are scattered at nuclei of the ambient matter. Recurring scattering processes slow down the neutrons and convert many of the fast neutrons into slow neutrons and thermal neutrons. The capture or scattering of a neutron with a nucleus generally is accompanied by spontaneous emission of gamma radiation. The remaining part of the gamma radiation is emitted by the fission products or by the neutron-induced radioactive isotopes when undergoing radioactive decay.

A fusion reaction in a nuclear weapon explosion must be triggered by a fission reaction, or, in other words, the nuclear explosion of a fission device is necessary to start the fusion reaction in a nuclear warhead. Fast neutrons with energies of 14 MeV are the only significant ionizing radiation produced directly in the thermonuclear reaction. Some of these fast neutrons will escape, but a significant part will react with the nuclei present in the exploding device in scattering or capture reactions. Those neutrons that hit remaining uranium or plutonium nuclei may cause additional fission reactions leading to the emission of further fast neutrons. The initial radiation from boosted fission devices or multistage warheads, where both fission and fusion processes occur, consists essentially of neutron and gamma radiation. The relative proportion of the two types of ionizing radiation emitted from the exploding device as initial radiation depends on the weapon design and the total yield of the explosion. Not only does the relative proportion of gamma radiation and neutron radiation change while traveling through the air to more distant regions but also the energy spectrum changes because of scattering and capture reactions as a result of the interaction of the ionizing radiation with the ambient air. The change in energy spectrum is typically more distinct with neutrons.

Because the main part of the initial radiation is emitted nearly instantaneously, not the dose rate but the dose itself is the significant quantity concerning initial radiation. As expected, the initial gamma dose decreases with increasing distance to the nuclear explosion. Basically, the dose is governed by the inverse square law, but absorption and scattering in air also has to be taken into account, which leads to a deviation from a simple functional dependence. Of course, the initial radiation effect radii depend strongly on the weapon yield, the HoB, device type, and meteorological conditions. The dependence of the slant range for several dose values on the weapon yield is shown in Figure 5.14a for the gamma part of the initial radiation. Values from 1 to 100 kt yield are given for a pure fission device and from 100 kt to 20 Mt for a thermonuclear device with 50% fission and 50% fusion. The given values are averaged and considered to be reliable within a factor of 0.5–2

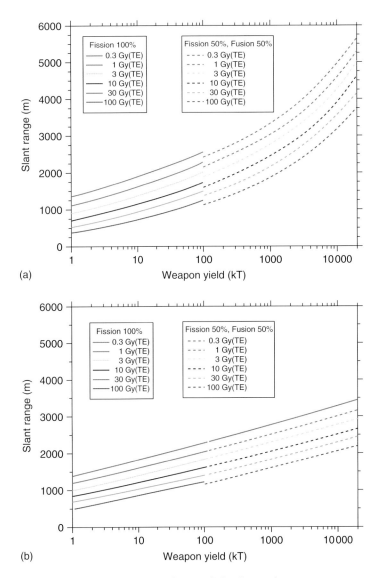

Figure 5.14 (a), (b) Slant ranges for specified radiation doses, given in tissue equivalent (TE) absorbed dose, for positions near ground as a function of the weapon yield; 1–100 kt for a pure fission weapon; 100 kt to 20 Mt for a thermonuclear weapon with 50% fission and 50% fusion: (a) initial gamma dose and (b) initial neutron dose.

for most fission weapons and within a factor of 0.25–1.5 for most thermonuclear weapons.

Strikingly, the distance for a given gamma dose increases more rapidly for high weapon yields. The reason for this effect is the fact that the density of the air following the shock wave is significantly lower, especially for high yield explosions. The delayed gamma radiation emitted by the fission products will experience that lower air density on its outward travel. This effect is called the *hydrodynamic enhancement* of the initial gamma dose. Hence the importance of the fission products to the initial gamma radiation increases with increasing weapon yield. Figure 5.15 shows the relative total initial gamma dose plotted against early time after explosion; this is exemplary for a low and a high yield explosion at the displayed ranges. Clearly, in the low yield case about 65%, and in the high yield case only about 5%, of the total initial gamma dose is received within the first second after explosion.

For the initial neutron radiation it is even more difficult to predict a functional correlation between weapon yield and neutron dose at a particular distance from ground zero when not considering the actual type of device. As mentioned before this is because neutrons change their energies over a much wider range than gamma quanta while scattering with surrounding nuclei. Figure 5.14b gives an estimate of the slant range for given initial neutron doses according to weapon yield. Again the low yield values are for pure fission devices and the high yield values for a 50% fission, 50% fusion thermonuclear device. Because the fast neutrons mostly travel in front of the shock wave and account much more for the absorbed dose than the slower neutrons, there is no hydrodynamic enhancement effect with the initial neutrons.

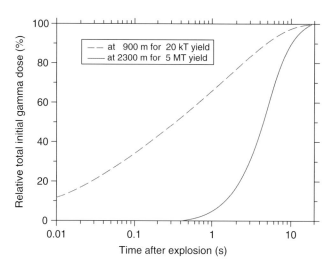

Figure 5.15 Time dependence of the relative initial gamma dose received for a 20-kt burst at about 900 m and a 5-Mt burst at about 2300 m distance.

Concerning biological effects of initial radiation, it has to be pointed out that the biological impact of neutron radiation is higher than for gamma radiation at the same *absorbed dose* measured in Gray (Gy). Fast neutrons are more effective by a factor of about ten. For low yield weapons, often called *tactical nuclear weapons*, the initial radiation generally is the dominant hazard compared to thermal and blast effects. Furthermore, people have nearly no chance of avoiding the initial radiation, because the effect is instantaneous and available shielding is often not effective. Therefore, initial radiation doses can easily exceed the threshold for acute radiation effects. Table 5.5 shows clearly that the dominant effects of high yield weapons, often called *strategic nuclear weapons*, are thermal and blast effects and therefore initial radiation effects are of minor importance in the not-totally-destroyed regions.

In radiation protection, doses are given in the biologically assessed quantity called *dose equivalent* and measured in sievert (Sv), expressing a kind of probability of the chance of developing radiation-induced cancer in the long term. Concerning acute radiation effects the absorbed dose by the different tissues in the body is the more meaningful physical quantity. The acute radiation effects for short-term whole body exposure are shown in Table 5.8 and are given as absorbed doses for human tissue, often designated as tissue equivalent (TE) absorbed dose. For further explanation of dose units see Chapter 9.

5.2.4
Residual Nuclear Radiation

All ionizing radiation that is emitted 1 min or later after the nuclear weapon burst is called *residual nuclear radiation* or just *residual radiation*. There are different contributory factors such as device type, yield, and HoB and also the nature of the surface at ground zero. First of all every nuclear device consists of fissible material, usually uranium or plutonium. In a pure fission device only about 20% of that fissible material undergoes fission and is converted into highly radioactive fission products, which in turn emit mainly beta and gamma radiation. The non-fissioned part of the uranium or plutonium is vaporized during the explosion process and disperses with the other weapon debris. Fissile material is radioactive itself and emits mainly alpha radiation. The tremendous neutron flux accompanying a nuclear burst produces further radioactive material by activation of the exploding weapon debris and dust. This activated material starts to emit ionizing radiation itself. The amount and form of neutron induced radioactivity varies a lot with surface and air bursts. The additional presence of high energy fusion neutrons from a thermonuclear device plays a significant role here. Last but not least, in the direct vicinity of ground zero the neutron radiation activates the surface of the ground and other material and objects around. The radioactivity in this area is called neutron induced gamma activity (NIGA).

The primary hazard of the residual radiation results from the creation of fallout caused by the radioactivity produced during the explosion process. These particles can be dispersed over large areas by the wind according to the meteorological conditions and their effects can be significant at distances well beyond the range of

Table 5.8 Radiation effects of acute whole-body exposure [21].

Radiation dose	Biological effects
Below 1 Gy	No obvious sickness occurs; detectable changes in blood cells begin to occur at 250 mGy, but occur consistently only above 500 mGy; changes set in over a period of some days and may require months to disappear; at 500 mGy atrophy of lymph glands becomes noticeable; depression of sperm production becomes noticeable at 200 mGy – an exposure of 800 mGy has a 50% chance of causing temporary sterility in males
1–2 Gy	Mild acute symptoms; tissues primarily affected are the hematopoietic and sperm-forming tissues; mild to moderate nausea (50% probability at 2 Gy), with occasional vomiting, setting in within 3–6 h after exposure and lasting for several hours or days; mild clinical symptoms return in 10–14 days; recovery from other injuries is impaired; enhanced risk of infection
2–4 Gy	Illness becomes increasingly severe and significant mortality sets in; hematopoietic tissues are still the major affected organ system; nausea becomes universal (100% at 3 Gy); onset of initial symptoms within 1–6 h; symptoms last 1–2 days and a 7–14 day latency period sets in; recurring symptoms may include hair loss (50% probability at 3 Gy) and hemorrhage of the mouth; at 3 Gy the mortality rate without medical treatment is about 10%; recovery takes one to several months
4–6 Gy	Mortality rises steeply from 50% at 4.5 Gy to about 90% at 6 Gy (without medical treatment); hematopoietic tissues are still the major affected organ system; initial symptoms appear in 0.5–2 h and last up to 2 days; latency period remains 7–14 days; when death occurs it is usually 2–12 weeks after exposure and results from infection and hemorrhage; recovery takes several months to a year
6–10 Gy	Bone marrow is nearly or completely destroyed; survival requires marrow transfusions; gastrointestinal tissues are increasingly affected; onset of initial symptoms is 15–30 min; symptoms last 1–2 days, followed by a latency period of 5–10 days; final phase lasts 1–4 weeks; death from infection or internal bleeding
10–50 Gy	Rapid cell death in the gastrointestinal system causes severe diarrhea, intestinal bleeding, loss of fluids, and disturbance of electrolyte balance; the onset time drops from 30 to 5 min; initial bout of severe nausea and weakness followed by a period of well-being lasting a few hours to a few days ("walking ghost" phase); the terminal phase lasts 2–10 days; death is certain, often preceded by delirium and coma; therapy only to relieve suffering
Above 50 Gy	Metabolic disruption is severe enough to interfere with the nervous system; immediate disorientation and coma; onset within seconds to minutes; convulsions occur; victim may linger for up to 48 h before dying; it is assumed that a dose of 80 Gy caused by fast neutrons will immediately and permanently incapacitate human beings

the before mentioned nuclear weapon effects. It is common to separate fallout into the early fallout (local fallout) and the delayed fallout (global or long-range fallout). Early fallout is defined as that which reaches the ground during the first 24 h after explosion and is produced predominantly with explosions where the fireball touches the ground. Thereby, soil and dust particles are sucked into the rising cloud. When the vaporized fission products cool down in the cloud, a large portion condenses on the soil and dust particles. This leads to relatively large radioactive fallout particles with diameters of 100 µm or more, which settle on a relatively short time scale, forming the local fallout.

The fission products of the local fallout have a very complex composition of more than 300 different isotopes of about 36 chemical elements, decaying mainly by emission of beta radiation, frequently accompanied by gamma radiation. The decay process of such a number of isotopes with predominantly multilevel decay steps leads to a significant variation of the isotope composition with time. Directly after the fission process there are numerous isotopes with short half-lives within the fission products. The starting radioactivity of the fission products of 1-kt fission yield is about 10^{21} disintegrations per second, which is an activity of 10^{21} Bq. Generally, the multilevel decay generates more and more long-living radionuclides, so that the dose rate decreases rapidly in the first hours after explosion. A rule of thumb is: for every sevenfold increase in time after the explosion the dose rate decreases by a factor of 10. This rule corresponds to the so-called $t^{-1.2}$-law. With reference dose rate $\dot{D}_{H+1}$ at 1 h after explosion the dose rate as a function of time is $\dot{D}(t) = \dot{D}_{H+1} \cdot t^{-1.2}$, where t is regarded as the time in hours after explosion. This decay formula is valid within about 25% up to two weeks and within a factor of two up to about six months after the nuclear explosion. After that time the dose rate decreases at a much more rapid rate than predicted by the $t^{-1.2}$-law. Figure 5.16 shows the approximate time dependant behavior for the decay of local fallout during the first 50 weeks after the nuclear burst.

Even in the first 24 h the local fallout can travel a fairly large distance of a few hundred kilometers and may cover a large area, with significant fallout leading to an immediate biological hazard. Of course the real fallout pattern depends strongly on the weapon yield, the type and HoB, the nature of the surface, and the meteorological conditions. Especially, the direction and speed of wind are important. When the wind changes during transportation of the radioactive cloud irregular shapes for fallout can occur. With relatively steady wind conditions the fallout pattern is a kind of cigar shaped. The higher the wind speed, the longer and thinner the cigar pattern will be. Much research has been carried out to model and simulate the dispersion by prediction codes since the beginning of the nuclear age. These codes mainly run on computers with high computational capacities and are often classified. The accuracy of such dispersion calculations and fallout prediction depends basically on two aspects. First, a realistic source term has to be given or generated by the code. The source term describes the starting distribution, the amount and composition of radioactive particles, and their size distribution. For nuclear weapon scenarios the cloud rising process has to be included in the starting part of the simulations. The source term modeling

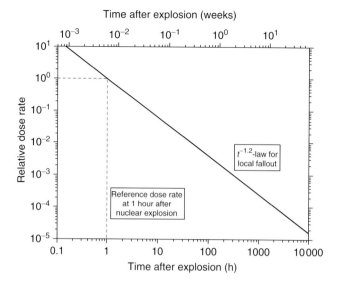

Figure 5.16 Relative dose rate versus time of the fission products in the local fallout approximated by the $t^{-1.2}$-law for the standard reference time of 1 h after explosion (H + 1).

needs a lot of knowledge about weapon design and nuclear processes. The second important aspect is the availability of preferably accurate and fine-spaced weather data for the actual dispersion calculation. One must keep in mind that these types of simulations will never be able to predict reality perfectly. But they do give a good understanding of the dimension and scale of the hazard. The better the input data, the more realistic the simulation will be.

Air bursts generally produce a negligible amount of local fallout. Essentially all of the residues in this case will contribute to the delayed fallout. When no soil and dust from the surface is sucked into the fireball and the rising cloud, radioactive particles will be smaller after condensation from the vaporized state than in the local fallout case. The sinking velocity of these smaller radioactive particles is significantly lower than that of local fallout particles. Because of the relatively large starting height in such an air burst combined with the enormous thermal energy liberated the weapon debris as well as the cloud will rise to even higher atmospheric layers where the particles are transported around the whole globe, often several times before they settle onto ground. Especially, when particles can enter the stratosphere they can remain there for several months or even years. Consequently, only the long-living radionuclides are of importance for the delayed fallout, mainly strontium-90 (half-life 28 years) and cesium-137 (half-life 30 years). In this aspect, the dispersed uranium or plutonium that was not fissioned also has to be mentioned because of the long radioactive half-lives and the predominant alpha activity. The major biological hazard of the delayed fallout is not the external irradiation by gamma and beta radiation but the risk of incorporation by ingestion and to a lesser extent inhalation. The delayed fallout will cause wide contaminated areas of relatively low values but all the crops and animals in this area will accumulate radioactivity.

5.3
Indirect Effects

> Indirect effects are impacts that are not direct caused in the ambient of the nuclear explosion or by a chain of physical mechanisms.

Indirect effects are:

- transient radiation effects on electronics (TREE);
- nuclear electromagnetic pulse (NEMP);
- ionization of the atmosphere – this can interfere with radar and radio-communications for short periods;
- residual nuclear radiation, by radioactive dust and ash created in the nuclear explosion;
- pumping of the radiation belt (e.g. inner Van Allen radiation belt) by charged particles produced in a nuclear explosion above the earth's atmosphere and captured by the geomagnetic field.

5.3.1
Transient Radiation Effects on Electronics (TREE)

TREE effects are mainly caused by initial nuclear radiation but also play a role in very high activity fallout shortly after a nuclear weapon burst. They are caused exclusively by ionizing radiation and therefore are different to the NEMP effects on electronics that are caused by a transient electromagnetic field of non-ionizing radiation (see next section). Because relatively high dose and dose rate levels of neutron and gamma radiation are necessary for significant degradation of electronic components, TREE is mainly restricted to tactical nuclear weapon scenarios. As stated in Section 5.2.3, medium and high yield nuclear bursts are dominated by blast and thermal effects by which the electronic components would be destroyed anyway in those regions were TREE effects would appear.

TREE effects are caused by ionization processes of gamma and neutron radiation in electronic components and by neutron induced lattice displacements mainly in semiconductor components. Figure 5.17 shows the chronological sequence of weapon effects for a typical tactical nuclear device of low yield for a distance where TREE effects might be of concern [6]. Clearly, the impact of the prompt gamma radiation happens first and is separated from the prompt neutron radiation pulse, so that the two disruptions can in principle be regarded independently. Basically there are three different effects that must be considered when talking about TREE:

1) lattice displacement by the neutrons of the initial radiation;
2) total dose ionization by gamma and neutron radiation of the initial radiation and the gamma radiation of the (early) residual radiation;
3) dose rate ionization by the prompt gamma part of the initial radiation.

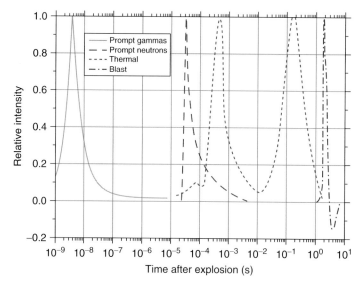

Figure 5.17 Chronological sequence of weapon effects for a typical tactical nuclear weapon scenario.

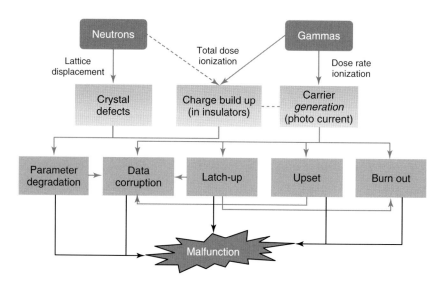

Figure 5.18 Different mechanisms of transient radiation effects on electronics (TREE).

As can be seen in Figure 5.18 these three effects can lead to various impacts on electronic components, mainly in semiconductors, which in the end will lead to a malfunction of the electronic device or the whole electronic system.

Before proceeding with a more detailed description of TREE, a few explanatory notes have to be made concerning dosimetry aspects. For TREE the significant interaction of the ionizing radiation takes place in electronic components. As the dominant radiation effects occur in semiconductors, silicon was chosen as the reference material. As absorbed doses are always related to a certain material, the TREE community uses the absorbed dose for silicon. This is measured in Gray and the material silicon is set in brackets, as Gy(Si). Notably, the absorbed dose for silicon in Gy(Si) is generally different to the absorbed dose in human tissue in Gy(TE). Usually in TREE, total dose ionization is measured in centigray, cGy(Si), and dose rate ionization in centigray per second, cGy(Si) s^{-1}, where 100 cGy(Si) = 1 Gy(Si). The lattice displacements are described best by the neutron fluence, measured in neutrons per cm^2. Because the neutrons impacting the electronic component have a wide energy spectrum, which changes significantly with weapon type and distance, and the displacement capability of neutrons of different energy varies significantly, it has been agreed to refer the TREE neutron fluence to the neutron energy of 1 mega-electronvolt (1 MeV equivalent damage in silicon). If an electronic component is irradiated with monoenergetic electrons of 14 MeV, the lattice displacement damage in silicon is significantly higher than for 1 MeV neutrons. Hence to get the same displacement effect as for the 14 MeV irradiation, one needs a higher neutron fluence with monoenergetic 1 MeV neutrons.

Because neutrons are uncharged particles they can easily approach and enter other nuclei near their flight path. In contrast, their ability to interact with the electron shell is limited. For TREE the significant neutron interactions are elastic and inelastic scattering of fast neutrons in the energy range 100 keV to 14 MeV. Figure 5.19 shows the principles of the different scattering processes.

The scattering events will lead to displacement production mainly in the semiconductor materials with accompanying ionization. Further ionization will be produced by the photons and other charged secondary particles that arise as a result of inelastic scatter and capture events. The recoil nucleus generally will have a much higher kinetic energy than the threshold for generating a lattice displacement in the semiconductor material. Therefore, much of the displacement

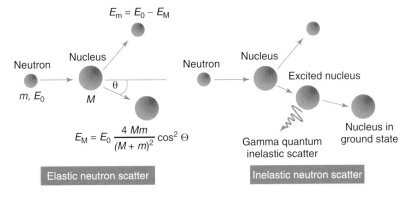

Figure 5.19 Schematic presentation of elastic and inelastic neutron scattering at nuclei.

damage will be produced in a displacement cascade. Displacement damage is especially significant in crystalline materials and can lead to changes in volume, elastic constants, thermal conductivity, and electrical resistivity. Materials that have the highest purity and regularity of crystal structure will be most vulnerable to neutron induced displacement damage. In tactical nuclear weapon scenarios the changes in electrical properties of semiconductors are the main concern. The vulnerability of electronic components due to neutron induced displacement damage generally occurs in the 1 MeV equivalent fluence range of $10^{12}-10^{16}$ neutrons per cm^2.

The total dose effect is governed mainly by gamma radiation. For tactical nuclear weapon scenarios the relevant energy range of the gamma radiation is between 100 keV and a few MeV and so the Compton effect is dominant for the interaction of the photons with the electronic components. The Compton effect is an inelastic scattering process of the incident photon, which knocks a Compton recoil electron out of the electron shell of an atom. While permanent chemical changes due to ionization are the main radiation hazard for human, at relevant radiation levels chemical changes are generally irrelevant in military equipment. The main problem due to total dose ionization is charge separation and permanent charge buildup, mainly for silicon dioxide dielectric material associated with silicon MOSFET device construction, in optical glasses and in fiber optics. Practically, total dose ionization effects in electronics play a relevant role in total dose ranges of 10^4-10^7 cGy(Si).

The main problem with semiconductors, concerning TREE, is the transient current production due to the prompt gamma radiation in tactical nuclear scenarios. If semiconductors are irradiated at high gamma dose rate and an electric field is applied, hole–electron pairs are generated causing the flow of a primary photocurrent. Often this process will also create secondary photocurrents in the semiconductor. These photocurrents may result in transient upsets that may lead to corruption, latch-up, and burn out in the electronic semiconductor devices (Fig. 5.18). Transient current carrier production also occurs in insulator materials and therefore capacitors also may show prompt and delayed conductivity changes, although they are of minor importance at tactical levels.

5.3.2
Nuclear Electromagnetic Pulse (NEMP)

Nuclear explosions of all kinds are accomplished by an electromagnetic pulse (EMP). The intensity and duration of this NEMP and the area over which it is effective varies considerably with the yield of the detonation and the height of the burst point. The strongest electromagnetic fields are produced near the burst by explosions at or near the earth's surface. Nuclear detonations at high altitudes produce fields that are strong enough to be of concern for electrical equipment over a huge area (largest effected area).

5.3.2.1 Generation of Electric Field
The NEMP is basically caused by (initial) gamma rays emitted in the nuclear detonation and those produced by neutron interaction with weapon residuals or

the surrounding medium. The gamma rays interact with air molecules and atoms by the Compton effect and produce an ionized (source) region that surrounds the burst point. The negative charged electrons move outwards faster than the heavier positively charged ions. As a result the region nearer to the burst point has a net positive charge whereas the region farther away has a net negative charge. This separation of charges produces an electric field. If the nuclear explosion occurs in a perfect homogeneous atmosphere and the gamma rays were emitted uniformly in all directions the electric field would be radial and spherically symmetric. Owing to the perfect symmetry there would be no electromagnetic energy radiated from the ionized deposition region.

However, in practice inhomogeneities of the surrounding media (differences in air density, variations of water vapor content, etc.) or nonuniform configurations of the nuclear warhead will interfere with the symmetry of the source region. This disturbance of symmetry of the ionized region generates a net electron current. This time varying current results in the emission of a short pulse of electromagnetic energy, which is strongest in the direction perpendicular to the current. The radiated electromagnetic field is called the nuclear electromagnetic pulse (NEMP).

The electric field strength of the NEMP will reach its maximum in about 10 ns and will fall off to quite small values in a few tens of microseconds. In contrast to the short duration of the pulse, it carries a considerable amount of energy, which is proportional to the yield of the exploding warhead. At a distance the energy of the radiated field can be picked up by antennas or any kind of conductors and converted into strong electric current and high voltages. Electric and electronic equipment may thus suffer malfunctions or in worse cases severe damage.

The Starfish Prime test was successfully detonated with a weapon yield of 1.4 Mt at an altitude of 400 km over Johnston Atoll in the Pacific Ocean on 8 July 1962. The Starfish Prime nuclear detonation caused an EMP that was far larger than expected. The experiment became known to the public as its EMP caused electrical damage in Hawaii, about 1445 km from the detonation point. The EMP knocked out about 300 streetlights, set off numerous burglar alarms and damaged a telephone company microwave link.

5.3.2.2 NEMP in High-Altitude Burst

One important type of NEMP is generated by nuclear explosions at high altitude above the atmosphere (e.g. above 50 km). This NEMP has three distinct time components that result from different physical phenomena, which are dominated by the inhomogeneity of air density (Figure 5.20). The three components of the NEMP, as defined in the IEC standard 61000-2-9 [1], are called:

1) early time (E1) component,
2) intermediate time (E2) component,

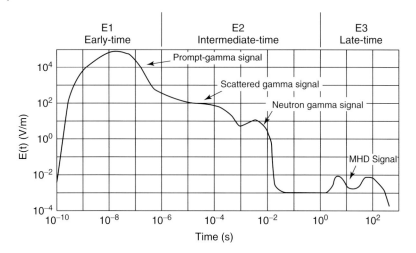

Figure 5.20 Schematic presentation of the NEMP generated by a high-altitude burst [1].

3) late time (E3) component.

5.3.2.3 Early Component of NEMP (E1)

For a high altitude nuclear explosion the prompt gamma output is the dominant radiation source. The gamma rays that are emitted from the nuclear explosion in an upward direction will travel through a region where the atmosphere has a low and decreasing density. As a result the gamma rays travel great distances before they are absorbed. In contrast, rays emitted from the explosion in a downward direction will encounter a region where the atmosphere density is increasing. At altitudes between 40 and 20 km these gamma rays will interact with the air molecules and atoms to form the source region for the early component of the NEMP (E1) (Figure 5.21). This ionization zone has a thickness of up to 80 km at the center and can extend all the way to the horizon (e.g. to 25 km for an explosion at an altitude of 500 km).

Initially, the electrons produced by the Compton effect move in the direction radial from the point of burst, the same direction as the gamma rays travel. The geomagnetic field forces the Compton electrons to turn to a transverse direction and follow a curved path along the field lines of the geomagnetic field. The turning of the Compton electrons in the earth's magnetic field results into the emission of electromagnetic radiation that adds coherently. The spectral range of the radiated electromagnetic field pulse (NEMP) is determined by the rise time of the Compton current. This is typically of the order of 10^{-8} s, which corresponds with frequencies around 100 MHz. Therefore, the resulting early time (E1) NEMP is characterized by a 10–90% rise time in the order of 2.5 ns, a duration of $T_{FWHM} = 24$ ns, and a spectrum up to several 100 MHz (Figure 5.22; Table 5.9).

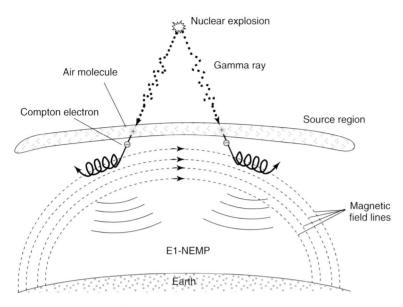

Figure 5.21 Principle of the generation of the early time (E1) NEMP in a high-altitude burst.

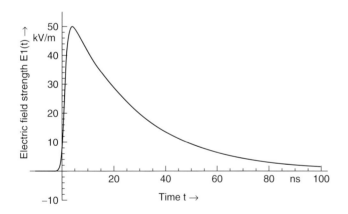

Figure 5.22 Early time (E1) waveform [1].

Table 5.9 Characteristics of the early time (E1) waveform.

Pulse amplitude of electrical field strength	Maximum rate of change of electric field strength	Rise time	Energy density
E_{max} 50×10^3 V m^{-1}	$\left.\frac{dE}{dt}\right|_{max}$ 16.2×10^{12} V s^{-1} m^{-1}	$\tau_{10 \to 90}$ 2.47×10^{-9} s	$\frac{1}{\Gamma_0} \int_0^\infty E^2 \times dt$ 0.114 J m^{-2}

As the NEMP of a high altitude nuclear explosion radiates from all parts of the source region, including its edges, and at various angles the effect of the NEMP extends to the tangent point on the earth's surface viewed from the burst.

A particular parameter of the NEMP (e.g. peak electric field, polarization) depends on the explosion yield, the HoB, the location of the observer on surface, and the orientation with respect to the geomagnetic field. Generally, one can expect that the peak electric field of a high altitude burst will be tens of kilovolts per meter (up to 75 kV m^{-1}) over most of the exposed area. Figure 5.23 shows a typical distribution of E1 amplitude on earth's surface for burst with an assumed yield of a few hundred kilotons at an attitude (HoB) from 100 to 500 km and a burst location (ground zero) between 30° and 60° north. The figure clearly shows that over most of the affected area the peak electric field exceeds 50% of the highest peak value (E_{max}).

5.3.2.4 Intermediate Component of NEMP (E2)

The intermediate (E2) component is generated by scattered gamma rays and inelastic gammas produced by weapon neutrons. This intermediate component is a pulse that lasts from about 1 μs to 1 s after the beginning of the EMP (Figure 5.24). The E2 component has many similarities to the EMPs produced by lightning, although the EMP induced by a nearby lightning strike may be considerably larger than the E2 component of a nuclear EMP. Because of the similarities to

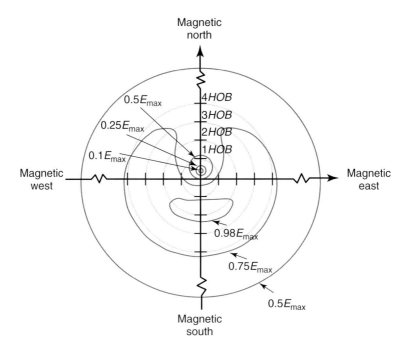

Figure 5.23 Typical distribution of E1 amplitude on earth's surface [high of burst (HoB) 100–500 km, ground zero between 30° and 60° north, yield higher than 100 kt].

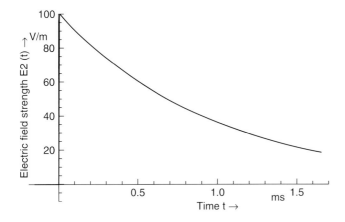

Figure 5.24 Intermediate time (E2) waveform [1].

Table 5.10 Characteristics of the intermediate time (E2) waveform.

Pulse amplitude of electrical field strength	Maximum rate of change of electric field strength	Rise time	Energy density
E_{max}	$\frac{dE}{dt}\big|_{max}$	$\tau_{10 \rightarrow 90}$	$\frac{1}{\Gamma_0} \int_0^\infty E^2 \times dt$
100 V m^{-1}	$20.6 \times 10^6 \text{ V s}^{-1} \text{ m}^{-1}$	3.66×10^{-6} s	$13.2 \times 10^{-3} \text{ J m}^{-2}$

lightning-caused pulses and the widespread use of lightning protection technology, the E2 pulse is generally considered to be the easiest to protect against. Table 5.10 shows the characteristics of the E2 component.

5.3.2.5 Late Time Component of NEMP (E3)

The physics of the late time or magnetohydrodynamic component (E3) of NEMP differs substantially from the other components (E1, E2). The E3 is formed by the interaction of the ionized source region of the burst with the geomagnetic field (Table 5.11).

Table 5.11 Characteristics of the late time (E3) waveform.

Pulse amplitude of electrical field strength	Maximum rate of change of electric field strength	Rise time	Energy density
E_{max}	$\frac{dE}{dt}\big|_{max}$	$\tau_{10 \rightarrow 90}$	$\frac{1}{\Gamma_0} \int_0^\infty E^2 \times dt$
$38.5 \times 10^{-3} \text{ V m}^{-1}$	$41.9 \times 10^{-3} \text{ V s}^{-1} \text{ m}^{-1}$	735×10^{-3} s	$71.1 \times 10^{-6} \text{ J m}^{-2}$

The high temperatures and extensive nuclear radiation of a nuclear explosion lead to vaporization and ionization of the weapon material. This moves out as an expanding plasma shell with a velocity in the range of 10^6 m s^{-1}. Owing to the conductivity of the plasma, currents are set up in this shell whose magnetic flux tends to cancel the earth's flux. As a result the geomagnetic flux is kept out of the plasma shell. As the shell expands it pushes away the earth's magnetic field, yielding a higher concentration of magnetic flux outside of it and nearly no flux inside it. The expansion of the plasma is slowed down by the increasing magnetic pressure and the viscosity of the surrounding media (e.g. air). Finally, when the magnetic pressure equals the kinetic pressure the expansion of the plasma stops and the magnetic field goes back inside, into its natural place (Figure 5.25).

The E3 component is a very slow pulse, lasting tens to hundreds of seconds with a maximum field of tens of millivolts per meter that occurs about 2–5 s after the nuclear explosion (Figure 5.26). The E3 component has similarities to a

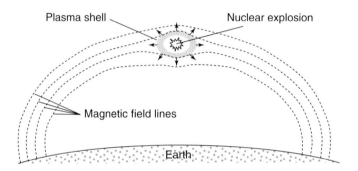

Figure 5.25 Distortion of the earth's geomagnetic field caused by an expanding plasma shell of a high-altitude nuclear explosion.

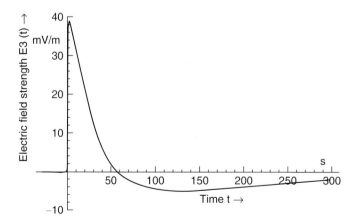

Figure 5.26 Late time (E3) waveform [1].

geomagnetic storm caused by a very severe solar flare. Like a geomagnetic storm, E3 can produce induced currents in long electrical conductors, which can then damage components such as power line transformers.

> In 1962, the Soviet Union conducted several high altitude nuclear tests in space over Kazakhstan to investigate EMP and collect damage data. Although these weapons were smaller than the Starfish Prime test, the EMP caused greater damage due to a stronger earth's magnetic field at the test. In particular, the late time component of NEMP (E3 component) induced an electric current surge in a long underground power line that caused a fire in a power plant [9, 10].

5.4 Radiological Weapons

It is essential to realize the basic difference between a nuclear and a radiological weapon. A nuclear weapon is characterized by the deliberate execution of an uncontrolled chain reaction of fissible nuclear material like highly enriched uranium or plutonium. The fission reaction may be succeeded by a nuclear fusion reaction in a second stage. The nuclear fission and the fusion reaction are the source of the tremendous energy release leading to the different nuclear weapon effects discussed in previous sections. In contrast, the energy release from an explosive type of a radiological weapon is created by conventional chemical explosives. What makes a radiological weapon hereby is the attempt to disperse radioactive material in the course of the detonation process of the conventional explosives. Theoretically, any radioactive isotope could be utilized for radiological weapons. In contrast to nuclear weapons, fissible nuclear material is not necessary for radiological weapons, but could of course be used. In such a case the nuclear material would simply get dispersed but not undergo fission reactions. Figure 5.27 shows a possible classification of nuclear and radiological weapons. While there are only two types of nuclear devices the variety of radiological devices is broader. In particular, it has to be pointed out that the deployment of a radiological weapon need not be accompanied by an explosion, as can be seen later. Nuclear devices can be divided into regular nuclear warheads and so-called improvised devices. The latter is a non-professional attempt to build a nuclear weapon, presumably employing the gun-method. Appropriate nuclear material is necessary for this attempt, typically highly enriched uranium (e.g. through proliferation channels). It is assumed that improvised nuclear devices will show a degraded explosion ("fizzle") because the chain reaction might break down too early due to the non-professional design.

There are two main types of radiological weapons, called the radiological exposure device (RED) and the radiological dispersal device (RDD) (Figure 5.27). A RED is

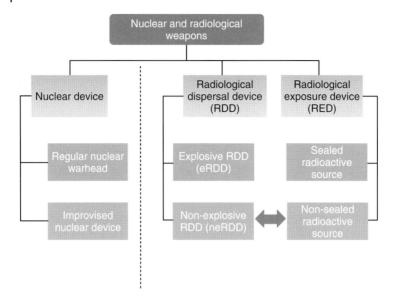

Figure 5.27 Classification of nuclear and radiological weapons.

a sealed radioactive source that is used to externally irradiate a single person or group of people with the intention of generating deterministic radiation damage (Table 5.8). Generally, a high activity RED is needed to achieve the necessary radiation dose over some hours, for example, while people are in a conference room. In addition, radioactive material appropriate for REDs will be limited to high yield gamma or neutron emitters because a long-range effect of the ionizing radiation is necessary. Non-sealed radioactive material might also be applied for REDs but that seems to be far more unlikely because it is fairly tricky to handle highly active gamma or neutron radiation emitting non-sealed radioactive material. When a non-sealed radioactive source is used for an RED nevertheless, there will be an additional risk of local radioactive contamination in theater. An RDD is intended to spread radioactive material to the environment, leading to contaminated areas. There are two types of RDDs: the explosive radiological dispersal device (eRDD) – known also as a *"dirty bomb"* – and the non-explosive radiological dispersal device (neRDD). Summarizing an eRDD is a conventional bomb spiked with radioactive material, while a neRDD is thought to be a device with radioactive liquid or powder that is dispersed by a sprayer system either on the ground or airborne.

5.4.1
Radioactive Material and Radiological Weapons

The most important aspect for radiological weapons is the radioactive material to be used. There are several requirements suitable radionuclides must fulfill to be an effective radiological weapon. First, the half-live of the radioactive substance must be high enough, so that there will be no significant decay during acquisition, preparation, and application of the radiological weapon. Furthermore, the emitting radiation of the radioactive nuclides should have sufficient potential to cause health effects. The most important requirement is the availability in the required relatively high quantities. Therefore, mainly radioactive material of widespread use should be of concern. Many studies and attempt at prioritization have been made in the past that has lead to a handful of radionuclides most specialists agree upon. Table 5.12 shows the most likely radionuclides for use in RDDs. Indicated is also the half-life of the isotopes, the primary radiation type, primary chemical form, high end source activity, and application forms for the mentioned activities. Although the official SI unit of the activity is the becquerel (Bq), which stands for one disintegration per second, the activity is given in curies (Ci) because of the more familiar numbers with high activities. The conversion factor is 37 GBq for 1 Ci. Notably, these activities are so immensely high that handling will be extremely dangerous and will lead to lethal radiation doses in very short periods of time. Consequently, potential culprits must have a profound knowledge of working procedures and must utilize special equipment.

Table 5.12 Most likely radionuclides for misuse in radiological weapons [3, 11].

Nuclide (half-life)	Primary radiation type	Primary form	Maximum source activity (Ci)	Application
^{90}Sr (28.6 yr)	Beta	Ceramic (SrTiO$_3$)	300 000	Radioisotopic thermal generators (RTGs)
^{137}Cs (30.2 yr)	Beta + gamma	Salt (CsCl)	200 000	Irradiator
^{60}Co (5.3 yr)	Gamma + beta	Metal	300 000	Irradiator
^{238}Pu (87.8 yr)	Alpha	Ceramic (PuO$_2$)	130 000	RTG for space probes
^{241}Am (432 yr)	Alpha + neutrons	Pressed ceramic within beryllium	20	Single well-logging source
^{252}Cf (2.6 yr)	Neutrons + alpha	Ceramic (Cf$_2$O$_3$)	20	Neutron radiography, well-logging
^{192}Ir (74 d)	Beta + gamma	Metal	1000	Industrial radiography
^{226}Ra (1600 yr)	Alpha + gamma	Salt (RaSO$_4$)	100	Old medical therapy sources

5.4.2
Impacts of Radiological Weapons

5.4.2.1 Radiological Exposure Device

Generally, for radiological exposure devices (REDs) radioactive sources of extremely high activity that emitt gamma or neutron of high energy are needed. For the following assessment the focus will be on gamma radiating sources. As shown in Table 5.8 a whole body dose of not less than 6 Gy is necessary to produce severe acute health effects in men. We shall assume furthermore the irradiation time to be 3 h (e.g. in a conference room) with a distance of 2 m between the potential victim and the radioactive source. To irradiate the projected dose of 6 Gy one would need a gamma dose rate of 2 Gy h^{-1}, which is approximately 2 Sv h^{-1} for ambient dose equivalent (for more details see Chapter 9, Measurement of Ionizing Radiation). For the highly gamma efficient radionuclide cobalt-60 this would imply a sealed source with the immense activity of 23 TBq (or 622 Ci). For preparation and transportation of an unshielded source, this would imply a gamma dose rate of about 8 Sv h^{-1} for a distance of 1 m to the source, which implies also an enormous acute radiation hazard for the culprits. One could shield the source with, for example, 20 cm of lead, which would reduce the dose rate to about 70 μSv h^{-1} outside the shielding at 1 m distance. This is a dose rate that is not dangerous to human health for a long time yet, but definitely far beyond the legal limits and easily detectable with standard radiation meters. Above all, such a lead container would have a mass of about 1200 kg and would therefore be easily noticeable in theater. This scenario should show that the deployment of an RED against selective people is possible, but is considerably restricted for practical reasons.

> **Gamma radiation** is electromagnetic radiation and consists of particles with no rest mass that travel with velocity of light and can interact with matter by ionization processes. These particles are called photons. Theoretically, gamma radiation cannot be shielded totally in matter but only attenuated. The ability of matter attenuating gamma radiation depends mainly on the density of the material but also on the energy of the photons. For example, 1.2 cm of iron reduces the dose rate for 662 keV photons (^{137}Cs) by a factor of $\frac{1}{2}$, and is, therefore, called the half-thickness of iron for this gamma energy. For practical aspects gamma radiation is said to be totally shielded when the dose rate falls below the ambient background gamma dose rate (about 0.7 μSv h^{-1} in most Western countries).

5.4.2.2 Radiological Dispersal Device

The impacts of an radiological dispersal device (RDD) are more complex than for an RED, as can be seen in Figure 5.28. The following description refers mainly to explosive radiological dispersal device (eRDDs), the so-called dirty bombs. Except

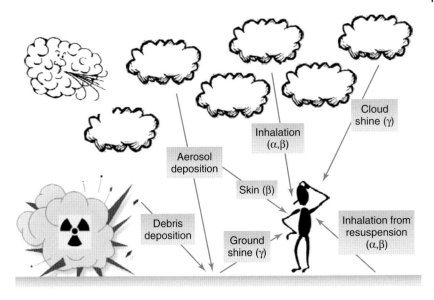

Figure 5.28 Impacts of an explosive radiological dispersal device event (dirty-bomb event).

for the explosion process itself the mechanisms of impact of neRDDs, however, is not much different, so that the main results are applicable for the latter, too. Figure 5.28 shows the different exposure paths caused by an eRDD event. External irradiation by gamma radiation[1] is induced by contamination of the ground. Radioactive aerosols will deposit out of the radioactive cloud traveling downwind from the spot of detonation. This is called the *ground shine*. During passage of the radioactive cloud there is also *cloud shine*, which is an external gamma radiation hazard. Additionally, in the vicinity of the explosion there will be fragments of the radioactive material that could not be aerosolized. Supplementary to the external gamma radiation hazard there might be contamination of people with beta radiating particles, leading mainly to irradiation of the skin. In addition, the contamination of clothing might lead to a significant skin dose, because beta radiation can penetrate clothing. A new aspect of RDD effects is the inhalation risk raised by aerosols containing mainly alpha and beta radiating radioactive particles. The inhalation hazard is acute due to the aerosol concentration in ambient air during passage of the radioactive cloud but also occurs to a certain extent by resuspension in contaminated areas [11, 12, 13].

The trickiest part in building an eRDD is finding a sophisticated design, so that as much radioactive material as possible can be aerosolized during the detonation process. The ability of aerosolization depends on the form of the radioactive source (sealed or non-sealed), the chemical form of the radioactive material (metal, salt, ceramics, liquid), and the bomb design. Intensive theoretical

---

1) Principally this is true also for neutron radiation, but appropriate neutron-emitting sources are relatively rare.

and experimental studies have been performed recently in the USA, Canada, and Germany, investigating the aerosolization process using principally non-radioactive surrogates for the trials. The most extensive experimental studies have been carried out by F. Harper *et al.* at the Sandia National Laboratories [11, 12]. It turned out that there is a stress-induced aerosolization mechanism that affects the aerosol particle size. The aerosolization processes are vapor phase change, solid fracture across grain boundaries, liquid phase change, solid fracture along grain boundaries, and energy-limited spall solid fracture (from small to large particle size).

> **Alpha radiation** is a particle radiation consisting of helium nuclei. It does not pose an external radiation hazard because it cannot penetrate human skin. However, it is a serious risk concerning incorporation because of the high ionization potential of the helium nuclei on their short travel through tissue. Inhalation of radioactive alpha emitting aerosols in the inhalable size distribution of about 0.1–10 µm is in particular of concern. Among the radioactive substances, alpha emitting radioisotopes exhibit the highest radiotoxicity.

There are three regions of interest concerning aerosol particle size. Small particles in the range 0.1–10 µm present mainly an inhalation risk. Intermediate particles in the range 10–100 µm will be blown into the air due to the detonation process and will be transported and diluted by the prevailing wind field. This will cause radioactive contamination in the far from the detonation. Large particles in the range 100–500 µm will also be transported in the air but deposit rapidly on the ground and lead to a near-field contamination and primarily pose a ground shine problem. In addition to real aerosols there may be fragments (>500 µm) that behave ballistically and will contribute to an enhanced ground shine around the spot of detonation.

> **Beta radiation** is a particle radiation consisting of electrons (β^- decay) or positrons (β^+ decay). In air, beta radiation can have a range of several centimeters to several meters, depending on the kinetic energy of the particles. Beta radiation is absorbed by thin sheets such as metals like aluminum or acrylic glass. The main radiation hazard of beta radiation is the skin dose provoked by radioactive contamination on clothes or directly on the skin. Beta radiation can cause secondary photon radiation, called *bremsstrahlung*, on being slowed down by shielding material (especially metal).

The real particle size distribution of the generated radioactive aerosols is essential for hazard and risk analysis and is the key input parameter for the source term in dispersion calculations and simulations. Though there is a comprehensive data base of experimental particle size distributions under controlled test conditions (mainly classified), the prediction for real situation RDD events is extremely vague. It turned out that the most critical parameter is the device design, which is usually not predictable in a terrorist scenario.

Nearly all the realized experiments and simulations result in the fact that an immediate and acute health effect caused by ionizing radiation from an RDD is very unlikely or often not possible for the population in the affected areas. This is a big difference between initial radiation and early fallout radiation from a nuclear explosion. The dominant first effect of an RDD event would certainly be more psychological in a large part of the local population. In that sense one cannot really name an RDD a "weapon of mass destruction" but rather as a "weapon of mass disruption." Even when the radiation and contamination levels predominantly would not provoke immediate health effects, they would exceed legal limit values in significant areas. Some areas would have to be evacuated at least temporarily. Especially in densely populated areas such as city centers with a complex commercial structure, this would have a devastating psychological and also economical effect. Access to the contaminated area would have to be prohibited or restricted at least temporarily. Extensive and expensive procedures would have to be established for access and contamination control of people and material. As described later in Chapter 15 it is extremely hard to decontaminate urban infrastructure, so that contaminated areas would remain radioactive at a non-acute but – concerning current radiation protection rules – considerable level for a significant period of time.

References

1. IEC (1996) 61000-2-9. *Electromagnetic Compatibility (EMC) – Part 2: Environment – Section 9: Description of HEMP Environment – Radiated Disturbance*, International Electrochemical Commission.
2. Zak, A. (2006) *Nonproliferation Rev.*, **13** (1), 143–150.
3. Sublette, C. (2007) Nuclear Weapons Frequently Asked Questions. http://www.nuclearweaponarchive.org/Nwfaq/Nfaq0.html (accessed on 18 April 2012).
4. Webblog; Declassified data on effects of nuclear weapons and countermeasures against them; http://glasstone.blogspot.com (accessed on 18 April 2012).
5. The nuclear information project; documenting nuclear policy and operations; http://www.nukestrat.com (accessed on 18 April 2012).
6. NATO (1991) NATO Allied Engineering Publication 22 (AEP-22) Edition 1, A Guide to Transient Radiation Effects on Electronics at the Tactical Level (Land Forces), November 1991, NATO Restricted.
7. Federation of American Scientists (1998) Nuclear Weapons. http://fas.org/nuke/intro/nuke/index.html (accessed on 18 April 2012).
8. Glasstone, S. and Dolan, P.J. (eds) (1977) *The Effects of Nuclear Weapons*, 3rd edn, United States Department of Defense and United States Department of Energy.
9. Greetsai, V.N. et al. (1998) *IEEE Trans. Electromagn. Compat.*, **40** (4), 348–354.
10. Loborev, V.M. (1995) Up to date state of the NEMP problems and topical research directions. *Proceedings of the EUROEM 94 International Symposium*, Bordeaux, France, 30 May – 3 June 1994, EUROEM, Gramat, pp. 15–21.
11. Musolino, S.V. and Harper, F.T. (2006) Emergency Response Guidance for the

First 48 Hours After the Outdoor Detonation of An Explosive Radiological Dispersal Device, *Health Phys.*, **90** (4), 377–385.
12. Harper, F.T., Musolino, S.V., and Wente, W.B. (2007) Realistic Radiological Dispersal Device Hazard Boundaries and Ramifications for Early Consequence Management Decisions, *Health Phys.*, **93** (1), 1–16.
13. Egger, E. and Münger, K. (2005) The Dirty Bomb: How Serious a Threat? Background Information on a Current Topic, Spiez Laboratory, Federal Office for Civil Protection, 10 p.
14. Rabinowitz, M. (1987) *IEEE Power Eng. Rev.*, **PER-7** (10), 60–61.
15. Karzas, W.J. and Latter, R. (1962) *J. Geophys. Res.*, **67** (12), 4635–4640.
16. Karzas, W.J. and Latter, R. (1965) *Phys. Rev.*, **137** (5B), B1369–B1378.
17. Final Report, Evaluation of nuclear criticality safety data and limits for actinides in transport, Republic of France, Institut de Radioprotection et de Sûreté Nucléaire, Département de Prévention et d'étude des Accidents (*http://ec.europa.eu/energy/nuclear/transport/doc/irsn_sect03_146.pdf*).
18. Restricted data declassification decisions 1946 to the present (RDD-7), U.S. Department of Energy, January 1, 2001 (*http://www.fas.org/sgp/othergov/doe/rdd-7.html*).
19. The Nuclear Weapon Archive – A Guide to Nuclear Weapons, *http://nuclearweaponarchive.org* (accessed on 17 July 2012).
20. Nuclear Explosions: Weapons, Improvised Nuclear Devices, U.S. Department of Health and Human Services, *http://www.remm.nlm.gov/nuclearexplosion.htm* (accessed on 17 July 2012).
21. NATO Handbook on the medical aspects of the NBC defensive operations, AMedP-6(B); Part I – Nuclear.

Part III
CBRN Sensors – Key Technology for an Effective CBRN Countermeasure Strategy

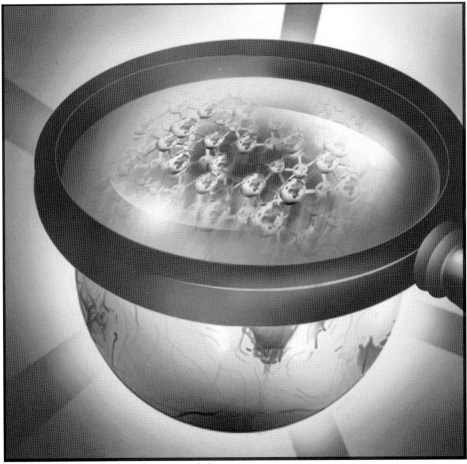

copyright by Jörg Pippirs, http://www.artesartwork.de

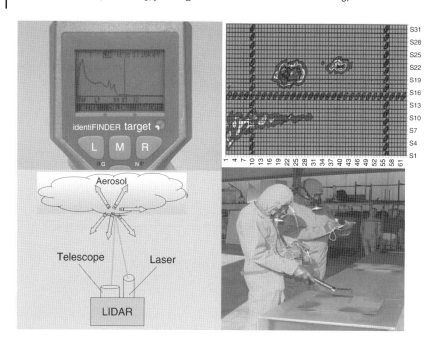

6
Why Are Reliable CBRN Detector Technologies Needed?
Birgit Hülseweh, Hans-Jürgen Marschall, Ronald Rambousky, and Andre Richardt

The potential use of weapons of mass destruction (WMD), such as chemical, biological, radiological, and nuclear (CBRN) weapons, is one of the most frightening present and future scenarios. Therefore, it is necessary to find an overall strategy for effective countermeasures. Developments recent decades have clearly shown that the immediate and reliable detection, identification, and monitoring of CBRN substances and its effects are increasingly a key technology in an international concept for CBRN risk management. State of the art technology should provide a reliable detection and identification capability that covers a wide range of different CBRN targets in multiple environmental matrices by deploying a field compatible system. Therefore, one of the most significant challenges for the future is to get micro-based detection systems into service as well as to implement an international standard for quality risk CBRN management.

6.1
Introduction

In the previous section you, hopefully, got an idea about the diversity of chemical, biological, nuclear, and radiological agents. However, before we begin to look at different technical countermeasures for chemical, biological, radiological, and nuclear (CBRN) warfare agents we have to address some general considerations about CBRN detectors as the key technology in any effective response strategy. For any successful technical CBRN protection strategy or concept effective detection mechanisms are a fundamental aspect. Reliable detection technologies are needed before, during, and after a CBRN incident (Figure 6.1).

Before a CBRN incident occurs detectors can be used for continuous monitoring to either prevent a CBRN incident (detect-to-protect) or for early warning in the event of an incident (detect-to-treat). During a CBRN incident detectors are needed to identify the precise nature and extent of the CBRN agent. After the CBRN incident detectors are essential for three main tasks:

6 Why Are Reliable CBRN Detector Technologies Needed?

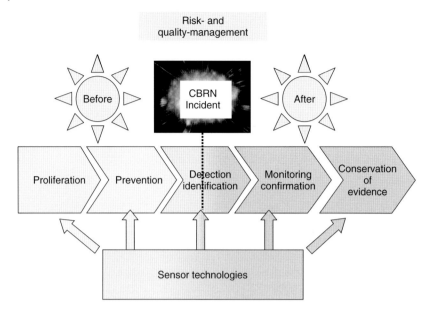

Figure 6.1 Sensor technologies as the key technology in the prevention, the case, and the aftermath of a CBRN incident.

1) to confirm the results of early identification,
2) to collect evidence for the use of international banned substances (forensic aspects [1]),
3) to confirm that the area is safe for reoccupation after decontamination.

Based on this variety of requirements for detector technologies, we should have in mind that the perfect detector does not exist and no single detector technology possesses all the desirable features and functions [2]. Furthermore, CBRN protective equipment is one of the most critical areas of a CBRN response strategy. For example, an integral consideration is that if the individual protective equipment (IPE) is too cumbersome then it affects the performance of first time responders and military operators. On the other hand, if the protective level is insufficient, then the health and the life of the wearer are put at risk. To make it even more complicated, decontamination procedures have to be adapted to the level of risk and the decontamination chemistry must not interfere with, for example, the protective layers in the IPE or with the detector equipment necessary for tracing the level of contamination.

6.2
A Concept to Track CBRN Substances

A general problem in discussing technologies and scientific developments for potential detector technologies is the often missing understanding of the meaningful use of detectors in a real CBRN scenario. Therefore, it is essential to understand that the capability to detect, identify, and monitor the full spectrum of CBRN agents at any operational phase is indispensable to mitigate the effects of CBRN substances at any phase of operation [3]. For this reason we will discuss some general aspects of detection, identification, and monitoring capabilities that we must have in mind not only to discover, characterize, and determine CBRN incidents rapidly but also to delineate areas of contamination and monitor changes over time. But what is a CBRN incident? We learned in Part I the history of the conventional use of chemical, biological, and nuclear weapons in World Wars I and II and in the cold war, but a modern definition of CBRN incident also has to include toxic industrial hazards or effects.

- **Definition CBRN incident**: Any occurrence involving the emergence of CBRN and toxic industrial hazards or effects, irrespective of source, cause, or intent [3].
  In a concept for tracking substances we have to discuss specific areas, for example, the required detection, identification, and monitoring capabilities and their related tasks. We must be aware that the process of hazard avoidance is a whole system. Obviously, the capability of one element will influence another and the different elements have to be combined to provide complete CBRN protection. For example, without efficient detector capabilities military and civilian personnel are not able to doff their IPE, to leave the collective protective equipment without IPE, to verify the decontamination procedures, and to reopen facilities after decontamination.
- **CBRN detection**: the discovery, by any means, of the presence of a chemical, biological agent, or radioactive material of potential military significance. The detection function also includes reconnaissance, survey, and surveillance.
- **CBRN hazards**: include the effects of nuclear attacks, the effects on man of biological or chemical attacks, or through the release of toxic industrial materials (TIMs).
- **CBRN detector**: device or system employed to recognize the emergence, presence, or absence of CBRN events or hazards, which is divided into (i) point detector: a detector that reacts to hazards at the point of interception; (ii) stand-off detector: a detector that reacts to distant events or hazards; and (iii) remote detector: a point (or stand-off) detector used at a distance from the protected force element [3].

Based on the definition of detection we can identify the primary objective of detection, which is to indicate the presence of health threatening levels of CBRN substances in seconds (Figure 6.2). Furthermore, detection has to provide timely information with generic results and appropriate sensitivity that will allow affected personnel to wear an appropriate level of physical protection (detect-to-warn). Based on these requirements for detection we can identify a need for close to

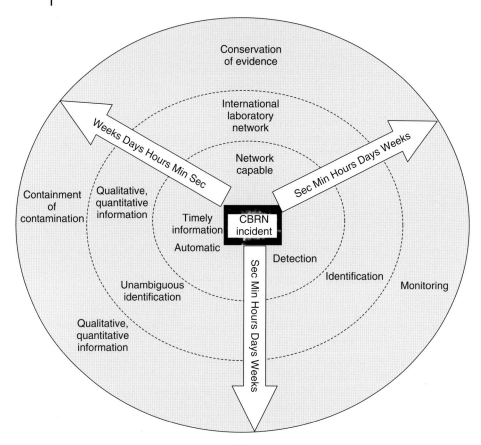

Figure 6.2 Time-dependent detection, identification, and monitoring of chemical, biological, radiological, and nuclear agents. Adapted from Reference [3].

real-time and automatic measurements [3]. In addition, detector technologies have to be sensitive enough to react to the most toxic hazards in seconds so that protective measures can be taken timely. After the generic detection of a potential hazard, identification of the hazard should provide qualitative and quantitative characterization. Such information on the hazard is indispensable for optimization of the countermeasures used to mitigate the CBRN effects. Over time, which can range from minutes to days, the quality of information will increase from provisional, through to confirmed, to unambiguous identification.

> **Levels of Identification of CBRN Substances**
>
> 1) **Provisional identification**: equipment required for provisional identification utilizes one or more generally accepted analysis method that can

quickly provide (within seconds up to minutes) the additional information required to confirm warnings caused by positive detection.
2) **Confirmed identification** requires advanced analysis equipment and knowledge about the suspected hazard, and is performed by CBRN-specialists or medical experts with scientific training. The analysis results may take more time to process (up to hours), will strongly reduce false results, and will be applied with higher sensitivity to cover lower quantities not measured before.
3) **Unambiguous identification** has to be performed by specialists in appropriate reference laboratories and may well take even more time than other identification levels (up to one or more days).

After identification of the hazard, continuous monitoring is necessary to determine if the hazard is still present.

We have to emphasize that detection limits and sensitivity standards can be different for the different types of detection, identification, and monitoring equipment. For example, the sensitivity is generally based on the hazardousness of individual CBRN agents. However, the selection of a threshold level for a detection system is also based on the needed level of personnel protection. Over recent decades we can observe an increased awareness of health effects from low-level exposures, which has resulted in lower sensitivity limits. We must be aware that not all detector systems need to have equal sensitivity and that the combination of different systems could be useful to provide the needed information for detection, identification, and monitoring. In general, for proposed sensor systems for the detection, identification, and monitoring of CBRN incidents a set of competing requirements can be identified based on the operational parameters and can be represented by spider charts (Figure 6.3) [4, 5]. The potential benefits of these spider charts are the possible comparison between sensors and obtaining an idea of the overall usefulness of a tested sensor [5]. Based on these general considerations we can characterize the detectors as follows:

- point detector,
- stand-off detector,
- remote detector.

A point detector reacts automatically to hazards at the location of the detector. The requirements for point detectors are (i) to react quickly and (ii) to give a signal automatically in the presence of a CBRN hazard. Furthermore, point detectors can be hand-held, portable, or fixed, and can vary from detection paper to advanced electronic equipment. The tasks of point detectors are, for example, to measure the exposure levels of personnel, to operate as a part of a network in a collective protection system, or to confirm the results of a decontamination process. In contrast to point detectors a stand-off detector allows the detection of CBRN substances by units from a position outside of the CBRN incident. The remote

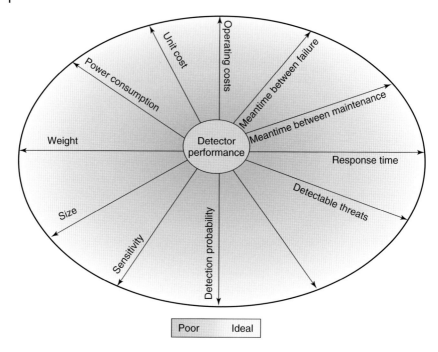

Figure 6.3 Spider chart as a useful tool for representing the most important parameters for sensor systems. For the different metrics the better performance is obtained on moving out from the center. Adapted from References [4, 5].

detector can be employed autonomously from civilian and military elements and can work unmanned at a reduced level of risk for civilian and military elements. In contrast to detectors, identification equipment enables the characterization and determination of the nature of a CBRN agent. Finally, there is one point we should not forget in the whole detection, identification, and monitoring process. Sampling equipment is necessary for the sampling of potential toxic materials in field or forensic analysis of a CBRN incident. To meet international standards for confirmed and unambiguous identification, Sampling Identification Biological Chemical Radiological Agents (SIBCRA) procedures have to be developed, standardized, and used. Under custody the qualified SIBCRA procedures are used for the collection and transportation of suspected CBRN materials to guarantee uniformity, safety, accountability, and evidence that will stand up in a court of law.

> **Sampling Identification Biological Chemical Radiological Agents (SIBCRA)**
>
> International agreed and confirmed standards are required to provide documented evidence of the produced laboratories results. This means that the results from the laboratory must meet a prescribed standard, for example:
>
> - good laboratory practice (GLP),

- quality assurance (QA),
- quality control (QC),
- standard operating procedures (SOPs).

Furthermore, it is necessary to demonstrate that the route from the raw data to the corresponding experimental results can be verified (Figure 6.4). For the provisional, confirmed, and unambiguous identification several methods can be used. Obviously, for unambiguous identification the highest level of certainty is required [6].

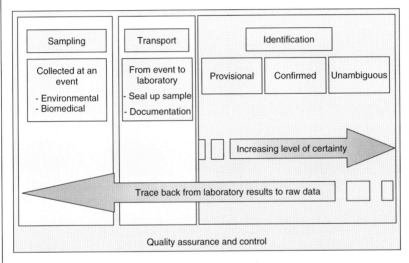

Figure 6.4 Generalized procedure for SIBCRA (Sampling identification biological chemical radiological agents) procedures as one possible step for a further forensic concept.

6.3
Low-Level Exposure and Operational Risk Management

To understand the difficulty of selecting the right detector strategy it is furthermore important to understand the complexity of a CBRN incident and thus some generalizations are needed. The severity of a health effect from CBRN agent exposure is the direct result of (i) the CBRN agent itself, (ii) its dose, (iii) the product of the CBRN agent concentration and duration of exposure, and (iv) the personal health situation of the affected individuals. Owing to the diversity of environmental conditions and possible scenarios it is not possible to have for each possible CBRN incident an adequate policy. It is common practice to use generic scenarios for estimating the severity of effects. These generic scenarios could be used as the basis for operational risk management (ORM) and furthermore to define low-level exposures and low-risk exposures and are therefore possible benchmarks for detectors [7]. However, an international accepted standard does not exist and therefore further standardization is needed. The term *CBRN risk management* is discussed further in Chapter 16.

Furthermore, we are not able to precisely estimate the time period that personnel can be exposed to CBRN agents. Therefore, only some generalized terms for duration of exposure are available. In particular, the consequences of the combination of exposure duration and low-level exposure concentrations are the primary focus

Table 6.1 Definition of Acute Exposure Guideline Levels (AEGLs).[a]

AEGL	Definition
General	AEGLs represent threshold exposure limits for the general public and are applicable to emergency exposure periods ranging from 10 min to 8 h. AEGL-2 and AEGL-3, and AEGL-1 values as appropriate, will be developed for each of five exposure periods (10 and 30 min, 1, 4, and 8 h) and will be distinguished by varying degrees of severity of toxic effects. It is believed that the recommended exposure levels are applicable to the general population, including infants and children, and other individuals who may be susceptible
AEGL-1	The airborne concentration, expressed as parts per million or milligrams per cubic meter (ppm or mg m^{-3}), of a substance above which it is predicted that the general population, including susceptible individuals, could experience notable discomfort, irritation, or certain asymptomatic nonsensory effects. However, the effects are not disabling and are transient and reversible upon cessation of exposure
AEGL-2	The airborne concentration (expressed as ppm or mg m^{-3}) of a substance above which it is predicted that the general population, including susceptible individuals, could experience irreversible or other serious, long-lasting adverse health effects or an impaired ability to escape
AEGL-3	The airborne concentration (expressed as ppm or mg m^{-3}) of a substance above which it is predicted that the general population, including susceptible individuals, could experience life-threatening health effects or death

[a] Airborne concentrations below the AEGL-1 represent exposure levels that can produce mild and progressively increasing but transient and non-disabling odor, taste, and sensory irritation or certain asymptomatic, nonsensory effects. With increasing airborne concentrations above each AEGL, there is a progressive increased likelihood of occurrence and the severity of effects described for each corresponding AEGL. Although the AEGL values represent threshold levels for the general public, including susceptible subpopulations, such as infants, children, the elderly, persons with asthma, and those with other illnesses, it is recognized that individuals, subject to unique or idiosyncratic responses, could experience the effects described at concentrations below the corresponding AEGL. Source: US Environmental Protection Agency (EPA).

Table 6.2 Important factors for detector selection [7–10].

Detection capability	Performance
Selectivity	Set-up and warm-up time
Sensitivity	Calibration requirements
Response time	Portability
Limit of detection	Power requirements
Response dynamic range	Resistance to environmental conditions
False alarm rate	

of medical research at the moment. A low-level exposure is exposure to a CBRN at a concentration at which there is low risk to human health in the short term. The most common terms for the range of low-level concentration and corresponding effects are the acute exposure guideline levels (AEGLs) for chemical substances. AEGLs represent temporary threshold exposure limits (10 min to 8 h) for three levels of severity and due to their general importance for employment protection the definition is given in detail in Table 6.1. This concept is useful for chemical warfare agents and is generally not appropriate and not applicable for assessing ionizing radiation and biological scenarios. Instead of AEGLs, for ionizing radiation scenarios a dosimetry concept has to be used that is based on the accumulated ionizing radiation dose a person receives from external and internal contributions. A more detailed description will be found in the appropriate chapters dealing with the radiological and nuclear aspects. For biological warfare agents a dosimetry concept is under discussion but not available yet.

As a result of the discussed parameters (technical and operational) it is understandable that during the selection process of a detector for a special task several technical and operational important factors should be considered (Table 6.2, Figure 6.3) [7, 8]. However, the ideal set of factors depends on the specific circumstances for the chosen type of operation and the ideal CBRN sensor for all possible scenarios does not exist [4].

6.4
Conclusions and Outlook

Sensor technologies can be assumed as the key technology before, in the case of, and in the aftermath of a CBRN incident. Reliable sensor technologies are needed for detection, identification, monitoring, and for collecting evidence. Furthermore, the process to prove the data resulting from different sensors becomes more and more important. Without on overall quality assurance (QA) the produced data cannot be traced back to the raw data and are therefore useless. For the future it can be expected that the international standards for collecting data will be tightened up.

References

1. Reutter, D. et al. (2010) *Biosecur. Bioterror.: Biodef. Strategy, Pract., Sci.*, **8** (4), 343–355.
2. Jopling, L. (2005) Chemical, Biological, Radiological or Nuclear (CBRN) Detection: A Technological Overview. NATO Report 167 CDS 05 E rev 2, NATO, p. 2.
3. European Defence Agency - EDA (2006) Concept for the Detection, Identification and Monitoring (DIM) of Chemical, Biological, Radiological and Nuclear (CBRN) and Toxic Industrial Hazards (TIH), Report of the Project Team on CBRN Detection, Identification and Monitoring.
4. Demirov, P.A., Feldman, A.B., and Lin, J.S. (2005) *Johns Hopkins APL Tech. Dig.*, **26** (4), 321–333.
5. Carrano J. (2004) Chemical and Biological Sensor Standards Study. CBS3_Final Report, Defense Advanced Research Projects Agency (DARPA), pp. 1–31.
6. Hancock, J.R. and Dragon, D.C. (2005) Sample Preparation and Identification of Biological, Chemical and Mid-Spectrum Agents. A General Survey for the Revised NATO AC/225 (LG/7) AEP-10 Edition 6 Handbook (DRDC Suffield TM 2005-135) Defense R&D Canada – Suffield.
7. Sun, Y. and Ong, K.Y. (2005) *Detection Technologies for Chemical Warfare Agents and Toxic Vapors*, 1st edn, CRC Press, Boca Raton, FL.
8. Sferopoulo, R. (2008) A review of Chemical Warfare Agent (CHEMICAL WARFARE AGENT) Detector Technologies and Commercial-off-the-shelf Items. Report No. DSTO-GD-0570, Human Protection and Performance Division, DSTO Defence Science and Technology Organization.
9. Fitch, J.P., Raber, E., and Imbro, D.R. (2003) *Science*, **302** (5649), 1350–1354.
10. Fatah, A.A., Barrett, J.A., Arcilesi, R.D., Ewing, K.J., and Helinski, M.S. (2000) Guide for the Selection of Chemical Agent and Toxic Industrial Material Detektion Equipment for Emergency First Responders, NIJ Guide 100-00, *Law Enforcement and Corrections Standards and Testing Program*, U.S. Department of Justice, Volume II, Section 3.0, pp. 7–13.

7
Analysis of Chemical Warfare Agents – Searching for Molecules

Andre Richardt, Martin Jung, and Bernd Niemeyer

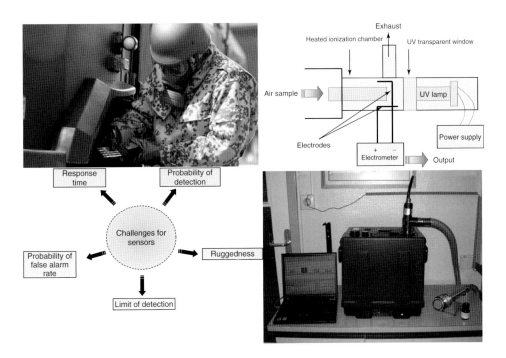

Since the first use of mustard gas and other gases on the battlefield during World War I it has been the aim to develop detection technologies to trigger countermeasures in time. One of the first detectors for chemical warfare agents was trained sniffers to detect the characteristic garlicky odor of mustard gas in time. Other detectors were based on color changes resulting from reactions between mustard and reagents. In the decades after World War I further detectors were developed. However, the development of detection technologies always lagged behind the discovery and synthesis of new chemical warfare agents. With the emergence of the first nerve agents the challenges for detection technologies were even greater and different methods, techniques, and instruments have been developed over the decades. Especially after the 9/11 incident in the year 2001 efforts were increased significantly to obtain better chemical warfare agent detectors for quicker reaction time in diverse scenarios.

7.1
Analytical Chemistry – the Scientific Basis for Searching Molecules

By choosing the appropriate detector technology for a defined scenario we have not only to understand the needed and required specifications but also the basics of sampling and analyzing probes and the scientific principles of different chemical warfare agent detector technologies [1–3]. Furthermore, the developed chemical warfare agent detectors and sensors have to be thoroughly tested to rigorous criteria. Changing environmental conditions such as temperature, humidity, additional compounds, and atmospheric pressure could lead to false results or misinterpretation and must therefore be understood, addressed, and corrected.

> - **Analytical chemistry**: a branch of chemistry that deals with the separation, identification, and quantification of compounds and mixtures.

> Analysis of substances: Before we start our tour through potential technologies for the detection of chemical warfare agents we have to set out fundamental aspects about the analyses of substances. Analytical chemistry is the study of methods for determining the composition and quantity of substances in environmental, clinical, forensic matrices, and quality control and thus consists of the qualitative ("what?") and quantitative ("how much?") questions (Figure 7.1).
>
>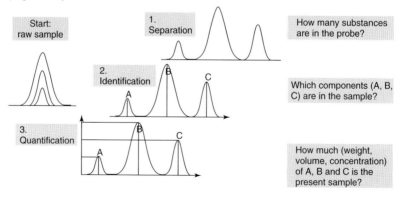
>
> **Figure 7.1** What is qualitative and what is quantitative? (1) Separation, (2) identification, and (3) quantification.

Figure 7.1 shows how to analyze raw samples ("start"). The various substances have to be **separated**, for example, mainly by extraction and chromatographic methods. None of the compounds have to be "lost," for example, irreversibly bound onto the chromatographic support or evaporated and thus lost through the gas phase. Thus we know: *How many substances are in the probe.*

The **identification** of the chemical species is realized by comparing known data, such as solubility in the extractant or the retention time within the chromatographic column under the actual flow condition, with relevant data from the literature or our own data set (e.g., calibration). We have to apply a detection method, characteristic for the different substances, for example, light absorption (Section 7.4.2). After this step we know: *Which compounds (A, B, C) are in the sample.*

The final procedure aims to **quantify** the substances present. Methods include evaporation of the solvent fractions and weighing of the residues (gravimetry) or comparing the peak areas in the chromatogram with calibration data. In the end we obtain full information on: *How much (weight, volume, concentration, etc.) of A, B, C is in the present sample.*

A typical correlation for analyzing the concentration by photometric methods is given by the Lambert–Beer law. It implies that the change in absorbance (A) is linear with the change in concentration (c):

$$A = c\varepsilon l = c\alpha$$

The extinction coefficient ε depends on the substance to be analyzed, the wavelength, environment parameters, such as temperature, and so on. The path-length l is the length the light beam passes through the analyte.

Therefore, determination of the reliability and significance of any result is a general analytical problem (Figure 7.2). Obviously, data of *unknown* quality are useless. Therefore, the evaluation of the analytical method and the determination of the possible systematic and random errors are contributing factors in obtaining verifiable results [4, 5].

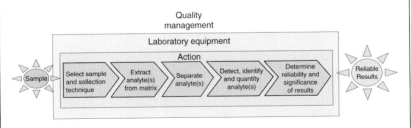

Figure 7.2 Process of analyzing a sample with reliable results.

Ruggedness against harsh environmental conditions and the rapid and reliable detection at very low concentrations are further factors that influence a decision for or against a detector technology. Therefore, this chapter aims to give an overview of (i) possible chemical warfare agent detection standards and criteria for deployment, (ii) several detector principles and technologies, (iii) test procedures of detector technologies for evaluation, and (iv) briefly, future technological developments. Finally the reader should first have an understanding of the scientific and technological basics of the most important detector technologies and their possible limitations. Based on this knowledge and the understanding of possible chemical warfare agent detection standards they should have a chance to choose the most suitable detector for a specific scenario. Some of the presented principles for chemical warfare agent detectors could be used by analogy for biological warfare agent detectors.

7.2
Standards for Chemical Warfare Agent Sensor Systems and Criteria for Deployment

Researchers, developers, and users have to adhere to different recommendation parameters for chemical warfare agent detectors capabilities, sensitivities, and response times for detector equipment. These parameters can be considered as guidelines and challenges for chemical warfare agent sensor systems (Figures 7.3 and 6.3) [1, 2, 6, 7].

7.2.1
Recommended Chemical Agent Concentration and Requirements for Chemical Warfare Agent Sensors

Because of the potential benefit to the general population the 10-min acute exposure guideline level (AEGL) level-1 concentration values for each agent can be a valuable

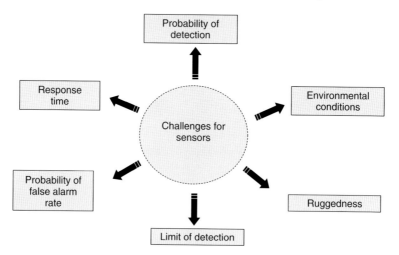

Figure 7.3 Some challenges for chemical warfare agent sensor systems.

Table 7.1 Joint Services Operational Requirements for point sampling detectors; maximum response time at immediately dangerous to life and health (IDLH) and higher concentrations [8].

Agent	Threshold exposure concentration (mg m^{-3})	Threshold exposure response time maximum (s)	Relative humidity (%RH) range	Temperature range (°C)
VX	1	10	5–100	−10 to +49
	0.1	30		
	0.04	90		
GA, GB, GD, GF*	1	10	5–100	−30 to +49
	0.1	30		
HD, L**	50	10	5–100	−18 to +49
	2	120		

*GA: Tabun; GB: Sarin; GD: Soman; GF: cyclohexyl sarin.
**HD: mustard; L: Lewisite.

proposal as alarm trigger points for new detectors. For the deployed military population AEGL-2 or AEGL-3 values might be set as objectives [1]. To enable the affected workers or population to don protective gear chemical warfare agent detectors should warn in time. For this purpose the Joint Services Operational Requirements (JSOR) were developed (Table 7.1). The JSOR are only one example how recommended chemical agent concentrations can be addressed and be used as benchmarks for new detectors. Other nations and organizations may have another basic level for their requirements for chemical warfare agent sensors.

7.2.2
Acute Exposure Guideline Levels (AEGLs) for Chemical Warfare Agents

Predicting chemical warfare agent dose–response effects over longer exposure times and lower concentrations is a medical challenge and not the aim of this book. For completeness, though, we will address this question. In a first step the conditions of a real-world chemical warfare agent incident have to be simulated in laboratory systems. Several factors such as the weather, chemical and physical state of the agent (see Part II), further chemicals present, and rate of exposure can be easily simulated, whereas other factors such as vegetation, soil, and handling by personnel can lead to greater uncertainties. These uncertainties have to be addressed when we want to compare laboratory data with a real chemical warfare agent incident. Nevertheless, some necessary standards exit for chemical warfare agent concentration levels [e.g., JSOR (Table 7.1)], their designated time period, and their designated application exit [1]. Exact data are necessary for any emergency planning and appropriate medical countermeasures.

For example, AEGLs (Table 6.1) are designed for the general public and are protective to the general population. Therefore, they can be used as an appropriate way to ensure that the deployed population is also protected. Thus AEGLs are also guidelines for medical investigations of long-term exposure. Further information about the medical aspects of AEGLs of chemical warfare agents is available in specialized literature [9–11].

7.3
False Alarm Rate and Limit of Sensitivity

Another important point that has to be considered is the choice of a detector system with characteristics that meet the desired capabilities. In Table 6.2 we have shown some of these characteristics for possible detector technologies. One possible tool is a method to relate statistical quantities to operational demands such as accepted false alarm rate and limit of sensitivity. Some important definitions have to be set before we discuss further details:

- **Probability of detection**: the probability that the system detects a given concentration, when the system takes a decision.
- **Probability of false (positive) alarms**: the probability of the system to trigger an alarm when no contamination is present.
- **Response time**: the time for the system to analyze the information and to take a decision.
- **False alarm rate**: the number of false alarms for the user within a fixed time scale.
- **Limit of detection (LOD)**: the lowest detectable amount of chemical warfare agent concentration that produces an alarm.

We do not want to discuss all these parameters in detail. However, if a detector system produces too many false (positive) alarms over a certain period of time it is understandable that further alarms will no longer be noticed by the user. Owing to the importance of false (positive) alarms and the limit of sensitivity as critical parameters for the suitability of given detector systems the reader needs to understand the complexity of the procedure to determine the full characteristics and performance of such a system.

False Alarms

If R is defined as the false alarm rate then the time between two false alarms is:

$$t = 1/R$$

The response time of the detector system T is the time needed for the system to take the decision for detection/no detection. Between two false alarms the system takes N decisions:

$$N = 1/RT$$

If one of these decisions is a false alarm, we have the equation for the probability of false alarm rates P_{fa} as:
$$P_{fa} = 1/N = RT$$
Two examples will highlight this equation a little:

1) Assume a detection system with a 1-s response time. A false alarm rate of 1 per month (30 days = 2 592 000 s) leads to a probability of false alarm P_{fa} of:
$$P_{fa} = 1/N = RT = 1/2\,592\,000\,\text{s}^{-1} \times 1\,\text{s} = 2.6 \times 10^{-7}$$
Interpretation of the results: this means in each 39 million decisions one is false. This would be a nearly perfect sensor.

2) Assume a detection system with a 60-s response time. A false alarm rate of 1 per 1 h leads to a probability of false alarm P_{fa} of:
$$P_{fa} = 1/N = RT = 1/3600\,\text{s}^{-1} \times 60\,\text{s} = 1.7 \times 10^{-2}$$
Interpretation of the result: in every 170 decisions one is false. This is a more realistic assumption for existing detector systems.

The limit of sensitivity is another important operational parameter and is given by the minimal concentration level that generates the user needed detection quality. Summarizing, the lower the probability of false (positive) alarms the better the detection probability. Furthermore, an ideal system should have a high sensitivity and in parallel a low false alarm rate (Figure 7.4).

The general idea of challenging the detector with different parameters will be discussed more as part of testing chemical warfare agent detectors.

7.4
Technologies for Chemical Warfare Agent Sensor Systems

We have covered some general basics for the deployment of chemical warfare agent sensors. Now we are able to discuss several technologies for the detection of chemical warfare agents (CWA). In general, most commercially available technologies are based on (i) mass spectrometry (MS), (ii) flame photometry (FP) or atomic absorption spectrometry (AAS), (iii) ion mobility spectrometry (IMS), (iv) colorimetric technology, (v) electrochemical (EC) technologies, (vi) photoionization (PI) detection techniques, and (vii) infrared (IR) spectrometry (Figure 7.5). Although MS, FP, and IMS can be considered as the main technologies for chemical warfare agent detectors today, we will give for all of these technologies a short overview of their principle of operation, their instrumentation, and technique specification. For more detailed descriptions the reader is referred to advanced literature [8, 12, 13].

7 Analysis of Chemical Warfare Agents – Searching for Molecules

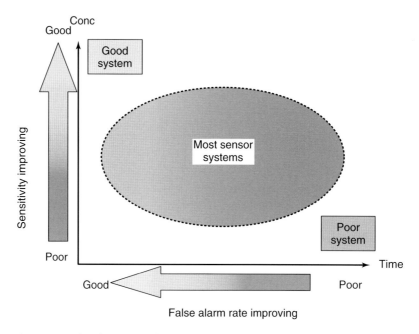

Figure 7.4 False alarm rate and sensitivity as important characteristics for detector systems.

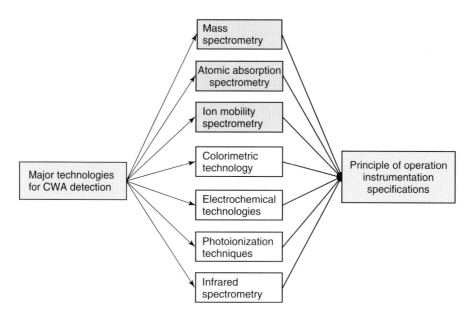

Figure 7.5 Major technologies for chemical warfare detection.

7.4.1
Mass Spectrometry

Mass spectrometry (MS) is one of the most important technologies for the detection of chemical warfare agents. It is an analytical technique that measures the mass-to-charge ratio of charged particles and can be used for:

- determining masses of molecules and fragments,
- determining the elemental composition of a sample or molecule,
- elucidating the chemical structures of molecules (Figure 7.6).

In Figure 7.6, MS as hyphenated analytical procedure is lined out; the detection method applied generally is an MS technique, while the separation process is any chromatographic procedure.

MS in different forms is a common, reliable, powerful, and sophisticated/expensive technology for the detection of chemical warfare agents, their degradation products, and possible adducts (Figure 7.6) [14]. Owing to the multitude of spectrometry technologies, not every form of MS technology can be explained, although all forms have their advantages and disadvantages. For the reader it is sufficient to understand the general principle of operation of MS sensors. This general MS principle (Figure 7.7) consists of ionizing chemical compounds to generate charged molecules or molecule fragments. The measurement of their mass-to-charge ratios leads to information about the unknown chemical substance. In a typical MS procedure a sample is injected or loaded into a high vacuum chamber. After vaporization the components are ionized by various methods. As

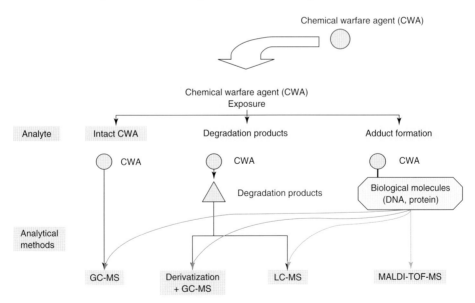

Figure 7.6 Schematic of changes of chemical warfare agents after exposure and possible corresponding mass spectrometry technologies for their detection [14].

188 | 7 Analysis of Chemical Warfare Agents – Searching for Molecules

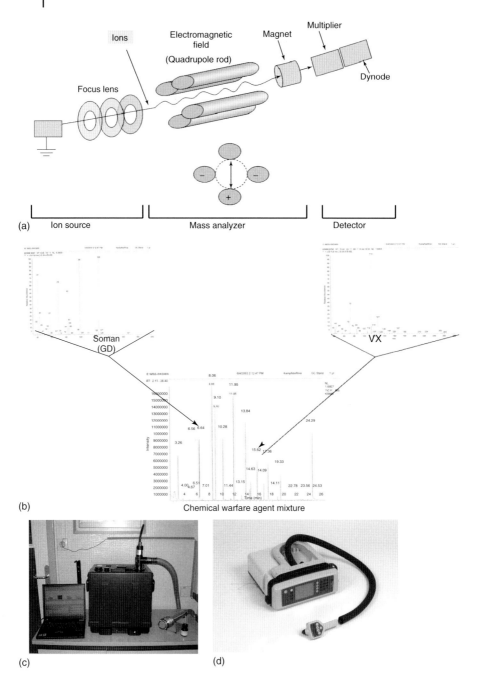

Figure 7.7 Principles and examples of mass spectrometry: (a) schematic view of a quadrupole mass spectrometer, (b) analysis of a chemical warfare agent mixture with the identification of VX and soman (GD), (c) Bruker MM2 GCMS, and (d) Inficon Hapsite GCMS.

charged particles the ions are separated according to their mass-to-charge ratio in an electromagnetic field analyzer. In a final step the ions are measured by a detector and the detector signal is processed into mass spectra (Figure 7.7). MS methods are very suitable for quantitative measurements.

All MS instruments consist of three parts (Figure 7.7):

1) In the ion source the material (the analyte) is ionized. After ionization the ions are then transported by magnetic or electric fields into a high vacuum atmosphere and to the mass-to-charge analyzer. Note that different techniques for ionization are available, which determine what types of samples can be analyzed by MS. Electron and chemical ionization are used for gases and vapors and are the main ionization technology for the ionization of chemical warfare agents. However, for completeness, ionization of liquid and solid samples by electrospray ionization and matrix-assisted laser desorption/ionization (MALDI) can be used. An inductively coupled plasma (ICP) is used for cation analysis. Other technologies include glow discharge, field desorption (FD), fast atom bombardment (FAB), and thermospray.

2) A mass-to-charge analyzer is the second part and separates the ions by applying electromagnetic fields (mass-to-charge ratio). The Lorentz force law and Newton's second law govern the dynamics of charged particles in electric and magnetic fields in vacuum. These differential equations are the classical equations used to describe the motion for charged particles. The particle's motion in space and time in terms of m/Q can be determined completely together with the particle's initial conditions. This is the reason why mass spectrometers could be thought of as "mass-to-charge spectrometers." It is common to use the dimensionless m/z number, where z is the number of elementary charges (e) on the ion ($z = Q/e$). This quantity represents the ratio of the mass number m and the charge number z.

There are different types of analyzers, applying static or dynamic fields, and electric or magnetic fields, but all of them use the two above-mentioned Lorentz and Newton equations. The simultaneous utilization of two or more analyzers is called *tandem mass spectrometry* (or mass spectrometry/mass spectrometry, MS/MS) and can be found in many mass spectrometers with improved sensitivity.

3) The final module of the mass spectrometer is the detector. The detector perceives the impact of an ion by generating a proportional electrical signal and provides the data for calculating the abundances of each ion present.

Another point is the data representation. MS produces various types of data, but the most common data representation is the mass spectrum. However, what we have to accentuate as the most complicated subject is MS data analysis. There are general subdivisions of data that are fundamental to understanding any data:

1) **Ion charge**: Many mass spectrometers work in either *negative ion mode* or *positive ion mode*. It is important to know whether the observed ions are negatively or positively charged. On the one hand, this knowledge is important

in determining the neutral mass but, on the other hand, it also indicates something about the nature of the molecules.
2) **Ion source**: The use of different types of ion sources results in different arrays of fragments produced from the original molecules. Data processing has to consider this.
3) **Origin of a sample**: By understanding the origin of a sample, certain expectations can be assumed as to the different molecules of the sample and their fragmentations. For example, a probe from a synthesis/manufacturing process will probably contain impurities chemically related to the target component. In contrast, a relatively crudely prepared environmental sample with biological materials will probably contain a certain amount of salt. This can result in the formation of adducts with the analyte molecules in certain analyses.
4) **Preparation of the sample**: Results can also depend strongly on the preparation procedure of the sample and how it was injected into the mass spectrometer.

Figure 7.7b exhibits the mass spectra of pure soman and VX, as well as a mixture, containing both substances. Commercially available handheld MS devices of different suppliers are illustrated in Figure 7.7c and d.

7.4.2
Atomic Absorption Spectrometry (AAS)

This technique, also called flame photometry is based on the light emission phenomenon and has long been successfully used for chemical warfare agent detection [1, 2, 15]. Because most chemical warfare agents contain the elements sulfur and/or phosphorus (see Chapter 3) these chemical warfare agents can be detected by the characteristic light emitted by excited sulfur and/or phosphorus atoms or clusters (Figure 7.8a). As seen from (Figure 7.8b) hand-held devices are available commercially.

The same principle can be used for other toxic chemicals containing sulfur and/or phosphorus atoms such as pesticides, insecticides, fungicides, and petroleum products. Flame photometry detectors (FPDs) are powerful tools in the analysis of samples containing toxic chemicals [3].

FPDs for chemical warfare agents are not selective for specific compounds because they respond to all sulfur/phosphorus containing molecules. To increase the selectivity and identification of chemical warfare agents the flame photometric technology is normally connected to a gas chromatography (GC) column for separation into single portions of pure substances and with a known retention time on the column for specific substances. Without the integration of a GC column the chemical warfare agent limit of detection (LOD) is in parts per billion to parts per million. These GC/FPD technologies were used to develop standard methods for the analysis of samples and to determine chemical warfare agent concentrations.

FPD is based on the principle that each element emits light with a particular wavelength when burned in an open flame. This technology is based on the fact that atoms possess a specific energy level based on the nucleus and the corresponding electron shell around the nucleus with a specific energy state. If the atom or

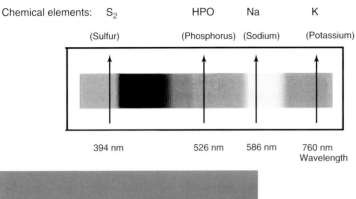

Figure 7.8 (a) Example of main emission line (or band) from sulfur, phosphorus, sodium, and potassium between 350 and 800 nm [2]; (b) Proengin_AP4C as an example of a portable chemical contamination control device used to detect directly chemical agents in the form of vapor, aerosols, and dust (Proengin, France).

molecule is in its lowest energy state ("ground state") it is in the most stable state. When it is stimulated by external energy (photons), electrons will be excited to higher states ("excited states"), which then become unstable and return to its ground state, and emit the gathered photon energy readily at specific wavelength. The emitted radiation of this electron transition is specific for an atom. Thus it is possible to identify the molecule by measuring the energy of the emitted radiation. Figure 7.9a gives the procedure and a set-up of FPD applying a hydrogen–air flame to provide a broad band of photon energy equivalents. When chemical warfare agents are broken down in an hydrogen–air flame the released sulfur and phosphorus are excited. When they return to lower energy levels, the excited sulfur and phosphorus in the form of S_2^* and HPO^* emit light at characteristic wavelengths. The main emission wavelengths of S_2^* (394 nm) and HPO^* (526 nm) are identified by the FPD with the use of wave-selective filters. Only photons with these specific wavelengths could pass through the filter and hit the photomultiplier tube (PMT), where the signal is transformed and amplified to a proportional electrical signal. The detection signal is then generated by monitoring the light intensity by the PMT. Because the intensity of the light released is proportional of the S_2^* and HPO^* numbers, a direct determination of the chemical concentration of the relevant chemical species is possible. This determination can be realized with the employment of calibration curves, where different concentrations are analyzed and the relevant signal intensity is correlated to substance concentration, as seen from Figure 7.9c for sarin.

FPDs consist of different system components (Figure 7.9b). These components are different inlets (hydrogen, air, sample), thermal and light filters, PMT, signal processor, and exhaust chimney. We can distinguish between the direct sample types of FPD and the indirect sample type of FPD where the flame photometer is combined with a GC-interface. The direct sample type of FPD mixes the sample in the combustion chamber with the hydrogen and burns at the tip of the burner. The detection of chemical warfare agents occurs quickly regarding the contents of sulfur or phosphorus. This is more or less "real time" detection. The combination of two independent methods (retention times determined by GC and the presence of sulfur and phosphorus) results in the identification of substances analyzed. However, the increase of information needs an increased analysis procedure time. This is an example of how getting more information requires longer detection or analysis time. The flame and hydrogen source we mention only briefly. The temperatures in the hydrogen–air flame of 2000–2100 °C are high enough to excite S_2 and HPO without ionizing them [1]. For field applications, small hydrogen alloy storage devices without a major safety hazard are used. The emitted characteristic radiation from the $S_2^* \rightarrow S_2$ and $HPO^* \rightarrow HPO$ transition passes through two quartz windows and optical filters before reaching the PMT. However, to protect the PMT from background influence and heat, thermal filters are applied. Furthermore, different wavelength filters are used to ensure that only radiation with 394 nm for $S_2^* \rightarrow S_2$ or 526 nm for $HPO^* \rightarrow HPO$ will pass through the filter. After the wavelength filters the photons reached the PMT, where the incoming light is converted into an electrical signal, which can be analyzed by a microprocessor. Specific compounds can only be indentified when the FP is coupled with a GC interface. By running standard samples under defined conditions (temperature, pressure, and type of column) the retention time and the presence of sulfur and/or phosphorus atoms are the key factors for matching the unknown compound with the standard.

FPs have been developed to detect sulfur or phosphorus atoms and therefore it is not possible to identify chemical warfare agents clearly (missing selectivity). In addition, interfering chemicals that emit photons at similar wavelengths as those of sulfur or phosphorus atoms could cause false positive alarms. The limits of sulfur and phosphorus detection are lower than the reported JSOR concentration levels. The response dynamic range for sulfur and phosphorus is over 10^4 µV and follows, normally, a linear concentration–response relationship. For quantitative analysis the background has to be corrected. FPD technology is very rugged and needs very low maintenance [1].

7.4.3
Ion Mobility Spectrometry (IMS)

Ion mobility spectrometry (IMS) is one of the most common technologies for today's units used to detect chemical warfare agents, toxic industrial chemicals

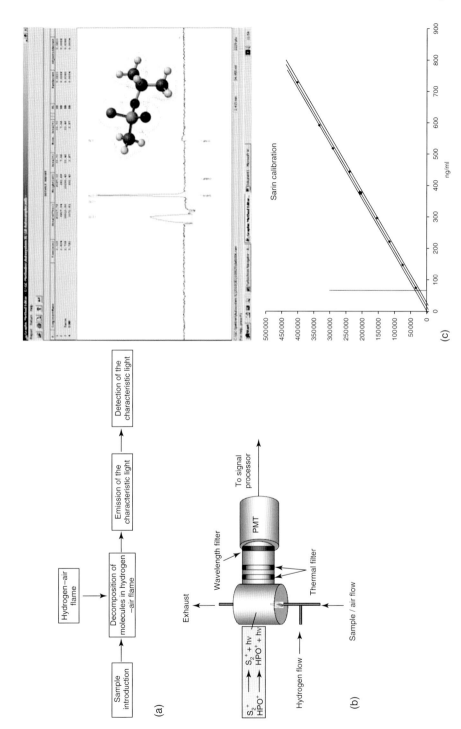

Figure 7.9 Flame photometry (FP): (a) block diagram of flame photometry analysis procedure; (b) typical configuration of flame photometry; (c) signal of three independent sarin calibration curves with the limit of detection by the indicated bar at around 70 ng ml^{-1} for sarin.

(TICs), explosives, and illicit drugs [1–3, 16–19]. IMS technology possesses same advantages that can be employed for the detection of chemical warfare agents:

- high sensitivity and selectivity,
- low LOD,
- rapid analysis,
- no or little sample preparation for gaseous samples,
- operation under atmospheric conditions.

The ability of IMS to separate different ions in a drift tube (Figure 7.10c) can be used to detect a wide range of chemicals and to identify them.

The shape, mass, and charge of an ion govern the time it takes to pass through an electric field within the counter flowing drift gas. The spectrum of an ion's mobility through the drift region under a constant strength and length of the electric field is the "signature" of the ion (Figure 7.10b). In general, a specific ion has a specific spectrum under defined conditions. Without setting specified conditions (e.g. air pressure, humidity, temperature, etc.) we will not obtain valid results. The characteristic fingerprint can be used to catalog different ions in libraries and to identify unknown samples through spectrum matching with the library within the detector unit.

The operational process of IMS detection (Figure 7.10a) includes:

- sample introduction,
- ionization,
- drift tube providing ionization and ion separation, applying mechanisms,
- ion collection,
- signal generation/microprocessor.

Via the inlet the sample vapor is drawn into the ionization region of the detector. According to the chemical and physical properties of the molecules the ionization source ionizes the molecules to negative or positive species. One important feature is that the ionization process occurs under atmospheric conditions and no carrier gas is needed. Especially in handheld detectors, a weak radiation source (e.g., ^{63}Ni) is common. In accordance with the used ionization source the ionization process and the charge competition occur. The charge competition is the process by which molecules compete for charges to become ionized. After the ionization process the ions are injected into the drift region via a very short pulse. The pulse is controlled by an electronic gate. When the gate is "open" the ions enter the drift region. It is important that the "open" time is kept as short as possible – otherwise the separation of the different ions will not be sharp. Furthermore, it is possible to control the type of ions entering the drift tube by regulating the gate. For handheld detectors, refer to (Figure 7.10d and e), the drift tube itself is several centimeters long and a homogeneous electric field can be created [1]. For detection of chemical warfare agents it is important (i) to keep the air flow dry and free from other organic matter, (ii) to maintain adequate control of the drift temperature, and (iii) retain pressure control.

After this process the charged ions enter the drift tube with an electrical field. In milliseconds the ions reach the ion collector at the other end of the drift tube. The drift of the ions can be explained by the force exerted on the ions from the electrical field as well as the interaction of the counter flowing drift gas. Additionally, a usually dried airflow is added in the opposite direction of the ion drift and the ions in the drift tube can collide with molecules in the drift flow. In general, smaller as well as higher charged ions reach the collector earlier than larger or lower charged ions. The collision process separates the ions on their individual basis of collision events and finally determines the mobility. Another important factor in the drift tube is

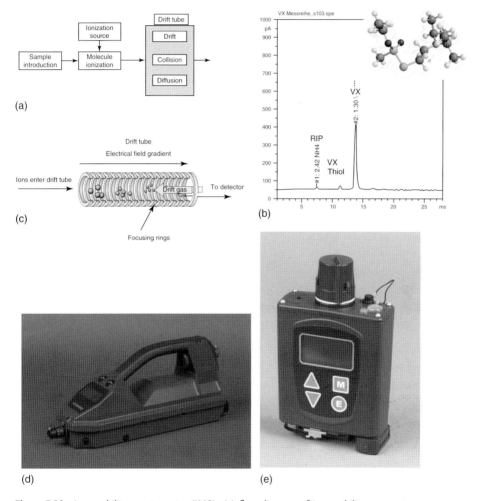

Figure 7.10 Ion mobility spectrometry (IMS): (a) flow diagram of ion mobility spectrometer analysis procedure; (b) typical analysis of VX with RIP (reactant ions positive), VX Thiol, and VX peaks; (c) typical configuration of IMS. Examples of ion mobility spectrometers: (d) Bruker Raid-M100 IMS; (e) Smith Detection LCD3.3 IMS.

the dispersion of the ions through diffusion. The diffusion depends on the drift temperature and on the nature of the ions and produces a concentration gradient of the ions through the drift tube. This concentration gradient results in broader peaks for longer drifting than for ions with a shorter residence time in the drift tube. The ion collector is located at the end of the drift tube. The various ions collide with the ion collector at different times due to the selection process in the drift tube.

After the whole way through in the drift tube with mechanisms of drift, collision, and diffusion the ions reach the ion collector and produce an electrical signal that is proportional to the number of the ions collided with the ion collector. This proportionality can be used as a strong indicator for the relative concentration of the chemical warfare agents. This signal is compared with the detector's library of signatures and an audible/visible alarm is generated when the electrical signal matches the criteria set for chemical warfare agents [1].

Most chemical warfare agents have relatively high vapor pressure and direct analysis of contaminated air sample has to be performed quickly. The sample inlet of IMS sensors is designed for this purpose. The unknown air sample is injected into the IMS without any sample preparation. The sample path is generally heated. In addition, to reduce moisture problems in the sample, thin heated membranes (e.g., dimethyl silicone membranes) with low permeation rates for water or other particles (e.g., dust) are integrated.

In principle, IMS sensors can be utilized to detect any ionized molecule. Current instruments can detect multiple chemicals of common interest. The nonselective identification properties of IMS detectors make them useful for detection of different vapors. One problem could be false positive alarms that occur when the mobility (K) of a non-hazardous substance is similar to one of the assigned peaks of a chemical warfare agent [1]:

$$K = v/E$$

where

v is the drift velocity of an ion (cm s^{-1}),
E is the electric field gradient (V cm^{-1}),
K is the mobility of ions (cm^2 V^{-1} s^{-1}).

To reduce the false alarm rates, several techniques have been developed, such as (i) adding a gas chromatographic column in front of the IMS or (ii) use of chemical additives. False negative alarms can occur on analyzing complex sample matrices. In general, the sensitivity and low LOD of IMS sensors mean they are capable of detecting chemical warfare agents at determined concentration levels. Nevertheless, the dynamic response range of IMS sensors is two to three orders of magnitude and the more exponential relationship between concentration and response could be a disadvantage for the detection of chemical warfare agents under field conditions. Furthermore, some environmental conditions like pressure, humidity, and temperature could cause some problems regarding the performance of IMS-based sensors. Therefore, changing environmental conditions have to be addressed very precisely and to be compensated for by calibration or technical counteractions.

7.4.4
Colorimetric Technology

The idea of using a color change resulting from the reaction between chemical warfare agents and reagents has been realized in different colorimetric detectors [1, 2, 20, 21]. The normal case with a colorimetric detector is the use of one reagent, which is designed for the detection of one class of chemicals; thus for different classes of chemical warfare agents several sensors are needed for field applications. However, one great advantage of the use of color change is the low false alarm rate because of the very high selectivity of this method (Figure 7.11). The detection can be performed by the same handheld device, which even can deliver an acoustic or optic alarm (Figure 7.11a).

One type of multi-compound and multi-chemical group analyses has been realized by Dräger (Lübeck, Germany) by placing different reagents within a transparent reaction tube. Only small bands react with different agents or groups of chemicals by color change (Figure 7.11b). The higher the concentration of the chemical to be analyzed the shorter the monitoring time (Figure 7.11c).

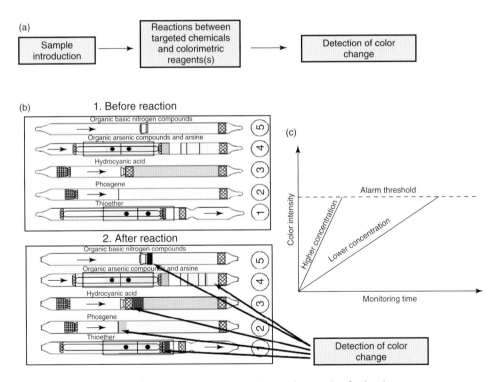

Figure 7.11 Principle of color-indicating detection: (a) general principle of color detection; (b) indication of color change (Fa. Dräger, Germany); (c) alarm threshold by different concentrations over time.

Use and Limits of Detection of Different Color Indicating Tubes

Figure 7.12 shows the colorimetric reaction device in operation (Table 7.2).

Figure 7.12 Example of the use of color indicating tubes (Fa. Dräger, Germany).

Table 7.2 Sensitivity of detection of the color-indicating tubes (a) CDS 1 and (b) CDS 2 (Dräger tubes).

Chemical warfare agent	Detection	Sensitivity
(a) CDS 1		
Hydrogen cyanide	Hydrogen cyanide	1 ppm
Phosgene	Phosgene	0.2 ppm
Lewisite	Organic arsenic compounds and arsine	3 mg m^{-3} (org. arsine) 0.1 ppm arsine
N-Mustard	Organic basic nitrogeneous compounds	1 mg m^{-3}
S-Mustard	Thioether	1 mg m^{-3}
(b) CDS 2		
Nerve agents	Phosphoric acid esters	0.025 ppm
Phosgene	Phosgene	0.2 ppm
Chlorcyan	Chlorcyan	0.25 ppm
Chlorine	Chlorine	0.2 ppm
Yperite	Thioether	1 mg m^{-3}

The sensitivity of the tube analyses is very low, ranging from 0.025 ppm (for nerve agents (e.g. VX) with CDS 2) to 1 ppm for hydrogen cyanide (HCN) with the tube CDS 1.

7.4.5
Photoionization Technology (PI)

Another common detection technique is PI technology, which is often employed in combination with for GC systems in laboratory environments. This technology

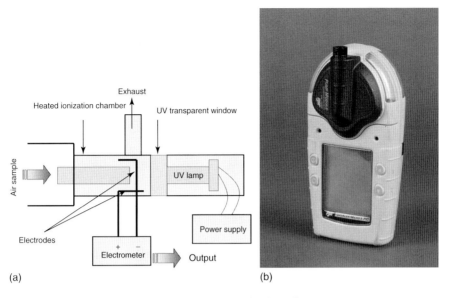

Figure 7.13 (a) Schematic view of photoionization technology; (b) BW GASALERTMicro5 PID.

is highly sensitive, allowing the detection of compounds in very low concentrations (parts per billion to parts per million) [1, 2]. Photoionization detectors (PIDs) can be used as first responders to obtain fast information about various chemicals, including inorganic compounds, and about the presence of volatile chemical constituents in the air. This can be achieved by the general principle of operation (Figure 7.13a). Any compound with an ionization potential lower than the UV-radiation source in the detector can be ionized and detected [2]. The UV lamp generates the ionization energy to ionize the sample molecules in the air.

The formed positive ions are attracted to the negatively charged electrode. After the charge is released the ions are neutralized and an electrical current can be measured. Because the produced current is related to the amount of ionized substances it is also proportional to the concentration of the target substance. The concentration of the analyte is displayed as parts per million or parts per billion on the LED of the instrument, shown in (Figure 7.13b) [2]. This technology is utilized routinely in the laboratory to analyze aromatic hydrocarbons or heteroatom-containing compounds such as organosulfur or organophosphorus species. As a general rule for the identification of substances with PI sensors we can state if the substances being detected contain a carbon atom then PI technology can be employed [2].

7.4.6
Electrochemical Technologies

Electrochemical (EC) sensors can detect toxic compounds in vapors at parts-per-million level. These sensors play a significant role in the detection of gaseous

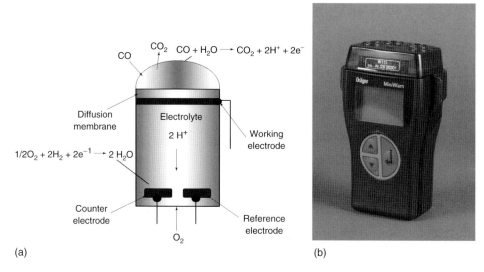

Figure 7.14 (a) Typical structure of an electrochemical sensor; (b) Dräger MiniWarn EC.

inorganic materials and their specific application is the detection of toxic industrial chemicals. The principle of EC sensors is based on the electric current between chemicals in the air and the sensor electrode (Figure 7.14a). The electric current is generated by oxidation–reduction reactions. Long-term monitoring, real-time measurement, and safety applications are some of the possible areas of use.

A typical EC sensor consists of three electrodes (Figure 7.14). The (i) working or sensing electrode (WE), (ii) counter electrode (CE), and (iii) reference electrode (RE) are immersed in an electrolyte. The different electrodes are responsible for the oxidation–reduction process (WE, CE) and to control the electric potential of the sensing electrode (WE) through a feedback mechanism (RE). The oxidation or reduction of the targeted chemical occurs during contact of the substance with the WE, which consists of a porous polymer membrane, and is one part of the reaction. The resulting part of the reaction occurs at the CE and completes the oxidation–reduction reaction. The lifetime of EC sensors can be reasonably long if the maintenance guidebook is followed. EC sensors react with targeted chemicals at the low parts-per-million level and are therefore relatively sensitive. The selectivity is not as high as other technologies due to the generation of an electric current for any reaction of a compound with the electrolyte. Furthermore, changing environmental conditions could affect the performance of EC sensors. An example of a commercial device is given in Figure 7.14b).

7.4.7
Infrared (IR) Spectroscopy

Infrared wavelengths range from 0.78 to 1000 μm. IR detectors can detect and identify a substance by its IR radiation absorption characteristics [3, 22]. The detectors can be divided in two types: (i) detection of IR radiation emitted from objects (e.g. IR cameras) and (ii) detection of IR radiation absorbed by objects (e.g. IR spectrometry), which is the subject of this chapter. This technique is like PI and can be used as a non-destructive detection method. It is based on the fact that functional groups in a molecule vibrate at specific frequencies within the IR region. A chemical warfare agent can be detected by determining the IR radiation frequencies absorbed by these functional groups (Figure 7.15b). Over the years large IR spectra libraries have been established. The characteristic frequencies of chemical warfare agents are well known and in field applications the detected chemicals can be compared with the characteristic IR spectra in the library. IR spectrometry shows fast detection with a high sensitivity, although changing environmental conditions may have greater impacts on its performance than on other technologies.

The principle of operation is based on the introduction of a sample into the detector, where it is exposed to IR irradiation (Figure 7.15a). IR spectroscopy is a pulsed sampling methodology, which means that analysis of the sample occurs non-continuously. Each sample has to be pumped into the detector for analysis. IR absorption at characteristic wavelengths occurs when the targeted chemical warfare agent is present. The concentration is proportional to the area of the spectrum peak. Signals are generated and sent to the microprocessor for identification. For detection purposes IR with wavelengths between 2.5 and 15 µm are the most common. Several important functional groups in chemical warfare agents have characteristic absorption bands in the IR spectrum, depending on their molecular environment and neighboring groups. Furthermore, less characteristic bands can be used for compound identification by supplying a specific molecular fingerprint [23]. IR spectrometry can be used as a sensitive quantitative tool:

$$A = c\varepsilon l = c\alpha$$

According the above-discussed Lambert–Beer law (Section 7.1), both the path length l and the extinction coefficient ε are unknown; thus the mathematical product, the quantity α, has to be obtained by a calibration curve using different substrate concentrations. However, it could be that at the low end of detectable limits the Lambert–Beer law is not followed the linearity perfectly for IR spectrometers. This indicates that field IR applications have qualitative up to semi-quantitative features. After the IR beam passes through the sample, IR transducers are applied to monitor the IR spectra. The photoacoustic effect can also be appropriated as a qualitative and quantitative IR detection technology. Another IR based method employs the Fourier-transform infrared (FTIR) spectromet. The Fourier transformation applied to data in FTIR results in a higher data precision than with the photoacoustic effect. More detailed information can be obtained in the literature [1].

The general instrumentation of an IR spectrometry based detector consists of (i) sample cell, (ii) an IR radiation source, (iii) specific wavelength filters, (iv) transducer, and (v) signal processor. Specific wavelength filters, so that only desired wavelengths reach the sample cell, can be used because the absorption spectrum of chemical warfare agents is very likely known. The sample cell is set

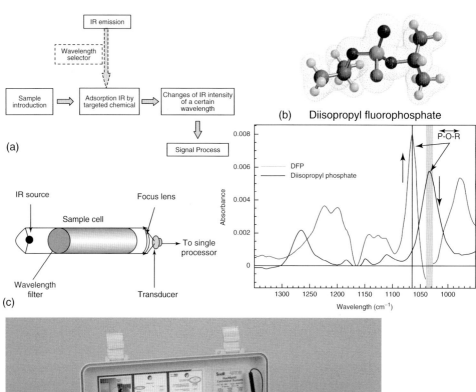

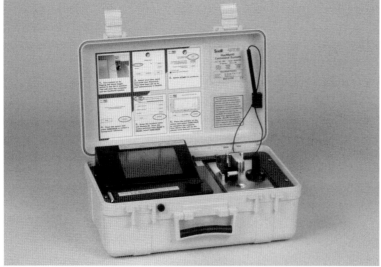

Figure 7.15 Infrared (IR) spectrometry: (a) flow diagram of the infrared technique; (b) infrared spectra of diisopropyl fluorophosphate (DFP) and diisopropyl phosphate recorded by an IR spectrometer – arrows indicate increasing and decreasing bands during the course of the reaction [22]; (c) typical configuration of signal wavelength infrared sensor; (d) Smith Heimann HazMatID FTIR.

in such a way that through a window the IR beam can penetrate into the cell. The window is set perpendicular to the radiation path for maximum IR radiation. The IR transducer can detect the IR absorption. As an example, the IR spectra of diisopropyl fluorophosphate and diisopropyl phosüphonate are presented in (Figure 7.15c). After exposure to IR radiation the sample remains chemically unchanged and can be analyzed with another method again, which leads to improved sample identification reliability. However, several environmental factors such as relative humidity can affect IR detector performance. In (Figure 7.15d) a commercial IR equipment is displayed.

7.5
Testing of Chemical Warfare Agent Detectors

We have learned that sampling and identification of chemical warfare agents have to follow strict quality controls. This applies also to the setting of detectors. For the whole testing procedure we have to establish a continuous quality system to verify the reliability of any result during testing [1, 19]. Before we start a testing procedure we have to establish an objective, which we should have in mind throughout. Therefore, for a future certification we can challenge the detector test-unit during the procedure with the following general questions (Figure 7.16):

- Can the detector establish how clean the environment is – What contaminations do we expect?

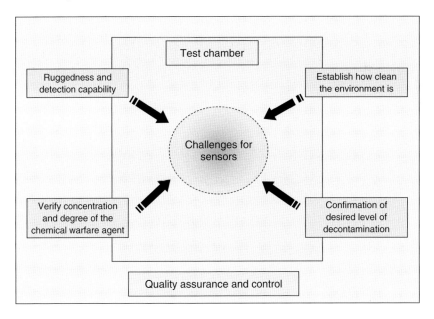

Figure 7.16 Major challenges for chemical warfare agent sensors during the testing process [1, 24].

- What are the ruggedness and detection capabilities?
- Can the detector measure the concentration and the nature and the degree of the chemical warfare agent?
- What level of decontamination is desired?
- Can the detector verify the decontamination process?

To achieve these ambitious stipulations for chemical warfare agent sensors a test methodology development (TMD) can be set by each sampling group, and institution, which may vary from nation to nation. For example, test methodology development addresses the chemical agent identification and detection requirements under the Joint Chemical Agent Detector (JCAD) performance specification. The emerged general questions can be further divided by the TMD into following areas:

- simultaneous generation of two agent vapors at two or more constant or dynamic concentration levels;
- simultaneous generation of one or more chemical interferences with the analytical procedure at two or more constant or dynamic concentration levels;
- characterization of chemical interferents at two or more concentration levels;
- quantification of agent concentrations and dosage;
- monitoring and recording of unit-under-test performance attribute data (Figure 7.17).

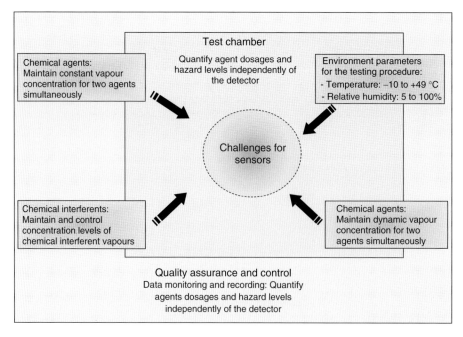

Figure 7.17 Test methodology development (TMD) for chemical warfare agent sensors [1, 24].

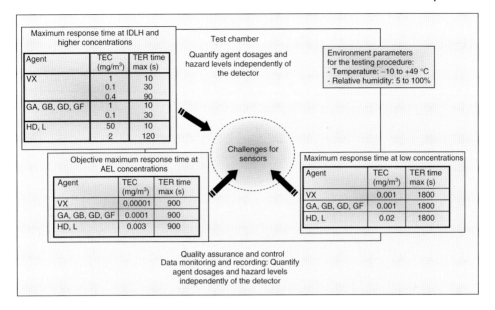

Figure 7.18 Example of how chemical warfare agent detectors can be tested. Maximum detection times from IDLH levels to desired AEL concentrations for selected chemical warfare agents; TEC: threshold exposure concentration, TER: threshold exposure time, AELs: airborne exposure limits, and IDLH: immediately dangerous to life and health [1, 24].

Before a detector can be evaluated the environmental parameters have to be defined. The environment parameters can be set by single nations or by organizations in a different way. In addition, the required/desired detection performance for a specific sensitivity range for the detection of chemical warfare agents has to be set (Figure 7.17). As one possible example, Figure 7.18 shows for respective chemical warfare agents the concentration levels from (AELs) immediately dangerous to life and health (IDLH) levels to airborne exposure limits levels (desired levels) versus desired maximum detection time.

The goal is now to characterize the behavior of the system in two different ways:

1) Run the system in a neutral environment (without any contamination). These measurements produce the background distribution under different environmental conditions.
2) Challenge the system with different incoming contaminations to determine the signal distribution.

The outcome from these evaluations gives evidence for the usefulness of the whole detector system, and can be placed in different quality categories. Furthermore, the results can be applied to compare different systems in terms of detection times, detector sensitivities, and concentrations, which are plotted along the horizontal and vertical axes (abscissa and ordinate axis). Variations of relative humidity (5–100%) and temperature effects (−10 to +49 °C) are investigated in the test

Table 7.3 Examples of potential field interferents and interference chemicals [1].

Field interferents	Interference chemicals
Diesel: exhaust, vapor, and on fire	Calcium hypochloride vapor
Kerosene: vapor, and on fire	Selected toxic industrial chemicals (TICs)
Gasoline: exhaust, vapor, and on fire	Pesticides, fungicides, and insecticides
Burning rubber, cotton, and wood-fire smoke	Selected decontaminants

chamber to determine the ability of the detector to perform measurements under various environmental conditions. When discussing the performance of a detector we should have in mind that proof of acceptable performance under controlled, clean laboratory conditions is not sufficient. Detectors also have to be tested under simulated operational conditions such as natural environment (temperature, rain, dust) and also induced environmental conditions (vibrations, engine exhaust, and other flammable chemicals). In addition, other chemicals have to be investigated to determine the cross reaction and to reduce the false alarm rate (Table 7.3). The challenge is to find a detector that operates sensitive and reliable under concentrations of the interfering substances. Therefore, the detectors are proved in combination with interfering vapors at the 0.1% and 1% headspace saturation concentration level of interferent vapor. If a false alarm occurs, the concentration of the applied interfering substances can be reduced stepwise to 0.1%. If the concentration of the interfering substances causes no further false alarms, the detector can be challenged with the target chemical warfare agent under the influence of the interference substance.

Summarizing the parameters for assaying detector systems and having in mind the complexity of the whole process, we have to consider that the highest priority of a detector is the detection of chemical warfare agents within a time-corridor and concentration limits to protect military and civilian first time responders and the population in general. If the detector meets this criterion, further examination is needed to meet the requirements of:

1) false alarm rate,
2) interference substances,
3) simplicity of operation,
4) achievement of operational stability (Figure 7.19).

7.6
Conclusions and Future Developments

We have discussed the challenges facing chemical warfare agent detectors in order to meet the required benchmarks in a chemical warfare agent incident. The quality aspect and the need to establish a quality system were also discussed. It seems

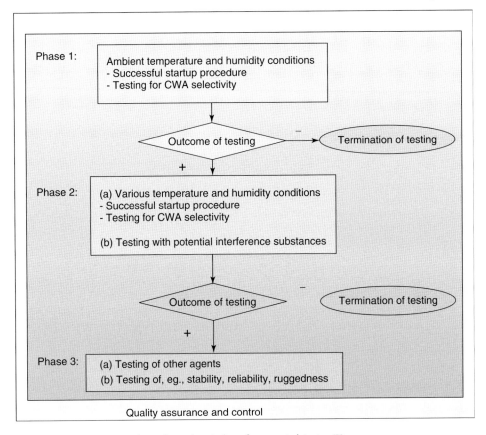

Figure 7.19 Testing procedures for a chemical warfare agent detector [1].

obvious that in the case of a chemical, biological, radiological, and nuclear (CBRN) incident laboratory data have to traced back to the raw data. Different technologies have been presented for potential application to detect possible chemical warfare agent contamination. Different detector systems are available and can be used for different scenarios. However, all of the discussed technologies have their advantages and disadvantages. We have seen that a major remaining challenge is to increase the detection reliability under field conditions, especially if different substances are present and if they interfere with the relevant agents or the analytical procedure intended to be applied. For the future, one focus for researchers and developers is to develop a capability for detection of chemical warfare agents in combination with a wide range of TICs [2, 3]. In addition, new technologies are under investigation for their potential as future chemical warfare agent detectors [24, 25]. Furthermore, because the properties of the chemicals influence the effectiveness of a particular technology, network sensor arrays have to be developed. A combination of technologies may offset problems posed by individual sensors and enable a more robust response [2, 3]. Another major challenge remains in the development of detectors with significantly improved specificity and sensitivity beyond currently available devices. The area of operation for these detectors could be in determining the extent of contamination for effective evacuation and to give evidence after contamination of when a site is safe for reoccupation.

References

1. Sun, Y. and Ong, K.Y. (2005) *Detection Technologies for Chemical Warfare Agents and Toxic Vapors*, 1st edn, CRC Press, Boca Raton, FL.
2. Sferopoulo, R. (2008) A review of Chemical Warfare Agent (CHEMICAL WARFARE AGENT) Detector Technologies and Commercial-off-the-shelf Items. Report No. DSTO-GD-0570, Human Protection and Performance Division, DSTO Defence Science and Technology Organization.
3. Janata, J. (2009) *Principles of Chemical Sensors*, 2nd edn, Springer. ISBN: 0387699309.
4. Skoog, D.A., West, D.M., and Holler, F.J. (1996) *Fundamentals of Analytical Chemistry*, 7th edn, Saunders College Publishing.
5. Hancock, J.R. and Dragon, D.C. (2005) Sample Preparation and Identification of Biological, Chemical and Mid-Spectrum Agents. A General Survey for the Revised NATO AC/225 (LG/7) AEP-10 Edition 6 Handbook (DRDC Suffield TM 2005-135) Defense R&D Canada – Suffield.
6. U.S. Department of the Army (2008) Pamphlet 385-61385-61 Toxic Chemical Agent Safety Standards, Headquarters Department of the Army, Washington, DC.
7. Carrano, J. (2004) Chemical and Biological Sensor Standards Study, CBS3_Final Report, Defense Advanced Research Projects Agency (DARPA), pp. 1–31.
8. U.S. Department of the Army (1998) Proposed Chemical Warfare Agent Detector Operational Requirements. Appendix A of JCAD Study Plan (Draft) 6/11/1998 modified for contractor distribution.
9. Kehe, K. and Szinicz, L. (2005) *Toxicology*, 214 (3), 198–209.
10. Katos, A.M. et al. (2007) *Toxicol. Ind. Health*, 23 (4), 231–240.
11. Russell, D. and Simpson, J. (2010) *Clin. Toxicol.*, 48 (3), 171–176.
12. Fraden, J. (2003) *Handbook of Modern Sensors: Physics, Designs, and Applications*, 4th edn, Springer, Berlin. ISBN: 978-1441964656.
13. Kealey, D. and Haines, P.J. (2002) *BIOS Instant Notes in Analytical Chemistry*, Taylor & Francis. ISBN: 978-1859961896.
14. Mesilaakso, M. (2005) *Chemical Weapons Convention Chemicals Analysis: Sample Collection, Preparation and Analytical Methods*, John Wiley & Sons, Ltd, Chichester. ISBN: 978-0470847565.
15. Tsuge, K. and Seto, J. (2009) *Minirev. J. Health Sci.*, 55 (6), 879–886.
16. Nieuwenhuizen, M.S. (2006) in *Encyclopedia of Analytical Chemistry* (ed. R.A. Meyers), John Wiley & Sons, Ltd, Chichester, pp. 1–17.
17. Murray, G.M. and Southard, G.E. (2002) *IEEE Instrum. Meas. Mag.*, 5 (4), 12–21.
18. Creaser, C.S., Griffiths, J.R., Bramwell, C.J., Noreen, S., Hill, C.A., and Thomas, C.L.P. (2004) *Analyst*, 129, 984–994.
19. Seto, Y., Kanamori-Kataoka, M., Tsuge, K., Ohsawa, I., Matsushita, K., Sekiguchi, H., Itoi, T., Iura, K., Sano, T., and Yamashiro, S. (2005) *Sens. Actuator B*, 108, 193–197.
20. Eiceman, G.A. and Karpas, Z. (1994) *Ion Mobility Spectrometry*, CRC Press, Boca Raton, FL.
21. Fatah, A.A., Arcilesi, R.D., Peterson, J.C., Lattin, C.H., and Wells. C.Y. (2005) Guide for the Selection of Chemical Agent and Toxic Industrial Material Detection Equipment for Emergency First Responders, Guide 100-04, Vols. I and II, Department of Homeland Security, pp. 1–5.

22. Kosal, M.E. (2003) The Basics of Chemical and Biological Weapons Detectors, Center for Nonproliferation Studies, Monterey.
23. Gäb, J., Melzer, M., Kehe, K., Richardt, A., and Blum, M.M. (2009) *Anal. Biochem.*, **385** (2), 187–193.
24. Mlsna, T.E., Cemalovic, S., Warburton, M., Hobson, S.T., Mlsna, D.A., and Patel, S.V. (2006) *Sens. Actuator B*: **116** (1-2), 198–201.
25. Nilles, M., Connell, T.R., and Durst, H.D. (2009) *Anal. Chem.*, **81** (16), 6744–6749.

8
Detection and Analysis of Biological Agents

Birgit Hülseweh and Hans-Jürgen Marschall

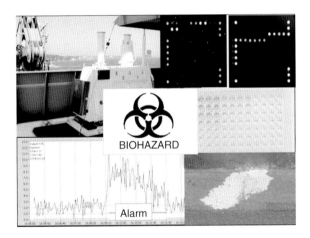

Compared to chemical warfare agents, biological pathogens and biotoxins are remarkably effective in very low doses, and they are colorless and odorless. Most of them are neither volatile nor dermally active. In extreme cases, biological agents are more effective than chemicals and a disorder can be triggered by a relatively low number of pathogens. In addition, some of the microbial pathogens have incubation periods ranging from several hours to several days or weeks. Symptoms of disease are often retarded. In addition, identification is commonly complicated by the natural biological background. Discrimination of living and dead bacteria is essential to estimate the biological threat and high sensitivity is indispensable for biological detection and identification systems. Compared to chemical and nuclear detection, there are currently only a few commercial solutions available but diverse technologies are under study.

CBRN Protection: Managing the Threat of Chemical, Biological, Radioactive and Nuclear Weapons,
First Edition. Edited by A. Richardt, B. Hülseweh, B. Niemeyer, and F. Sabath.
© 2013 Wiley-VCH Verlag GmbH & Co. KGaA. Published 2013 by Wiley-VCH Verlag GmbH & Co. KGaA.

8.1
What Makes the Difference?

Analysis of biology is inherently different from analysis of chemicals and one of the major problems during the analytical process of biological warfare agents is the naturally occurring biological background. This background is further enhanced during grain harvest or flowering of grasses by high numbers of harmless spores and generic particles in the air, a situation that could increase the probability of false alarms during detection.

Unspecific detection of biology molecules in the air provides rather rare and non-specific information in nearly real-time, but it does not yield any identification of a biological agent or threat. It only indicates its possible presence. In contrast, classification categorizes the biological agents into rough and major groups without having an idea about the exact nature of the agent. As introduced in Chapter 4, the most relevant groups for detection are bacteria and their spores, fungi, viruses, and toxins. First the process of biological identification specifies the B-agent. The process itself can last from several seconds to several minutes to several hours or days. Consequently, more reliable detection and identification is wanted the longer it takes to identify the biological threat (Figure 8.1).

> Rapid generic **biological detection** is rather non-specific and does not provide any identification of the agent. It only gives an indication of the presence of a biological agent.
>
> In contrast, **identification** specifies the B-agent and reveals its nature.

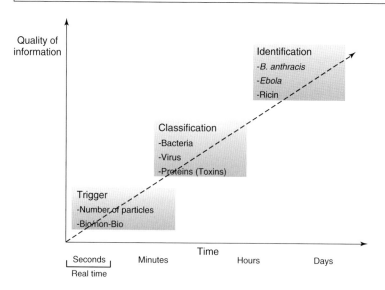

Figure 8.1 Correlation of time requirement and information content.

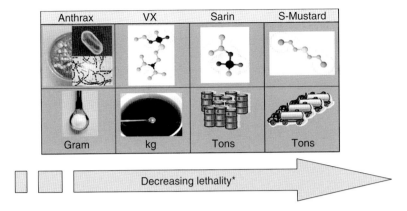

Figure 8.2 Comparative lethal doses of anthrax spores and some chemical warfare agents for 10 million people.

A further characteristic that makes biological agents attractive as bioweapons is their remarkably low effective doses, particularly if the agent is transmissible from person to person. For example, under ideal conditions a lethal human dose of anthrax is probably in the range of about 1×10^{-6} g. Figure 8.2 estimates the required amount of *Bacillus anthracis* spores necessary to kill about 10 million people in comparison to a few chemical weapons. However, the estimation neither considers how the agents are dispensed nor how the agents are taken up by the human body.

Moreover, the main interest of biological detection is to measure augmentable B-agents because only these can initiate and spread an infection. In a simple sense or preliminary test this is often equated with culturability of an unknown sample. Not all viable microorganisms are cultivable.

Whereas rapid and preliminary detection of B-agents is mainly based on physical and chemical properties like the particle size, particle distribution, and their biological fluorescence, identification focuses prevalently on the immunological and molecular characteristics of the B-agents (Figure 8.3). Moreover, *in vitro* culturing, specific staining, and microscopic analysis can reveal the agents' nature.

The process of analysis for biological agents is broadly comparable to the chemical analysis process as illustrated in Figure 8.4. The flowchart is quite general and all steps are of equal importance.

The analysis starts with the detection of an aerosol, its collection, and sample preparation, if necessary. The preliminary physical characterization of the aerosols' content could occur either before or after sample collection and is then followed by more specific investigations based on different technologies. In any case, the importance of valid and reproducible identification techniques cannot be overestimated for reliable results.

As explained in the introduction to chapter 6, there are two main detection strategies applied in the management of bio-surveillance – "detect-to-protect" and "detect-to-treat." The first is used in situations where there is sufficient time

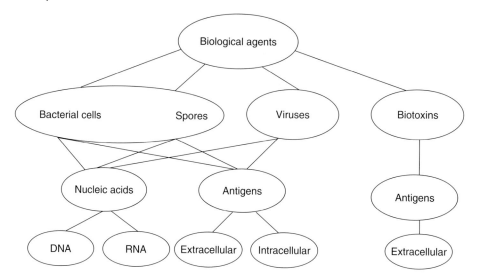

Figure 8.3 Biological agents and their integral components that can be used at different stages of detection and identification. Antigens are biological molecules or structures detected by antibodies.

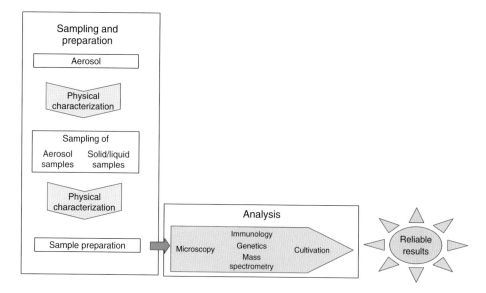

Figure 8.4 Generalized biological agent analysis process.

detection for personnel to be protected or even evacuated, and the second in situations where exposure has already occurred, or is imminent, and treatment is the only remaining option.

Although several technologies and tools exist, "detect-to-protect" detection is currently not possible for biological threat agents because early symptoms of a BW agent caused infection are often unspecific. In addition, specific symptoms of a bacterial or viral infection are often retarded. Furthermore, the available devices for B-detection are too large and slow to give a real-time answer. Therefore, biodefense strategies of many nations still tend to combine several stages of detection.

8.2
The Ideal Detection and Identification Platform

The prospects of biological warfare and terrorism present many challenges for the ideal detection and identification platform. It has to fulfill a host of criteria and the ultimate approach is to obtain an integrated system containing all components necessary for the detection as well as the identification of biological agents.

The device itself should be inexpensive and easy to use. In addition, the instrument should be portable and rapidly deployable. At the same time the methods and assays for detection and identification should be rapid, sensitive, specific, and inexpensive. Moreover, it means that the instrument is capable of detecting low concentrations of target agents without interference from background materials and with no false results. In particular for B-agents this means that the device is able to discriminate live from dead agents (Figure 8.5). Ideally, it should be capable of identifying various biothreat agents in one sample (multiplex capability) and be able to detect modified or previously uncharacterized agents. In addition, reliable reproducibility is an important requirement for a detection platform and demands the stability and consistency of reagents as well as assay conditions.

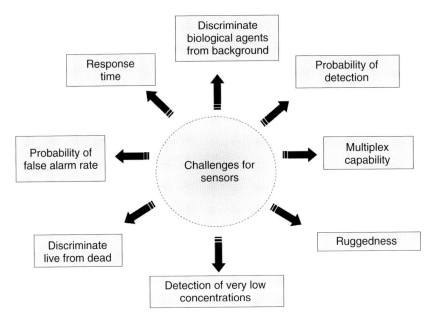

Figure 8.5 Challenges for biological warfare agent sensor systems.

8.3
Bioaerosols: Particulate and Biological Background

Our environment and atmosphere is filled with particulates that originate from numerous sources and include dust, dirt, and fog as well as microorganisms (Figure 8.6) [1, 2].

The particulate background in an aerosol can be either natural or man-made and changes depending on the local and meteorological conditions, as indicated in Figure 8.7 for airborne bacterial concentration within 24 h.

The particulate background in an urban setting will dramatically depend on whether there is traffic or not and will significantly differ from the background in a rural landscape, desert, or coast. Additionally, the biological background is dependent on relative humidity, wind speed, temperature, and daylight.

The physical detection of an aerosol cloud differs significantly from the identification of the contents of the cloud. Depending on the level of sophistication, aerosol detectors can provide rapid but only nonspecific information about the aerosol particles in a cloud, such as:

- number,
- size,
- distribution of particles.

The challenge for an aerosol detection system is to discriminate between biological and non-biological agents. It has to pick out specific signals from the biological

Figure 8.6 Particles that make up an aerosol.

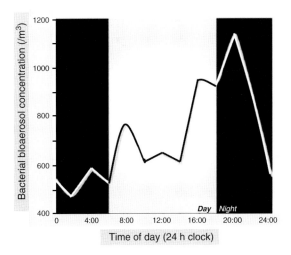

Figure 8.7 Fluctuation of airborne bacterial concentration on a single day and response of a biological agent real-time sensor (BARTS) during field trials according to Reference [3].

agents while differentiating them from those originating from the non-biological background. The major group of particles of interest in an aerosol is in the size range 0.5–5 μm in diameter since these are efficiently deposited in the alveoli of the human respiratory tract (see Figure 4.12). Moreover, it is desirable to distinguish between biologically viable and non-viable bacteria and viruses because only the viable organisms contribute to the spread of infection.

8.4
Aerosol Detection – A Tool for Threat Monitoring

Pathogens are most likely to be disseminated as aerosol while toxins, on the other hand, may be disseminated either as aerosol or liquid drops.

As mentioned in Section 8.1 the composition of different aerosols differs significantly and each is subjected to a strong fluctuation. In any case, biological aerosol detection in nearly real time is an urgent civilian and military requirement for threat monitoring and the perfect aerosol detection system should be able to:

- give real-time results,
- operate continuously, 24 h a day and 7 days a week,
- trigger no false alarms,
- require only minimal maintenance [4].

8.4.1
Cloud Detection

In its simplest form a detector acts as a "gateway" or trigger for further analysis. If the sample displays characteristics of biological particles, it is passed to the next level of analysis, namely, its classification and identification.

8 Detection and Analysis of Biological Agents

Several detection technologies have been developed in the past and despite the fact that diverse devices are commercially available most technologies still have their flaws. Some of the current and available technologies are depicted here, and their assets and drawbacks are discussed (Figure 8.8).

Like chemical agents, biological compounds are known to have infrared (IR) spectra, and radiation yields spectral fingerprints that result from a superposition of the individual spectra of all molecules of the target.

These spectra can be exploited to detect a threat plume from a distance of a few hundred meters up to several kilometers.

Principally, we distinguish two types of sensor systems for standoff-detection, **passive** and **active** ones, and whereas chemical agents can be detected by both sensor types small amounts of airborne biological agents can be only reliably detected by active systems. Moreover, these active systems allow at best a distinction between bio- and non-bio aerosols, which is due to the fact that excitation by radiation is limited to a few molecular bonds. IR spectra generated by standoff-detection from biological samples all look very similar and do not allow a discrimination of microbial species.

Passive detectors record the infrared spectrum emitted or absorbed relative to interference with the surrounding background and the only source of excitation energy results from the ambient background, which could be either the sun or artificial equivalents. In contrast, **active detectors** operate with transmitted infrared laser radiation that is scattered back to a co-located receiver from aerosols in the atmosphere or from a topographic target. The laser energy is attenuated by the natural atmosphere, the chemical and biological agents, and other aerosols and gases in the environment.

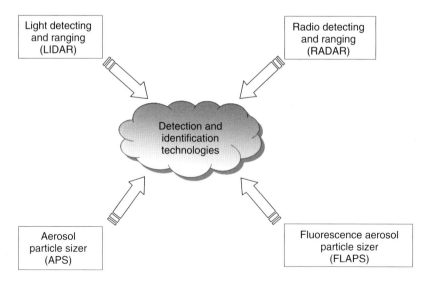

Figure 8.8 Technologies for cloud detection.

Both sensing techniques extract the signal from the temporal spectrum by sophisticated algorithms but there is a fundamental difference regarding the ability to predict the signal source term for the two approaches. For an active detector the signal can be mathematically predicted from the detector system characteristics and from the scattering and absorption cross sections of the aerosols and gases as well as the topographic target characteristics that are present, allowing a complete model to be developed of the measurement process and uncertainties. In contrast, the source term for passive detection, like radiation from a widely varying background, cannot be accurately predicted to develop such an end-to-end model. Passive infrared detection technology compares their excitation pattern against a library of known chemical signatures.

The passive detector will work best when chemical agents are in the vapor phase, whereas the active detector has the potential of working better when the organic matter of chemical or biological origin is in the form of an aerosol [5].

8.4.2
Radio Detecting and Ranging (RADAR) and Light Detection and Ranging (LIDAR)

Radio detecting and ranging (RADAR) is a recently upcoming technique that is long known from weather reporting. The technique uses reflecting radio waves to characterize and monitor the shape, size, directionality, and speed of a cloud. The radiation is scattered by the diverse particles of the aerosol and the elapsed time before the radio waves return to a receiver and the change in the radio waves energy upon return provide information about an aerosol cloud. An approved and promising tool for first line cloud detection and recognition is light detection and ranging (LIDAR) (Figure 8.9). This laser active technique is based on the same physical principles as RADAR, except that instead of the reflection of radio waves higher energy light waves are used. Additionally, by using high energy ultraviolet (UV) lasers the signature (UV spectra) of the aerosol can be identified as biological or non-biological. As shown in Figure 8.9 LIDAR is an optical remote sensing technology that measures the distance to the cloud (object) by determination of the time delay between transmission of a pulse and detection of the reflected signal.

Today, different LIDAR technologies attempt to exploit different signatures of an aerosol. The results depend upon factors such as the:

- wavelength,
- laser power,
- ambient conditions, for example, water vapor in the atmosphere.

In addition, the various physical properties of particles in an aerosol affect the extinction and backscatter intensity of the signal. When UV radiation is used as the excitation source in a LIDAR system the laser-induced fluorescence (LIF) can indicate at short distances that a cloud is biological in nature (Figure 8.9). Moreover, notably, detectors based on UV-LIF are most effective at night or during low light periods because of the relative opacity of the air to these wavelengths and to the high UV background during the day.

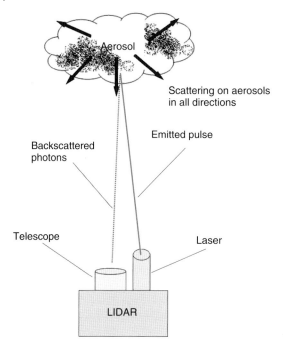

Figure 8.9 Schematic description of a LIDAR system.

The limitations in detection are mainly the distance and the resolution of the detector. The more specific the level of data is desired, the closer the instruments must be located to the cloud. In summary, LIDAR technology could be used to provide early warning for field personnel. However, it should be taken into consideration that this technology still delivers only limited information about biological particles in an aerosol and cannot detect suspicious biological warfare agents.

8.4.3
Aerosol Particle Sizer (APS), Flame Photometry, and Fluorescence Aerosol Particle Sizer (FLAPS)

Aerosol particle sizers (APSs) are a simple form of sensor that can detect higher than normal concentrations of airborne particles of a predefined hazardous size range. In these systems, particles are drawn through a nozzle into a constant and high-speed air flow and the introduced particles accelerate at rates proportional to their size. The particles pass through a laser beam that characterizes the so-called aerodynamic diameter by velocity measurement. Because of the universal high aerosol background these two parameters normally are not sufficient for warning and protection. To reduce the false alarm rates of a detector a distinction between viable and nonviable organisms is desirable, which requires the application of additional physical techniques.

One such technology is flame photometry, a technology that is also applied for the detection of chemical agents as explained in Chapter 7. As in conventional systems the aerosol is drawn through the detection chamber, in which the analytical hydrogen flame detector burns the particles. From the absorption lines in the resulting spectra the elemental composition can be generated. Bioparticles typically have different element ratios than those of inorganic origin. So at least it is possible to survey the rise in the amount of organic particles in the surrounding air. In view of the above-mentioned problems and the ever-changing aerosol content of the atmosphere, these systems do not give any reliable biothreat warnings. As an alternative **fluorescence particle-sizing technique** was introduced. Commercial solutions are available from several companies and the method provides enhanced discrimination capabilities compared to the relatively simple APS systems already mentioned. Biological particles can be distinguished from non-biological material based on the excitation of the biospecific intracellular fluorescing compounds. Common target molecules in cell material of biological organisms and the cell wall of bacterial spores are:

- the amino acid tryptophan,
- the coenzyme nicotinamide adenine dinucleotide (NADH),
- the cellular energy storage molecule adenosine triphosphate (ATP),
- the vitamin riboflavin as central component of the flavoprotein cofactors FAD (flavin-adenine dinucleotide) and FMN (flavin mononucleotide). (Figure 8.10).

Fluorescence of one of these compounds in an aerosol cloud indicates that the sample is of biological origin, a feature that also is important for reducing false alarms. Drawbacks mainly emerge from the large amount of bioaerosols in the surrounding atmosphere, which prevents the exclusive detection of real biothreats.

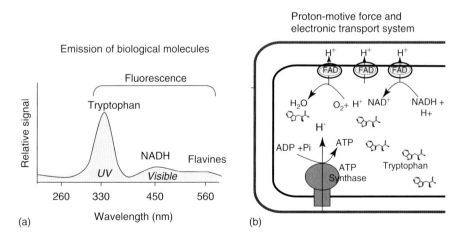

Figure 8.10 Excitation and emission spectra of different biomolecules (a) and their cellular localization (b).

So a change of the concentration of fluorescing particles can be taken as a severe warning, preferably in combination with separate detection technologies.

8.4.4
Detector Layout Topology, Sensitivity, and Response

Aerosol detectors are employed to discover the emergence, presence, or absence of oncoming aerosol clouds, and as explained already in Section 6.2 we distinguish in regard to their location, mode of action, and deployment, two distinct but complementary types of detectors (Figure 8.11):

- stand-off detectors
- point detectors.

Point detectors and stand-off detectors could be either stationary systems or mobile units. They aim in particular to detect the threat before the target population is exposed to the floating aerosol cloud. To assess the risk of a biological attack properly it is important to know how many biological individual organisms are contained in an aerosol.

The standard description of the sensitivity of a biological aerosol detection equipment is given in agent containing particles per liter of air (ACPLA) rather than in particles per liter of air (PPL). In general each ACPLA independent of size will form one colony-forming unit (CFU) or plaque forming unit (PFU) if seeded or cultured, even if there are several infectious agents in this particle. However, ACPLA only defines the number of colonies resulting from the single particles in 1 l of air and does not determine whether a health threat exists. As illustrated in Figure 8.12 collection procedures as well as identification methods significantly influence the result of an existing or not-existing biological threat [6].

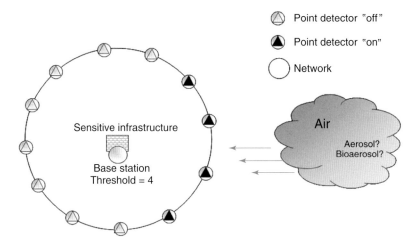

Figure 8.11 Example of a detector layout topology.

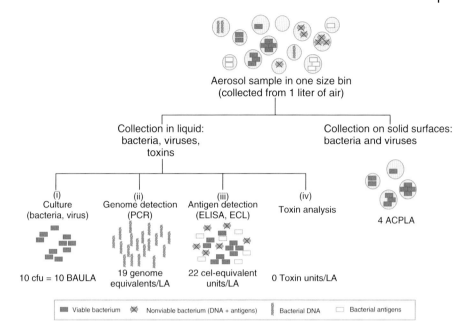

Figure 8.12 Analysis a microorganism-containing aerosol. Adapted from Reference [6].

Therefore, today other and perhaps more appropriate units like biologically active units per liter of air (BAULA) and biologically active particles per liter of air (BAPLA) are discussed and investigated [6]. Both measures take only viable and therefore CFUs into consideration. This means, if the sample is collected into liquid, the unit of measure will differ depending on the subsequent analysis. While signal contribution from genome and antigen identification comes from viable and nonviable bacteria, culturing determines only the viable ones.

8.5
Sampling of Biological Agents

Sampling of biological agents is the most critical step for detection and identification, since the effective or lethal dose for some agents can be extremely small (see Chapter 4 for further explanation). Moreover, sampling presents a challenge for the viability of the microorganisms and the efficiency of recovery and concentration can significantly affect detection as well as identification limits.

Monitoring the efficiency and success of sampling should be an essential part of each initial investigation and could be assessed on the basis of international and standardized protocols. However, international and standardized sampling protocols do not currently exist, although the NATO subgroup Sampling Identification Biological Chemical Radiological Agents (SIBCRA) tries to develop a sampling expertise for chemical, biological, radiological, and nuclear (CBRN). Their stated aim is to harmonize quantitative as well as qualitative sampling procedures among nations in terms of particle size and concentration.

In general sampling is performed with the following principles in mind:

- sterile handling techniques must be used to avoid cross-contamination of samples;
- samples must be preserved during the entire mission and kept cool in containers to avoid inactivation of live agents;
- two sets of samples are necessary for confirmatory analysis;
- clear and unique labeling of all samples is essential.

In case of a reported aerosol threat the biological agent might be collected by either surface sampling or aerosol sampling. The various collection options are explained in the following.

8.5.1
Aerosol Sampling

For the direct collection of microorganisms from aerosol highly efficient collection devices must be employed, and as the biological threat agents are present at only low concentrations it might be necessary to pre-concentrate the air sample. It is advisable to use high volume samplers that have pre-concentration stages and can work with air throughputs of up to several cubic meters per minute.

For aerosol sampling two different strategies are common:

1) continuous sampling for several hours;
2) sampling triggered by aerosol detectors.

Both strategies face several difficulties and can harm the collected biological agents. Continuous sampling, for example, requires sufficient laboratory capacity and resources for round-the-clock analysis. Triggered aerosol sampling needs extremely sensitive and reliable triggering instruments. The collection of microorganisms from aerosols is usually performed by impaction, impingement, or centrifugal samplers with subsequent concentration on filters or into liquids (Figure 8.13).

Impaction samplers collect the biological particles on sticky surfaces, agar plates, membrane, or fibrous filters that are subsequently analyzed by several methods including cultivation (Figure 8.14).

Basically an impactor is a jet below which an impaction plate or surface is located. The aerosol is sucked through the jet and as a general rule larger particles cannot follow the 90° or more deflection of the streamline. Therefore, they impact on the impaction surface or plate. In contrast, particles smaller than the cut-off of the impactor remain airborne and follow the streamline. Thereby, several different stages can eliminate unwanted particle sizes and characterize the aerosol of interest.

For the use of filtration for microbial air collection the air is drawn by a pump or vacuum line through the system. Then the filters are either eluted or incubated onto the surface of culture media. However, filtration is less convenient than

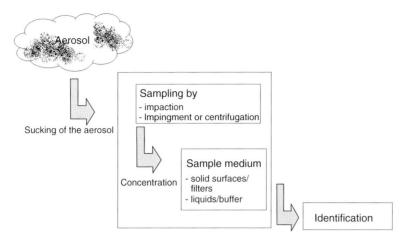

Figure 8.13 Sampling process.

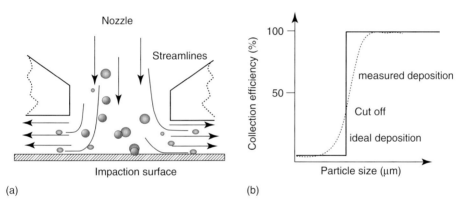

Figure 8.14 Physical principle of air collection of microorganisms by impaction (a) and comparison of the ideal and measured deposition of an impactor (b).

impaction-based sampling, causes considerable losses of antigens, and may cause drastic inactivation, for example, by sampling and dehydration stress in the trapped microorganisms and viruses. In addition, there are considerable difficulties for elution.

Impinging samplers collect the bioaerosols into liquid through an inlet tube, as illustrated in Figure 8.15, and trap the particles as well by the principle of high-speed impaction. The jet, positioned just above the bottom of the impinger, consists of a short capillary tube that operates as critical orifice, so that particles impinge at or near sonic velocity.

The collection of particles by an impinger is normally gentler and allows the direct analysis of samples by all conventional immunological and molecular methods. At present several high-performance impingers are commercially available. A selected list of bioaerosol samplers is summarized by Loh [8] and Verreault et al. [2].

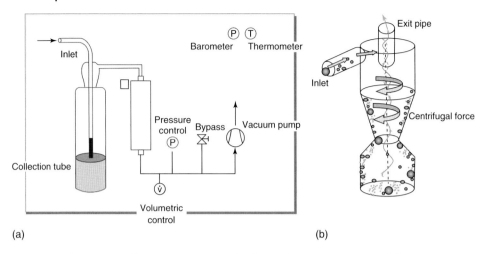

Figure 8.15 Schematic presentation of collection by an impinger (a) and flow pattern in a tangential flow cyclone (b) modified according to Reference [7].

Finally, centrifugal samplers employ the principles of a cyclone to extract biological particles from air. Most cyclones fall into one of the following types based on different inlets (tangential and axial) and outflow (returned flow and uniflow); the most common type is the tangential flow cyclone.

The samplers work by taking in air at a tangential angle to the wall of the cyclone or micro-centrifuge tube, thus creating centrifugal force and pushing the denser particles to the walls of the tube. Biological cyclones have wetted walls, so that the particles of interest are gently impacted tangentially into the circling liquid.

Table 8.1 summarizes the most commonly used aerosol sampling techniques.

Table 8.1 Features of the main air sampling methods.

Method	Principle	Suitable for the collection of	Collection media or surface
Impingement in liquids	Air drawn through small jet, directed against liquid surface	Viable particles, liquid aerosols	Buffered gelatin, peptone, nutrient broth, tryptone saline liquid media
Impaction on solid surface	Air drawn into sampler, particles deposited on dry surface	Viable particles	Dry surfaces, coated surfaces, sticky surfaces, agar
Cyclone	Centrifugal force	Viable particles	Liquid/buffer

8.5.1.1 Surface Sampling

Particles suspended in air normally sediment according to their aerodynamic diameter. Larger particles settle more quickly than smaller ones (Figure 8.16).

To get as many agents as possible surface samples for biological detection should be taken preferentially from places of most probability for aerosol settlement. This means, for example, from shrubby plants or protruding edges of obstacles downwind. Alternatively, and if not otherwise possible, samples could be also taken from soil or liquids.

Currently the most promising methods are:

- swab sampling,
- wipe sampling,
- replicate organism detection and counting (RODAC),
- vacuum spray extraction sampling (Table 8.2).

Swab surface sampling represents the primarily applied technique. However, the method is limited with respect to the level of contamination and the surface area. The same is true for wipe sampling and RODAC, where an agar plate is placed upon the area to be tested and pressed on with moderate pressure. Subsequently, the plates are incubated for 48 h.

Moreover, in the case of swab and wipe sampling the material has to be carefully chosen, otherwise extensive losses by irreversible absorption can occur.

> In summary, swap and wipe sampling as well as RODAC are particularly suitable for screening of small nonporous surfaces but are ineffective for larger and porous areas.

In contrast, vacuum spray extraction sampling, for example, with the M-Vac® from Microbial-Vac Systems Inc., Jerome, USA, also allows us to screen large

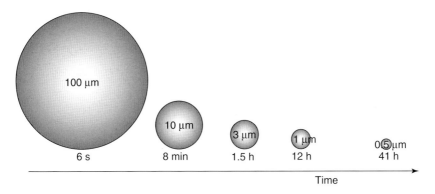

Figure 8.16 Correlation of the particle aerodynamic diameter with the time needed to settle 1.5 m in still air.

Table 8.2 Different types of surface sampling methods.

Type of sampling	Description	Application and target	Suitable for biological agents like
Wipe sampling	Sterile non-cotton gauze, moistened; wipe area of known size	Nonporous surfaces, usually small in area	Bacteria, viruses, fungi, and biological toxins
Swap sampling	Sterile non-cotton swab, individually wrapped, then moistened with sterile solution; wipe area of known size	Nonporous surfaces, usually very small in area, complex surfaces with crevices, corners	Bacteria, viruses, fungi, and biological toxins
RODAC[a]	Agar surface in culture dish, press onto surface, incubate	Nonporous surfaces, relatively small area, porous surfaces with limitations	Bacteria and fungi
Vacuum spray extraction	Liquid is sprayed onto a surface by a nozzle, collection is performed by vacuum	Porous and nonporous surfaces, relatively large areas	Bacteria, viruses, fungi, and biological toxins

[a] Replicate organism detection and counting.

non-porous surfaces like concrete plaster, floors, carpets, and wooden surfaces. For this purpose liquid is sprayed onto a surface by a nozzle located in the center of the sampling head while the agent containing liquid is collected into the sample bottle by vacuum (Figure 8.17).

In addition, high efficiency spray extraction sampling enables field teams to determine the decontamination effectiveness and reduces the possibility of false negative results.

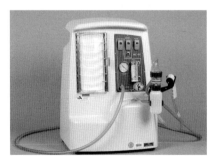

Figure 8.17 M-Vac™ surface sampler with single-use sampling head and liquid delivery bag from Microbial-Vac Systems Inc., Jerome, USA.

8.6
Identification of Biological Warfare Agents

Once the presence of biological aerosolized particles has been established, the next level of awareness relevant to detection of a potential biological warfare agent threat is to seek specificity with regard to the agent, that is, what exactly is it? Current technologies that aim to specify an agent rely on analysis of specific immunological biomarkers, the identification of unique nucleic acid fragments, the specific cultivation and classical analysis of the microorganisms, or the detection of unique physical and chemical properties.

8.6.1
Immunological Methods Based on Enzyme-Linked Immunosorbent Assay (ELISA)

Technologies based on this type of identification utilize the high specificity and affinity of antibodies to target bacteria, viruses, and toxins of interest. Since the 1970s, the enzyme-linked immunosorbent assay (ELISA) has been the standard method against which the performance of new identification technologies is measured [9–11].

> - **Enzyme-linked immunosorbent assay (ELISA)**: a highly sensitive immunoassay that uses an enzyme linked to an antibody or antigen as a marker for the detection of a specific protein.

All ELISAs form an antigen–antibody complex (Figure 8.18) and the binding event is analyzed either colorimetric, optically, electrochemically, or electrically. For this kind of identification the antibodies are labeled with enzymes, dyes, or markers that initiate the detection reaction.

Therefore, the antibody is the most critical reagent of the assay, and in the course of the last 25 years both polyclonal and monoclonal antibodies have become increasingly available for bio-detection. Whereas **polyclonal antibodies** are obtained from immunized animals and produced by different B cell resources, recognizing all antigens the individual ever had contact with, **monoclonal antibodies** are derived from a single cell line by genetic engineering and identify only one single and unique epitope of the antigen (Figure 8.19).

Typical antigens used for the specific immunological identification and differentiation of B-agents are:

- the envelope and nucleocapsid (core) proteins of viruses,
- the cell wall and outer membrane proteins of bacteria,
- polysaccharide components of the bacterial outer membrane,
- toxic protein subunits.

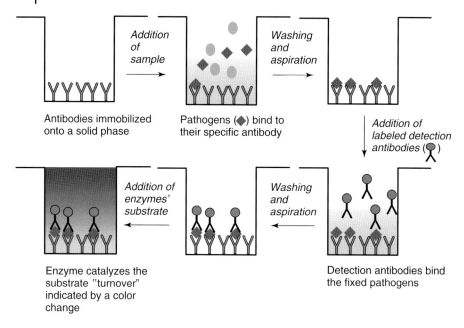

Figure 8.18 General principle of enzyme-linked immuno sorbent assay (ELISA) based identification.

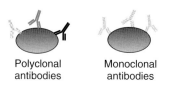

Figure 8.19 Binding strategy of poly- and monoclonal antibodies.

If antibodies with the desired specificity and affinity have been developed, they can be utilized in a broad range of immunoassays, but only a few technologies seem to be relevant for the rapid identification in biodefense. There are, for example, different categories of disposable immunoassays, assays with multiplexing capacity, and technologies that allow us to capture specific antigens from either highly complex mixtures of biomolecules (e.g., blood or sera) or extremely diluted samples. Each of these categories, along with examples of the corresponding technologies, is discussed below.

Immunochromatographic assays were first described in the late 1960s and were originally developed to assess the presence of serum proteins. Other early assays that used an immunochromatographic technique include those for the quantification of drugs in biological fluids [12]. Figure 8.20 illustrates the general principle of an immunochromatographic assay. The sample is applied onto an adsorbent pad that contains antibodies specific to the target analyte conjugated to

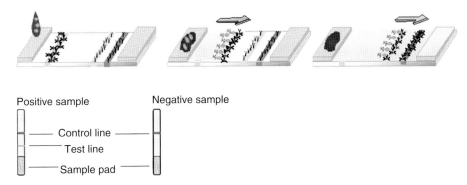

Figure 8.20 General principle of a lateral flow immunochromatographic assay (handheld assay) and example of a positive and a negative result.

either colored particles or gold. The immobilized antibodies bind their cognate antigen and by capillary action the antibody–antigen complexes as well as the excess unbound antibodies migrate down the membrane. The antibody–antigen complex is then captured by a specific secondary antibody immobilized at the test "T" line forming an antibody sandwich. The excess of the conjugated but unbound antibody continues migration and interacts with immobilized antibodies specific for the conjugate antibodies at the control "C" line.

This means in the case of a positive biological sample two lines will appear, the test line (T) and the control line (C) whereas in a negative sample only the control line will be visible. Many assays use nitrocellulose as membrane, and gold is the prevalently used conjugate for the first antibody. Frequently, the components of the membrane are fixed to an inert backing material integrated into a plastic cartridge. Today, several handheld assays recognizing biological warfare agents are commercially available, but only a few exhibit sufficient sensitivity, due to a lack of high-affinity, specific antibodies. Therefore, many nations have started their own developments in this special field of rapid identification.

Magnetic immunocapture assays are an especially effective form of an immunoassay. In general the conjugated capture antibodies are coated to the surface of magnetic microparticles and added to a complex sample. The specific antibodies bind their cognate antigen and by applying magnetic force the particle–antigen complex is captured and concentrated in relatively pure form. Compared to other immunological assays, this method affords high yields, especially from environmental samples.

Electrochemiluminescence (ECL) assays are one of the most promising new immunological technologies [13]. The method makes use of antigen-capture assays and the chemiluminescent label ruthenium. A cyclic redox reaction driven by an electric current between a ruthenium-containing molecule and a corresponding substrate leaves ruthenium in an electronically excited state, which then relaxes with emission of photons that are detected. Depending on the photo-amplifier used, this method is extremely sensitive as no adverse background problems occur.

Immunochip protein arrays are emerging to follow DNA chips as possible identification tool for biological agents, and diverse detection principles like fluorescence, colorimetry, and ECL have been already exploited [14, 15]. Although many technologies are still under study most biochip assays offer a highly effective method to analyze agents in parallel with little consumption of antibodies and minimal volumes of necessary reagents. The simultaneous identification from three up to seven biowarfare agents has been reported [16–18]. The degree of automatization varies from array system to system and they work with or without fluidic and liquid waste. Problems with this kind of technology mainly arise from antibody stability upon longer and repeated use in automated biosensors.

Most immunoassays today have sensitivity thresholds that are 100- to 1000-fold above the minimal infectious dose, a fact that still limits their utility. In addition, unspecific cross-reactivity and lack of long-term stability of reagents often cause problems. Therefore, today other forms than conventional monoclonal and polyclonal antibodies are also investigated for use in immunological methods (Figure 8.21). For example, recombinant antibody fragments like monovalent single chain antibodies (scFv) and bivalent diabodies are under investigation for several B-agents. Several reviews have been published that discuss the advantages of these new ligands and their applications [19, 20].

The primary advantage of these new binders is that they can be easily modified and improved by recombinant technologies. A substantial amount of research is concentrated on amplifying the binding kinetics, specificity, and durability of recombinant antibodies, thus making them more favorable as probes for biological warfare agents [21, 22]. Moreover, they can be produced in comparatively large

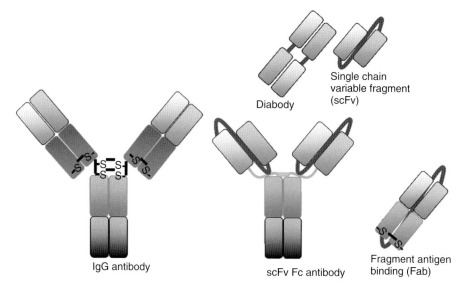

Figure 8.21 Different forms of recombinant antibodies in comparison to a classical IgG antibody.

quantities in biofermenters. Other alternatives to conventional antibodies are aptamers and peptide ligands. Aptamers react like antibodies, but consist of short nucleic acid stretches that recognize a target antigen by shape and morphology and not by sequence like in the polymerase chain reaction (PCR). Peptide ligands are short chains of amino acids. Both alternative binders can be chemically synthesized and are usually generated and selected by combinatorial methods.

8.6.2
Molecular Methods

In general, nucleic acid-based detection systems are more sensitive than antibody-based detection systems and several nations have already introduced selected assays and identification systems into their army. Undoubtedly, these nucleic acid-based detection systems will play a large role in future technologies. The PCR assay can detect ten or less than ten microorganisms in a short period of time and real time polymerase chain reaction (rtPCR) is today the most frequently applied assay for the identification of microbial agents.

> - **PCR (polymerase chain reaction)**: a molecular technique that allows the production of large quantities of a specific DNA from a DNA template using an enzymatic reaction without a living organism.

The method is restricted to nucleic acid containing agents. Whereas gene fragments from DNA carrying microorganisms can be amplified directly by PCR, RNA containing microorganisms require an additional step that transcribes the RNA genome to copy DNA (cDNA), a process that is called the reverse transcription. PCR uses the nucleic acid of an agent as template, from which a defined part of the sequence is amplified by the use of specific short nucleotide sequences (primers) and the addition of nucleotides and the amplification enzyme in an optimized buffer solution. The amplification occurs in an automatic cycler that repeats in its simplest form three major temperature steps, as illustrated in Figure 8.22a:

1) denaturation at 92–95 °C,
2) annealing at 50–60 °C,
3) extension at 60–72 °C.

During the denaturation step DNA fragments are heated to a temperature that reduces the DNA double helix to single strands. Thereby, these strands become accessible to the primers. The reaction mixture is then cooled to the annealing temperature. Primers anneal to the complementary regions in the DNA template strands and double strands are formed again between primers and complementary sequences.

In the third step (extension) the DNA polymerase synthesizes a complementary strand beginning at the primers. The enzyme reads the opposing strand sequence

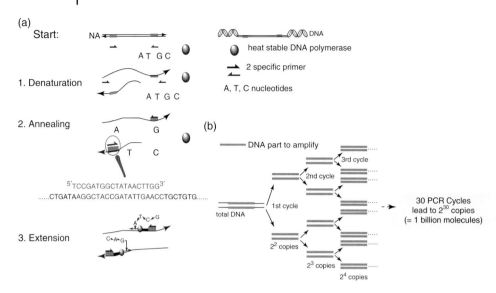

Figure 8.22 Specific amplification of DNA fragments by PCR: (a) general temperature steps; (b) illustration of the exponential amplification of DNA fragments.

and extends the primers by adding nucleotides in the order in which they can pair. The whole process is repeated over and over and therefore specific DNA fragments are copied on a logarithmic scale from minute quantities of source DNA material to huge amounts of DNA fragments (Figure 8.22b).

In rtPCR the production of a specific gene amplicon is monitored in nearly real-time. Diverse possible chemistries have been employed to detect the specific amplification of DNA. However, most techniques monitor an increase in fluorescence (Figure 8.23). In addition, there is an inverse linear relationship between the log of the starting amount of template and the corresponding Ct-value during rtPCR (Figure 8.23b)

A further and improved tool for the molecular identification of BW agents is DNA-based microarrays, which today are also referred to as *"gene chips."* Simply speaking these microarrays combine the specific amplification of a nucleic acid sequence by PCR with the specific hybridization or binding of the PCR product to the so-called nucleic acid probes, short specific but complementary single-stranded nucleotide sequences, immobilized onto a solid phase. DNA arrays offer like protein arrays a high degree of multiplex capability but most present day systems lack the necessary automation. Only a few systems combine at present the PCR amplification and hybridization on one device in one reaction without interrupting the whole assay. However, besides their general potential for identification purposes DNA-based microarrays offer a high potential for diagnostic objectives and medical treatment.

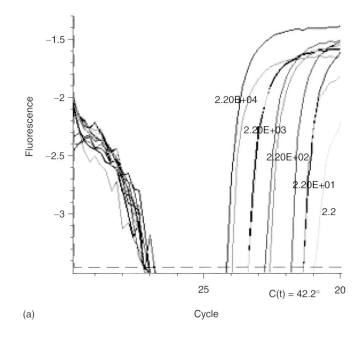

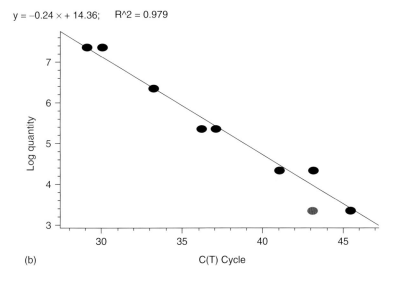

Figure 8.23 Specific amplification of a vaccinia virus (VacV) specific DNA fragment by PCR. (a) The point at which the fluorescence rises significantly above background is called the *threshold cycle* (Ct); (b) there is an inverse linear relationship between the log of the starting amount of template and the corresponding Ct-value.

8.6.3
Chemical and Physical Identification

As explained in Section 8.2 chemical and physical identification is a powerful tool for the detection and identification of chemical agents. Attempts to use these technologies also for direct identification of biological pathogens have long been hindered by the complexity of the generated spectra. In any event, today current and developing biosensors try to employ the physical and chemical characteristics (size, mass, or chemical composition) intrinsic to biological agents for identification. Examples of such technologies include:

- mass spectrometry,
- multiwavelength UV/visible spectroscopy
- Raman spectrometry.

Mass spectrometry selectively identifies components of a biological sample by molecular weight analysis and today many different analytical techniques are based on mass spectrometric principles that were already developed in the 1970s. Common to all mass spectrometric applications are the following steps:

- the production of electrically charged ions in a gas phase,
- the separation of these ions according to their mass (in daltons) to charge ratio in an electromagnetic field,
- determination of the number of individual ion species (ion detection).

It seems that mass spectrometry has found a niche in the analysis of proteinaceous toxins like ricin, SEB (staphylococcal enterotoxin B), and T-2 toxin [23], but the technique has also been used to identify bacterial and viral proteins as well as intact bacterial cells [24, 25]. Most current approaches utilize matrix-assisted laser desorption/ionization time-of-flight (**MALDI-TOF**), which is explained schematically in Figure 8.24.

The biological sample is mixed with a larger quantity of an organic molecule (the matrix) that efficiently absorbs the laser energy. After ionization of the sample all the molecules are accelerated in an evacuated tube. By measuring/calculating how long it takes for each ion to reach the detector at a fixed position, and

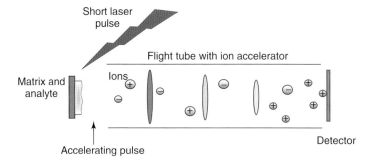

Figure 8.24 Simplified scheme of MALDI-TOF mass spectrometry.

MALDI spectra from *Bacillus* spore species

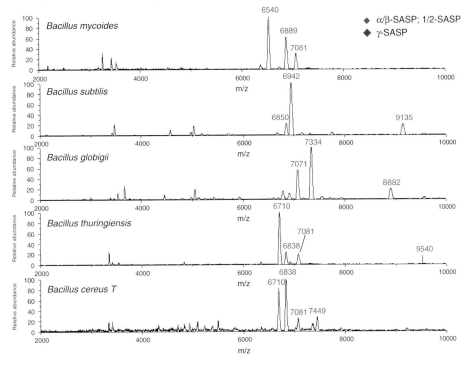

Figure 8.25 Mass spectra from *Bacillus* spores. Reproduced with permission from C. Fenselau, personal communication.

taking into account the length of the flight tube and the charge on the ions, one obtains the mass spectrum. In general, mass spectrometry is rapid and potentially requires little sample preparation but the main drawbacks are still its sophisticated instrumentation and high cost. Theoretically, mass spectrometry is capable of identifying all types of biological agents, as illustrated for different *Bacillus* spores [26], but the method requires small and highly concentrated samples of, for example, 10^5-10^7 cells ml^{-1} for whole-cell analysis (Figure 8.25).

Moreover, the method might lack specificity in complex matrices or mixtures of targets since today nearly all encouraging results have been achieved under conditions in which the target organism is present in much higher concentrations than other potentially interfering background organisms.

The multiwavelength UV/visible spectroscopy method uses the light-scattering and absorbing properties of suspended biological agents to generate a complete ultraviolet/visible (UV/Vis) spectrum.

The method is under study but currently requires extensive sample preparation to produce a pure suspension of target in a non-absorbing medium. Whether this kind of spectroscopy is sufficiently specific for biological identification has yet to be evaluated. For more details on this topic please refer to Reference [27, 28].

Raman spectroscopy and surface-enhanced Raman scattering (SERS) are currently under study as a means of identifying nucleic acids, pathogens, and toxins. Compared to Raman spectroscopy SERS attempts to overcome the complex biological spectra by using special tags for the indirect biological identification, which enhances the sensitivity of the technology by several orders of magnitude. Most of the tags are based on metals and use copper, silver, or gold. Owing to their roughened surfaces they act as carrier for the biological molecule as well as enhancer for the intensity of the Raman signals. For further information on this special kind of spectroscopy please see Reference [29]. In the first instance the advantage of all these methods and technologies is that they do not require additional biological reagents but affinity probes may be used to aid in target capture and increase specificity. To discriminate biological agents from each other all the subsequently introduced methods require the generation of complex spectral fingerprints. Therefore, successful identification by these methods clearly depends on the complexity of the associated data bank. Moreover, complex space-filling as well as non-portable devices limit currently the application of these technologies for biological identification.

8.7
Developing and Upcoming Technologies

Recent use of biological warfare agents has led to a growing interest in methods and technologies for the rapid detection and sensitive identification of pathogens and toxins. Only now is technology becoming available that permits identification of biological organisms within minutes.

For the future we expect improvements from new optical spectroscopic techniques and other label-free approaches that are all under study today to allow real-time or nearly real-time biological monitoring.

Whereas several years ago there was only one label-free specific detection method available, there are now many techniques available that monitor specific molecular interactions by principles like surface-plasmon resonance (SRP), changes in electrical impedance, and flow of ions [30]. Other physical properties that are currently exploited and pioneered for biological sensing are sound, generated by the vibration of a quartz crystal, and change in surface stress and resonance frequency by cantilever technology. For further information please refer to the current scientific publications. However, all these techniques will be only useful if the devices are engineered to the point that the systems are deployable and can be operated in the field. Moreover, all label-free sensing technologies have to prove that in comparison to the traditional immunological and molecular assays they achieve similar or better limits of detections (LODs), as they lack the amplification step of labeling. Applications that are currently limited by assay sensitivity might benefit from technologies like up-converting phosphor technology or time-resolved fluorescent technology. Both techniques make use of the very specific photophysical and spectral properties of lanthanides. When these lanthanides are used, for example, as label or reporter in immunochromatographic assays the background fluorescence signal or autofluorescence is dramatically reduced. Furthermore, these labels do not photobleach compared to classical fluorochrome-based assays.

We suppose that within the next ten years the focus and emphasis will be on techniques that offer a high or higher degree of miniaturization, portability, automation as well as multiplexing capability and expect significant improvements from progress in micro- and nanotechnology. The attraction of these relatively new technologies correlates with a reduction of suitable materials from bulk to micro- or nanometer size, a decrease in consumption of biological reagents, and less production of waste. Standard laboratory operations could be performed on small scales and become therefore more field-usable. Moreover, we expect shorter response times and faster analysis from these miniaturized devices due to shorter diffusion distances and smaller heating capacities. Their compactness might allow a massive parallelization of analytical processes. **Lab-on-a-chip** (LOC) technologies, for example, offer the possibility to integrate several laboratory functions on a single chip that is only a few square centimeters in size. Instruments capable of performing total and automated nucleic acid preparation as well as biochemical and immunological assays have been introduced [31–34]. Technologies like dielectrophoresis and patch clamping have been successfully integrated on a chip. Moreover, **nanopores** might have the potential capability to be used as a device for single-particle or possibly even single-organism detection. Although today their nearly real-time application is mainly limited by the fragility of the lipid membrane these deficiencies might be overcome by the fabrication of synthetic nanopores. Compared to their biological analogs synthetic nanopores are more stable and provide flexible pore sizes. They might be used as immunological sensor as well as a novel genotyping device.

Along with nanotechnology, technologies concerning nanoparticles, fluid dynamics and surface chemistry might also be of future interest [34]. Molecular imprinting, for example, could be prospectively applied to concentrate and specifically identify microorganisms and toxins from liquids [35]. In general this technology creates template-shaped cavities in polymer matrices and mimics the recognition and binding capabilities of natural biomolecules such as antibodies. In remembrance of the corresponding template the antigen is recognized sterically and selectively bound, while an incorrectly shaped molecule is not identified.

Which novel technology will settle the race within ten years cannot be decided without explicit practical experience, but certainly all the technologies have to be evaluated carefully in terms of sensitivity, specificity, cross-reactivity, portability, and degree of automatization.

8.8 Conclusions

In this chapter we have summarized state of the art technology of biological threat detection and identification. The advantages and limitations of most current methods in pathogen and toxin detection have been presented. Importantly, there is currently no real "real-time detection and identification" system but novel technologies are under study. Therefore, we hope to be closer to "detect-to-protect" detection and identification in a few years. Most current biological agent sensing approaches require point detection or point sample extraction, followed by biological discriminant detection or identification. To overcome the currently existing shortcuts a promising approach might be the use of different detector levels. For example, a nonspecific detector capable of detecting any and all biological agents could be combined with a rapid, structure-based identifier capable of identifying 10–20 of the leading threat agents. In addition, the independent use of two or more different detection and identification techniques would result in low false alarm rates and a high level of robustness against potential countermeasures. Critical needs still include rapid and autonomous sample preparation from diverse environmental matrices and a better characterization of the ambient bioaerosol backgrounds. Rapid detection of aerosols is under investigation and further developments in, for example, LIDAR technologies could lead to a breakthrough in this vital field. In addition, the development of highly specific detection molecules is a crucial point that should not be ignored.

References

1. Cox, C.S. (1987) *The Aerobiological Pathway of Microorganisms*, John Wiley & Sons, Inc., New York.
2. Verreault, D., Moineau, S., and Duchaine, C. (2008) *Microbiol. Mol. Biol. Rev.*, **72**, 413–444.
3. Fatah, A.A., Barrett, J.A., Arcilesi, R.D. Jr., Ewing, K.J., Lattin, C.H., and Moshier, T.F. (2001) An Introduction to Biological Agent Detection Equipment for Emergency First Responders, NIJ Guide 101-00, NCJ 190747.
4. Jim, H. (2002) *Anal. Chim. Acta*, **457**, 125–148.
5. Board on Chemical Sciences and Technology, Division on Earth and Life Studies, National Research Council of the National Academies (2003) *Testing and Evaluation of Standoff Chemical Agent Detectors*, The National Academies Press, Washington, DC.
6. Committee on Determining a Standard Unit of Measure for Biological Aerosols, Board on Chemical Sciences and Technology, Board on Life Sciences, Division on Earth and Life Studies, National Research Council of the National Academies (2008) *A Framework for Assessing the Health Hazard Posed by Bioaerosols*, The National Academies Press, Washington, DC.
7. Ogawa, A. (1984) *Separation of Particles from Air and Gases*, CRC Press.
8. Loh, W.L. (2007) in *CBRNe World* (ed. G. Winfiled and P. Baterman), Falcon Communications Limited, Winchester, pp. 78–80.
9. Voller, A., Bartlett, A., and Bidwell, D.E. (1978) *J. Clin. Pathol.*, **31**, 507–520.
10. Engvall, E. (1977) *Med. Biol.*, **55** (4), 193–200.
11. Voller, A., Bidwell, D.E., and Bartlett, A. (1982) *Lab. Res. Methods Biol. Med.*, **5**, 59–81.
12. Peruski, A.H. and Peruski, L.F. (2003) *Clin. Diagn. Lab. Immunol.*, **10** (4), 506–513.
13. Leland, J.K. and Powell, M.J.J. (1990) *Electrochem. Soc.*, **137**, 3127–3131.
14. Griffiths, J. (2007) *Anal. Chem.*, **79** (23), 8833–8837.
15. Kusnezow, W., Jacob, A., Walijew, A., Diehl, F., and Hoheisel, J.D. (2003) *Proteomics*, **3**, 254–264.

16. Huelseweh, B., Ehricht, R., and Marschall, H.J. (2006) *Proteomics*, **6**, 2972–2981.
17. Ehricht, R., Adelhelm, K., Monecke, S., and Huelseweh, B. (2009) *Methods Mol. Biol.*, **50**, 85–105 (Review).
18. Wang, J., Yang, Y., Zhou, L., Wang, J., Jiang, Y., Hu, K., Sun, X., Hou, Y., Zhu, Z., Guo, Z., Ding, Y., and Yang, R. (2009) *Immunopharmacol. Immunotoxicol.*, **31**, 417–427.
19. Morrison, S.L. (1992) *Annu. Rev. Immunol.*, **10**, 239–265.
20. Nelson, A.L. (2010) *Monoclonal Antibodies*, **2** (1), 77–83.
21. Honegger, A. (2008) *Handb. Exp. Pharmacol.*, **181**, 47–68.
22. Thie, H., Voedisch, B., Dübel, S., Hust, M., and Schirrmann, T. (2009) *Methods Mol. Biol.*, **525**, 309–322.
23. Ler, S.G., Lee, F.K., and Gopalakrishnakone, P. (2006) *J. Chromatogr. Appl.*, **1133**, 1–12.
24. Sauer, S., Freiwald, A., Maier, T., Kube, M., Reinhardt, R., Kostrzewa, M., and Geider, K. (2008) *PLoS ONE*, **3** (7), e2843.
25. Mazzeo, M.F., Sorrentino, A., Gaita, M., Cacace, G., Di Stasio, M., Facchiano, A., Comi, G., Malorni, A., and Siciliano, R.A. (2006) *Microbiology*, **72** (2), 1180–1189.
26. Ryzhov, V., Hathout, Y., and Fenselau, C. (2000) *J. Mass Spectrom. Soc. Jpn.*, **51** (1), 108–113.
27. Alupoaei, C.E. and García-Rubio, L. H. (2004) *Biotechnol. Bioeng.*, **86**, 163–167 [PubMed].
28. Alupoaei, C.E., Olivares, J.A., and Garcia-Rubio, L.H. (2004) *Biosensor. Bioelect.*, **19**, 893–903.
29. Smith, W.E. (2008) *Chem. Soc. Rev.*, **37** (5), 955–964.
30. Lim, D.V., Simpson, J.M., Kearns, E.A., and Kramer, M.F. (2005) *Clin Microbiol Rev.*, **18** (4), 583–607.
31. Shafiee, H., Sano, M.B., Henslee, E.A., Caldwell, J.L., and Davalos, R.V. (2010) *Lab Chip*, **10** (4), 438–445.
32. Tia, S. and Herr, A.E. (2009) *Lab Chip*, **9** (17), 2524–2536.
33. Myers, F.B. and Lee, L.P. (2008) *Lab Chip*, **8** (12), 2015–2031.
34. Cheng, M.M.-C, Cuda, G., Bunimovich, Y.L., Gaspari, M., Heath, J.R., Hill, H.D., Mirkin, C.A., Nijdam, A.J., Terracciano, R., Thundat, T., and Ferrari, M. (2006) *Curr. Opin. Chem. Biol.*, **10**, 11–19.
35. Alexander, C., Andersson, H.S., Andersson, L.I., Ansell, R.J., Kirsch, N., Nicholls, I.A., O'Mahony, J., and Whitcombe, M.J. (2006) *J. Mol. Recognit.*, **19** (2), 106–180.
36. Daniels, J.S. and Pourmand, N. (2007) Label-Free Impedance Biosensors: Opportunities and Challenges Electroanalysis, **19** (12), 1239–1257.

9
Measurement of Ionizing Radiation

Ronald Rambousky

Radiation detection differs totally from the analysis of chemical molecules and biology. Generally, ionizing radiation is the result of radioactivity of matter and is emitted by a radioactive atomic nucleus in consequence of a nuclear decay or other particular nuclear reactions. Furthermore, ionizing radiation can be produced in man-made nuclear accelerators and reactors. Humans do not have any sense organs to detect ionizing radiation – we cannot see, taste, smell, hear, or feel it. Certainly

CBRN Protection: Managing the Threat of Chemical, Biological, Radioactive and Nuclear Weapons,
First Edition. Edited by A. Richardt, B. Hülseweh, B. Niemeyer, and F. Sabath.
© 2013 Wiley-VCH Verlag GmbH & Co. KGaA. Published 2013 by Wiley-VCH Verlag GmbH & Co. KGaA.

this is a main reason why radioactivity and ionizing radiation is often regarded with skepticism by the general public, at least in most Western countries. Although ionizing radiation has been present in nature (naturally occurring radioactive materials and cosmic radiation) throughout man's history, it remained unnoticed until less than 120 years ago, when W.C. Röntgen discovered X-ray radiation (1895) and A.H. Becquerel (1896) discovered the radioactivity of matter. These discoveries initiated the development of the first coarse radiation meters. Especially, the nuclear weapons program, starting in the 1940s, later the civil nuclear power industry, and last but not least the safeguard efforts of the International Atomic Energy Agency (IAEA) boosted the development of a wide variety of radiation detectors and radioactivity measuring systems for many different kinds of measuring tasks in the different nuclear and radiological scenarios.

9.1
Why Is Detection of Ionizing Radiation So Important?

Because humans have no sense organs to detect ionizing radiation it is obvious that precautions are only possible with technical devices. The ozone that is produced by the interaction of photons with air molecules in irradiation facilities with extremely high gamma dose rate can be recognized by its characteristic smell. But this is an indirect recognition of ionizing radiation and not at all reliable. To understand the variety of measuring tasks and the need for different detector types the example of natural background radiation is examined. Figure 9.1 shows the different contributions to natural background radiation [1] together with some practical dose values for low and high dose radiation scenarios [2].

The data in Figure 9.1 shows impressively that there are orders of magnitude between dose and dose rate measurements necessary in low and high dose or dose rate radiation scenarios. While the natural gamma background radiation in most Western countries is about $0.07\ \mu\text{Sv h}^{-1}$, the dose rate in the vicinity of highly radioactive sources can reach hundreds of Sv h^{-1}. The figure also shows the different types of measurements necessary for evaluating the different components of natural background radiation. The terrestrial radiation consists of gamma radiation caused by radioactive nuclides of the primordial radioactive series, the radioactive gaseous radon, which is inhaled mainly by people living in buildings, and the ionizing radiation originating from the natural radioactive nuclides inside our body, mainly potassium-40 and carbon-14. For the different contributing dose paths generally different types of radiation measuring equipment is used [3], as will be explained later.

The other part of the natural background radiation is cosmic radiation [4]. On the earth's surface, cosmic radiation consists mainly of gamma radiation with an average annual dose of 0.3 mSv. But on going to higher altitudes the dose rate increases and other radiation particles (e.g., neutrons) begin to play a more important role. To assess the radiation dose in aircraft, radiation meters for

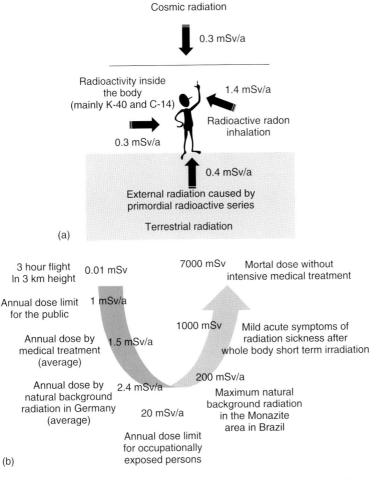

Figure 9.1 Natural background of ionization radiation in relation to some radiation scenarios: (a) different contributions of natural background radiation; (b) dose values of low and high dose radiation scenarios.

different types of radiation therefore have to be used. Correct measurements of cosmic radiation are especially important for health purposes of aircrew [5].

Cosmic radiation is generally influenced by the following four factors [6]:

1) **Altitude:** The earth's atmospheric layer provides significant shielding from cosmic radiation. As a rule, cosmic radiation levels rise with increasing altitude (up to about 20 km above ground). The radiation exposure at conventional aircraft flight altitudes of about 10 km is about 100 times higher than at ground level.

2) **Geographic latitude:** The earth's magnetic field deflects many cosmic radiation particles. The magnetic shield is most effective at the equator, decreases at higher latitudes and essentially disappears at the poles. This leads approximately to a doubling of cosmic radiation exposure from the equator to the magnetic poles.
3) **Solar activity:** the solar activity varies in a predictable way with a cycle of approximately eleven years. Lower solar activity leads to higher cosmic radiation levels and vice versa.
4) **Solar proton events (SPEs):** Sometimes large explosive ejections of charged particles occur from the sun. These explosions are not predictable but can lead to sudden increases in radiation levels in the atmosphere and on earth, the SPEs.

Several studies have been conducted to investigate the overall radiation exposure effect for aircrew and travelers. In general an increased radiation exposure during flights as compared to staying on the ground can be observed (Table 9.1). Most aircrews, with up to 1000 h per year, have an annual effective radiation dose in the range of 2–5 mSv. Occasional travelers obtain a fraction of this value.

This is only one example – beside home security questions – of why it is all the more important to track and measure enhanced radiation levels with appropriate measuring devices. The commanders and the users of radiation measuring equipment must be aware of the fact that, generally, for the different kinds of ionizing radiation devices with different detector types have to be used. In addition, the energy or energy range of the radiation plays an important role in the choice of the appropriate detection or measurement device. Last but not least, the purpose of the desired radiological analysis also determines the radiological equipment. Measuring the dose rate of a gamma radiation field, the contamination of a surface, the amount of a certain radionuclide in a food sample, or identifying

Table 9.1 Cosmic radiation dose on selected flights for November 2010.

From	To	Duration (h)	Estimated radiation dose (μSv)[a]
Sydney	Singapore	7.50	19
London	Tokyo	12.00	72
New York	Paris	7.00	45
Frankfurt	Los Angeles	9.50	63

[a] Estimated radiation doses based on a cruise altitude of 10 000 m and the shown flight duration. Calculations obtained from EPCARD portal *http://www.helmholtz-muenchen.de/en/epcard-portal/epcard-home/index.html* (accessed 29 November 2010). Other software may give slightly different values.

the radioactive nuclides in a sample will require the designated detectors and analysis equipment [7].

In the scope of military related operations and civil defense we focus on the following four types of ionizing radiation [8]:

- **Alpha radiation** is a particle radiation consisting of positively charged helium nuclei. The ionization capability is very high but the range is only some centimeters in air and some micrometers in human tissue.
- **Beta radiation** is also a particle radiation, consisting of electrons or their anti-particles, the positrons. The range of beta radiation in air is from some centimeters to some meters, depending on the energy of the particles.
- **Gamma radiation** is electromagnetic radiation of very high frequency (very short wavelength). According to the physical wave–particle duality, gamma radiation can also be described as photons (particles with no rest mass and no charge, but energy and impulse). X-Ray radiation is electromagnetic radiation, too. Photons are far-ranging, up to some hundreds of meters, depending on the activity of the source for practical use.
- **Neutron radiation** is a particle radiation consisting of neutrons, which are uncharged particles. Free neutrons are not stable, with a half-life of about 15 min. The energy range of neutrons is very large, ranging from 25 meV[1]) (thermal neutrons) to 14 MeV (fusion neutrons) for most applications. Neutrons also belong to the far-ranging radiation.

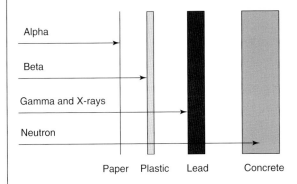

Figure 9.2 Shielding and attenuation effects for the different types of ionizing radiation relevant for military operations and civil defense.

Figure 9.2 displays the ability of materials to shield or attenuate the different kinds of ionizing radiation. While both alpha and beta radiation can be shielded by a sheet of paper or a disk of plastic, respectively, gamma radiation

---

1) Electron volt (eV) is a measure of energy in nuclear physics. 1 eV is the same energy as 1.6022×10^{-19} J; meV stands for 10^{-3} eV, MeV for 10^{6} eV.

theoretically cannot be shielded but only attenuated by material with a high mass index as, for example, lead, tungsten, or uranium. Neutron radiation must be thermalized by collisions with light atoms like hydrogen. The thermal neutrons are then captured by the surrounding material. This thermal neutron capture generally is accompanied by the emission of prompt gamma radiation.

9.2
Physical Quantities used to Describe Radioactivity and Ionizing Radiation

Before we start our tour through different ionizing radiation detector technologies we should recall some important physical quantities about ionizing radiation (Figure 9.3) [2, 8–10].

9.2.1
Activity

The activity (A) of a radioactive source describes the number of disintegrations of radioactive nuclides in 1 s. The physical unit for the activity is the becquerel (Bq),

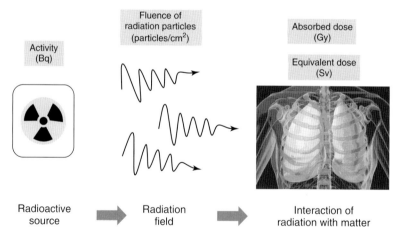

Figure 9.3 Illustration of the most important physical quantities describing radioactivity and ionizing radiation. The activity describes the amount of radioactive nuclides in the source. The decay of radioactive nuclides is related to the emission of ionizing radiation, creating a radiation field around the source. The interaction of the radiation with matter (e.g., human tissue) causes energy absorption by ionization processes. The absorbed dose is a measure of the deposited radiation energy and is measured in gray (Gy). The equivalent dose, measured in sievert (Sv), is a biologically assessed absorbed dose, taking into account that different radiation types have different biological effects for the same absorbed dose.

which is simply the decays per second. Radioactive decay is a statistical process that generally follows Poisson statistics. Therefore it is not possible to predict the disintegration of a single radioactive nuclide, but a large number of nuclides follow the well-known differential equation for radioactive decay $A = -dN/dt = \lambda N$, where A is the activity, N is the number of radioactive nuclides, λ is the decay constant, and t the time.

- **Activity** of a radioactive substance is the physical unit of the number of disintegrations per second.
- **Radioactivity** is the property of some kind of matter to undergo disintegrations on a statistical basis.

Simply speaking, the activity of a source is the amount of radioactive material. Because the specific activities (activity per mass) of most of the in-use radionuclides are extremely small, the mass of even highly active sources is often only in the milligram (mg) to gram (g) range.

Care must be taken not to mix up the count rate of a radiation detector with the activity of a source. Both can be expressed in the unit s^{-1} but there is a fundamental difference. There are disintegrations of radionuclides that statistically produce more or less than one radiation particle, perhaps actually of different types. Caesium-137, for example, emits one electron per decay and one photon in only 84% of all disintegrations. Cobalt-60 emits two photons per decay. The difference between activity and a measured count rate is also related to detector efficiency, sample geometry.

9.2.2
Absorbed Dose

The absorbed dose (D) – also known as *total ionizing dose* – is a measure of the energy deposited in a medium by ionizing radiation. It describes the energy deposited per unit mass of the medium. Therefore, the absorbed dose has the physical unit joule per kilogram ($J\,kg^{-1}$), which is given the special name gray (Gy) or radiation absorbed dose (rad) as a non-SI unit. The absorbed dose depends on the chemical composition of the related medium and strictly speaking the medium must be indicated in brackets after the Gy unit. For example, the absorbed dose of 5 Gy in silicon would be indicated as $D = 5$ Gy(Si). In most applications – especially in health physics – the related medium is human soft tissue. When high doses and deterministic radiation effects are concerned the gray is usually used to describe the relevant doses. For dealing with stochastic effects in the realm of radiation protection issues the absorbed dose is not the appropriate physical unit, because it is not a good indicator of the stochastic biological effects.

9.2.3
Equivalent Dose

The equivalent dose (H_T) is a measure of the radiation dose to the tissue type (T) – for example, muscle tissue, brain tissue, and lung tissue – where the biological effectiveness of the different types of radiation has been taken into account. X-Ray, gamma-, and beta radiation have a relative effectiveness of 1, alpha particles of 20, and neutrons between 2 and 20, depending on their energy. Equivalent dose is therefore a less fundamental quantity than absorbed dose, but is more biologically significant. The unit of the equivalent dose is the sievert (Sv). An old unit is called Röntgen equivalent man (rem). Both the Gy and the Sv can be written as J kg^{-1} but the meaning is quite different.

> The **radiation dose** is a measure for the "amount" of radiation a person was exposed to. The dose rate describes the "strength" of the ionizing radiation. There is an analogy with pouring water through a pipe in a barrel. The amount of water (volume) in the barrel corresponds to the radiation dose, while the flow rate of the water corresponds to the dose rate.

9.2.4
Effective Dose Equivalent

While the equivalent dose is strictly speaking defined only for each of the different organs in the human body, the effective dose equivalent is an estimate of the stochastic effect that a non-uniform radiation dose has on the whole human body. This is necessary because it turned out that the same equivalent doses lead to stochastic effects with different probabilities in the different organ tissues. The effective dose is the most important dose value because it is used for legal radiation protection limits and most of the radiation detectors, like personal dosimeters and dose-rate meters, are calibrated for an operational quantity describing the effective dose equivalent.

9.2.5
Operational Dose Quantities

The equivalent dose for an organ and the effective dose equivalent belong, strictly speaking, to the so-called protection quantities, which are not directly measurable. The operational doses and dose rates allow a practically easy determination by measurement and are a good estimate for the protection quantities like the effective dose.

The operational quantities are $H^*(10)$ (the ambient dose equivalent) for penetrating radiation and $H'(0.07)$ (the maximum directional dose equivalent) for non-penetrating radiation. They are measured with appropriate dose-rate meters by integrating the dose rate over a certain time interval. Other operational quantities are $H_p(10)$ (the personal deep dose equivalent) for penetrating radiation and $H_p(0.07)$ (the personal surface dose equivalent) for non-penetrating radiation. Personal dose equivalents are measured with personal dosimeters, which must be worn at representative positions on the body (e.g., chest, finger) and take into account the backscatter of the body itself.

9.3
Different Measuring Tasks Concerning Ionizing Radiation

As mentioned above, not only the different types of radiation (alpha, beta, gamma, and neutron radiation) demand special metrology but so do the different measuring tasks and purposes for radiation measurement and radiological analysis. The measurement of gamma radiation to determine the ambient dose rate of the present radiation field demands measurement technology other than nuclide identification on the spot [3]. Even the instrumentation for accurate gamma dose rate measurements according to legal specifications needs to be different to instrumentation used for a sensitive search of lost gamma sources in rapid motion (Figure 9.4).

Therefore, detection and measurement of ionizing radiation can be grouped into the following tasks and purposes:

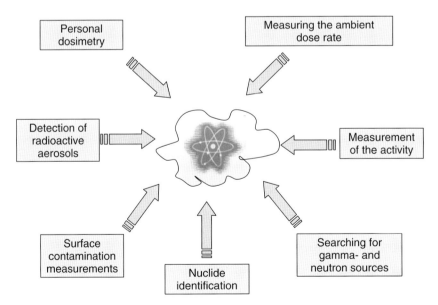

Figure 9.4 Different measuring tasks and purposes for radiation measurement and radiological analysis.

9.3.1
Personal Dosimetry

In **personal dosimetry**, the radiation dose is measured at a representative position on the surface of the human body for the radiation field present. The measured operational quantity is a good estimate of the effective dose the person was exposed to in the radiation field. Personal dosimeters are available for penetrating radiation (gamma and neutron radiation), measuring $H_p(10)$, and for non-penetrating radiation (X-ray and beta radiation), measuring $H_p(0.07)$. There are no dosimeters for alpha radiation available, because alpha radiation cannot penetrate intact human skin and is therefore not an external radiation hazard. The incorporation of radionuclides into the human body cannot be measured with personal dosimeters. The internal dose must be calculated according to complex formulas based on advanced radiological analysis of excrements or whole body counting [10].

9.3.2
Measuring the Ambient Dose Rate

Measuring the ambient dose rate in gamma-, X-ray, or neutron radiation fields is one of the commonest tasks concerning both radiation protection issues and also CBRN-issues on the battlefield. The dose rate in the different scenarios can cover many orders of magnitude. The natural background radiation in most Western countries is about 60–100 nSv h^{-1}.[2] On the nuclear battlefield dose rates in the range of 10 Sv h^{-1} might occur. Usually, a dose-rate meter that is used for accurate measurements in the range of natural background radiation cannot be used in high dose rate fields. The detector would become overloaded and damaged and in consequence would display wrong data. Radiation in most scenarios is usually not mono-energetic but will cover a certain range of the energy spectrum. In choosing the appropriate dose-rate meter, care must be taken to cover the dominant energy range of the radiation fields present. Dose rate range, linearity and energy compensation are the most critical technical requirements for choosing the right dose-rate meter. The dose-rate meters should be calibrated in $H^*(10)$ for penetrating radiation and in $H'(0.07)$ for non-penetrating radiation [7]. For military use on the nuclear battlefield a calibration in absorbed dose for soft tissue "free in air" is recommended by NATO [11].

9.3.3
Searching for Gamma- and Neutron-Sources

While **searching for gamma- and neutron-sources**, the respective dose rate is measured, too. But in this application high-volume detectors with a large efficiency are needed to obtain the necessary time resolution, especially when the detectors are used in fast vehicles or helicopters. In these applications requirements of energy

---

2) nSv h^{-1} stands for 10^{-9} Sv h^{-1} (nano-Sievert per hour).

compensation and dose rate linearity are not of such relevance as they are for "real" dose-rate meters. Therefore, in these kinds of application we are talking more about quantitative detection and not about a proper measurement of the gamma- or neutron dose rate. If the detection devices display a dose rate value it should be understood more as a good guess than a real statistically provable value [7].

9.3.4
Surface Contamination Measurements

Surface contamination measurements are carried out to detect radioactive particles on surfaces of, for example, vehicles, equipment, clothing, and personnel. Generally, two-dimensional probes with a large active probe area of about 100–300 cm^2 are used for this purpose. The devices are called *contamination monitors* and are built mainly for hand-held use either with separate or integrated probe. The detector generally can discriminate the alpha radiation but not between gamma and beta radiation. For gamma/beta discrimination usually a manual shielding element has to be placed in front of the probe window. The measured value is basically displayed as a simple count rate. If the radionuclide present and the corresponding nuclide specific calibration factor are known, the count rate can be easily converted into a surface activity in Bq cm^{-2}. Many of the newer contamination monitors provide an internal database with the calibration factors for various commonly used radionuclides [3].

9.3.5
Nuclide Identification

Measurement procedures and equipment for **nuclide identification** depend mainly on the type of radiation. The simplest way is to analyze the gamma radiation of a radionuclide. This identification by photons is called *gamma spectroscopy* [12]. There are already hand-held gamma spectrometers available for identification on the spot (Figure 9.5) but for accurate sample analysis you have to use a laboratory

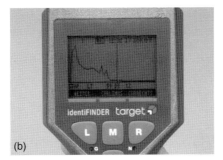

Figure 9.5 identiFINDER hand-held spectrometer manufactured by ICx Radiation Inc.: (a) identiFINDER in the dose rate scanning mode; (b) identiFINDER shows a typical NaI pulse height spectrum of ^{60}Co gamma radiation.

environment. Essential for the quality of identification is the energy resolution of the detector used. Hand-held gamma spectrometers are generally equipped with NaI(Tl)[3]-detectors (energy resolution not smaller than 6%). Laboratory gamma spectroscopy systems mainly use high-resolution HPGe (high purified germanium crystal) detectors (energy resolution about 0.15%).

Because of the broad energy distribution of beta radiation a classical beta spectroscopy is not possible. Rough nuclide identification is sometimes achieved by determining the so-called end point energy of the broad beta spectrum.

Alpha radiation emitted from decaying radionuclides – as well as gamma radiation – is characteristic and therefore suited for alpha spectroscopy [3]. A big problem concerning alpha spectroscopy in practical application is the high amount of absorption in the sample itself and in the air between sample and detector. This absorption causes a peak shift to smaller energies and a broadening of the peak that generally makes identification impossible. Therefore, a vacuum measuring chamber is essential for alpha spectroscopy. Additionally, complex radiochemical sample preparation is generally inevitable when the radionuclides are embedded in a dense matrix (e.g., soil and water samples).

9.3.6
Measurement of Activity

Measurement of the activity of a sample depends on the type of radiation and the property of the sample. With the latter, the main difference is between two-dimensional samples (e.g., swipe samples and filter samples) and three-dimensional samples (e.g., soil, water, or material samples). To determine the total alpha and beta activity of swipe or filter samples, so-called gross alpha/beta counting is performed (Figure 9.6). Nuclide identification is not possible with this measurement procedure, but often the total alpha + beta activity is a sufficient result. The sample (swipe or filter) is analyzed in a shielded chamber with a large area detector (like the probe of a contamination monitor) on top of – but not in contact with – the sample. The alpha counts are separated by a discriminator threshold. When the nuclide present and the corresponding calibration factor are known, one can calculate the total alpha and the total beta activity of the sample. For radionuclides in a liquid solution – or when a liquid solution can be prepared from the original sample – the activity can be determined by liquid scintillation counting (LSC) [13]. The alpha and beta particles emitted from the radionuclides in the solution produce weak flashes of visible light that are converted into electrons by a photosensitive detector and amplified by a photomultiplier in the LSC device. To determine the activity of gamma-radiating radioactive samples, generally, a gamma spectroscopic activity measurement is chosen [12]. Therefore, the sample must be transferred into a reference geometry for which an efficiency calibration of the used detector exists. An analysis laboratory should hold a minimum set of radioactive calibration standards for filter/swipe geometry and three-dimensional geometries

3) NaI(Tl) stands for thallium-doped sodium iodide crystal.

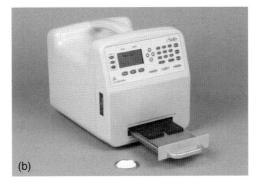

Figure 9.6 Examples of professional gross alpha/beta counters for activity measurement of filter and swipe samples: (a) Hand-E-Count (ThermoFisher Scientific) with accessories; (b) iSolo (CANBERRA) with sample holder and swipe sample.

(e.g., Marinelli beaker, Kautex bottle). Nowadays efficiency calibrations for more exotic geometries can be performed by calculation with specialized software tools when the detector is specially characterized (using radiation transport calculation tools) at the manufacturer beforehand.

9.3.7
Detection of Radioactive Aerosols

The **detection of radioactive aerosols** and the **measurement of the activity concentration**[4] **in the air** are often of special interest in radiation hazards. Fundamentally, there are two different methods. In the classical method, in a first step the radionuclides are sampled on a filter and then in the second step the filter is measured directly on the spot or somewhere in a designated laboratory. In contrast, in the second method – which is relatively new – an *in situ* measurement is performed (Figure 9.7). The air is also sucked over a filter by a pump, but on top of the filter there is already a large-area semiconductor detector [e.g., passivated implanted planar silicon (PIPS)-detector], which measures the alpha and beta count rate continuously. With this procedure the current aerosol concentration can be calculated and displayed. By peak height discrimination such detectors can distinguish between alpha and beta radiation [14].

There are a set of other special measuring tasks and procedures that are of minor importance concerning the realm of military and civil CBRN-defence. Examples are the determination of the whole body activity or incorporation measurements and excretion measurements to determine doses caused by the incorporation of radionuclides in the human body. These aspects are outside the scope of this introductory book.

---

4) The activity concentration is the activity per volume of air and is generally given in Bq m^{-3} (becquerel per cubic metre).

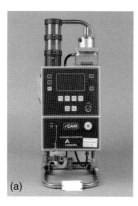 

Figure 9.7 Examples of *in situ* air monitors for detecting radioactivity in air and measuring the activity concentration: (a) iCAM (CANBERRA) is a portable device; (b) MyRIAM (SARAD GmbH) is a personal aerosol dosimeter.

9.4
Basics of Radiation Detectors

The basic concept for electrical measurement techniques in nuclear science is to let the ionizing radiation interact with the active detector volume and to detect the thereby produced charged particles (e.g., secondary electrons). Basically, the detector and the adjacent electronics (e.g., preamplifier, amplifier) produce impulses that are evaluated in the device electronics. Generally, the primary measured value is a count rate (pulses per second). In some cases the device electronics is able to convert the primary count rate into more meaningful physical quantities (e.g., dose or dose-rate meters). Often it is the task of the user to interpret the primary result of the count rate (e.g., contamination monitor). Beside the electrical based radiation detection meters there are several passive radiation detectors in use. However, we can define four primary detector categories (see also Figure 9.8)[7, 15]:

- gas-filled detectors,
- scintillators and light conversion devices,
- semiconductors,
- miscellaneous detectors.

Below we give a brief overview of the most common detector types and their basic functional principle [7].

9.4.1
Gas-Filled Detectors

Although gas-filled detectors are the ancestors of ionizing radiation detectors they continue to play an important role in routine survey meters [16]. Typical gas-filled detectors have a cylindrical form with the cathode as outer wall and a thin wire in

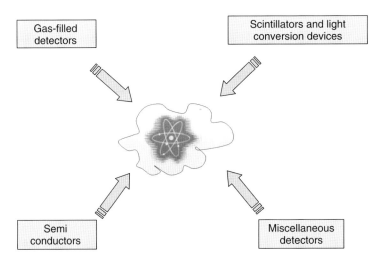

Figure 9.8 Primary detector categories.

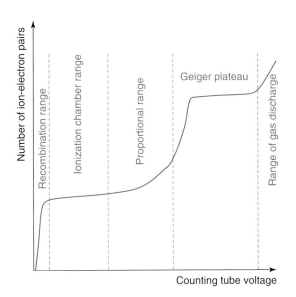

Figure 9.9 Characteristics of gas-filled counter tubes.

the axial center position as the anode. The detector is gas-tight sealed and filled with the so-called counting gas[5] (Figure 9.9). Depending on the high-voltage supply, gas-filled detectors work in three different modes, as an ionization chamber, a proportional counter, or a Geiger–Müller (GM) counter [7, 15].

5) An exception are the so-called gas-flow detectors (e.g., large area gas-flow proportional counters), which are not sealed and work with a steady flow of the counting gas.

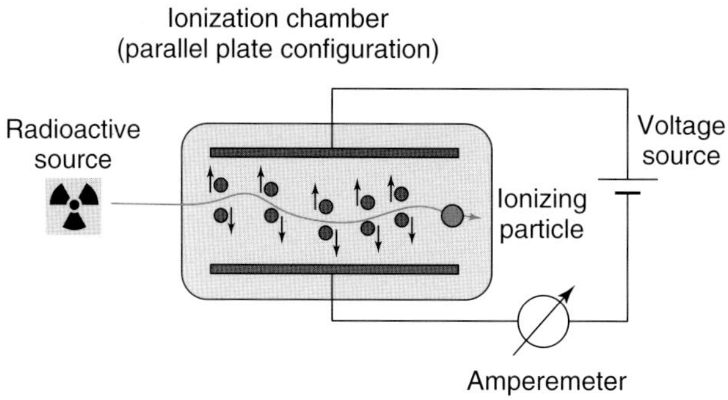

Figure 9.10 Schematic diagram of an ionization chamber.

The voltage on **ionization chambers** is just sufficient to separate the positive (ions) and negative charges (electrons) produced as a consequence of the interaction of the radiation particles with the detector gas and to avoid their recombination. Ionization chambers are mainly operated in the current mode, where they can be used for reliable and accurate dosimetry measurements (e.g., for calibrating radiation fields) and for the measurement of pulsed radiation. Figure 9.10 shows schematically the principle of an ionization chamber in parallel plate configuration, which is also a possible construction as well as the cylindrical form.

According to Figure 9.9 the voltage applied to a **proportional counter** is of such magnitude that the produced negative and positive charged particles are accelerated toward the anode and the cathode respectively. On their way through the counting gas of the tube the primary charged particles are able to ionize further gas atoms or molecules, which results in an amplification of the separated charge. In practice, amplification factors of 100 to 10 000 are possible. Inert gases like argon and xenon (with additives), carbon hydrates like methane and butane, as well as argon-methane mixtures at pressures of about 1 bar are used as counting gases. Leakages and aging processes of the gas filling can lead to a change of the electrical characteristics of gas-filled counter tubes and therefore to a change in radiological properties. The number of produced (separated) charges in a proportional counter is nearly proportional to the energy of the radiation particle. Therefore, proportional counters are well suited for the discrimination between alpha and beta particles, for example, in contamination monitors. Proportional counters (Figure 9.11) are also used successfully in neutron spectroscopy applications. To ensure that beta particles, low energy X-ray radiation, and above all alpha particles can reach the interior of the counter tube, counter tubes for those particle types have to be built with thin radiation entrance windows.

The best known radiation detector might be the **Geiger–Müller counter** (GM-counter is a common abbreviation) or **Geiger–Müller tube**. GM-counters work at even higher voltages than proportional counters, in the so-called Geiger plateau of the counter tube (Figure 9.9). Independent of the energy of the penetrating radiation particle a charge avalanche is generated in the tube because of the high voltage conditions. Therefore, amplification factors in the range of

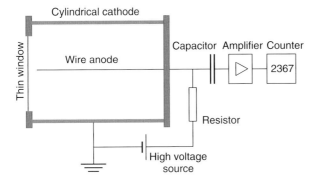

Figure 9.11 Schematic diagram of a cylindrical proportional counter.

10^6–10^8 are possible. Inert gases such as argon or neon are used as counting gases but, additionally, in GM-tube applications a quenching gas (ethanol and halogens) has also to be present to achieve a temporary reduction of voltage at the anode wire. This quenching process leads to the so-called dead time of the GM-counter. This dead time usually is in the range of 10–100 μs and describes the time period the detector is not able to process another radiation event after a successful avalanche generation. The dead time must be considered in the evaluation of the raw detector data. The effect of dead time increases with count rate. The ongoing damage of the counting and quenching gas during detector operation is the cause of the limited life of GM-counters. Therefore, performance checks are necessary on a regular basis with GM-counters.

9.4.2
Luminescence Detectors

Luminescence detectors are passive radiation detectors and have to be read out (analyzed) after irradiation in a separate process. They are mainly used to record radiation doses in environmental and personal dosimetry applications and possess the great advantage of also recording pulsed radiation. Therefore, in military applications luminescence detectors were intended to record the initial radiation of a nuclear weapon burst on the battlefield. In radiation protection applications luminescence detectors have been established for dose measurements in photon and beta radiation fields. Luminescence detectors are particular solids in which the ionizing radiation produces lattice defects during the irradiation process. In a second step, the so-called annealing process, the lattice defects are reversed by heating (thermo-luminescence detectors, TLDs) or ultraviolet irradiation (radiophoto-luminescence detectors, RPL). In this annealing process the luminescence elements emit small portions of visible light that are recorded by appropriate photomultipliers. The amount of emitted light is nearly proportional to the recorded radiation dose. TLDs usually are made of lithium fluoride (LiF) with manganese or titanium doping. For RPLs special phosphate glasses are common [15].

9.4.3
Photo-emulsion

A photosensitive film is also a passive radiation detector and is used for the classical film dosimetry in photon and beta radiation fields. In Germany, film dosimetry is still the most common way of legal dosimetry for occupationally exposed persons. The advantage is that film badge dosimeters (film in a special casing) are very robust and the data is not manually erasable without destroying the whole film (Figure 9.12). When ionizing particles penetrate the film, they cause structural changes in the emulsion layer, which leads to the development of silver particles on developing the film, which in turn results in blackening of the film. The blackening of the film depends on the irradiated dose and also on the energy of the radiation. To get the same blackening one needs 20 times the dose of ^{60}Co photon radiation (high energy) than with 60 kV X-ray radiation (low energy). Therefore, energy compensation with appropriate metal filter elements in the casing is necessary. With the appropriate energy compensation the efficiency of the film is dose equivalent over a wide energy range. In Germany the so-called gliding shadow film dosimeters have been used for some years. They have a dose range from 0.1 mSv to 1 Sv according to $H_p(10)$ and an energy range of 13–1400 keV [7].

9.4.4
Scintillators

Ionizing radiation can cause immediate emission of visible light in some kinds of material as consequence of excitation processes. The amount of emitted light, however, is very small. In a scintillation detector the emitted light is guided to a light-sensitive photocathode that absorbs the light by emission of electrons (Figure 9.13). These electrons are amplified in a photomultiplier so that an

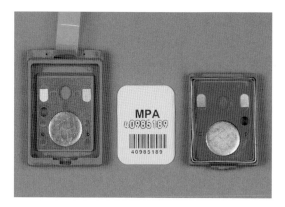

Figure 9.12 Gliding shadow film dosimeter used in Germany for legal dosimetry of occupationally exposed persons; the dosimeter is opened to show the integrated filter elements and the actual film badge (in the middle).

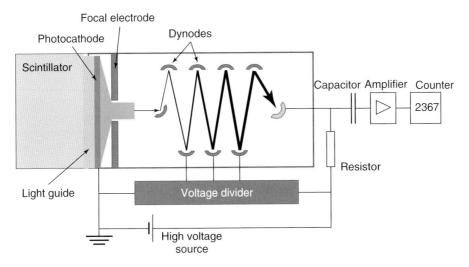

Figure 9.13 Schematic diagram of a scintillation counter.

appropriate current pulse can be produced. For sensitive detection of gamma and X-ray radiation high volume plastic scintillators or NaI(Tl)[6] crystals are used. For gamma-spectroscopic applications the most common detector material is also NaI(Tl), but new detector materials with specific properties have been developed, like CsI(Tl), LaBr$_3$, or BGO (Bi$_4$Ge$_3$O$_{12}$) [17, 18].

An interesting application of scintillator material is zinc sulfide. Polycrystalline zinc sulfide doped with silver atoms ZnS(Ag) can be put on a planar photoconducting carrier substance, so that contamination monitors with active detector areas of more than 100 cm^2 can be produced. Because of the difference in ionization capability of alpha and beta particles, scintillation detectors can discriminate the two types of radiation in two different measuring channels (Figure 9.14).

In LSC applications liquid organic and non-organic scintillators are used that are mixed into the liquid sample (Figure 9.15). There are also measurement procedures where swipes can be put in a special measuring vial without any further sample preparation together with the liquid scintillator. When the radionuclide present and the associated calibration factor for the used geometry (type of vial) is known, one can determine the activity in the sample [13].

Figure 9.15a shows the Triathler LSC device from the manufacturer HIDEX. It is a tabletop device that can be easily transported in a suitable casing and therefore is appropriate for use in mobile laboratories or even in the field. The two-dimensional pulse height spectrum shown in Figure 9.15b gives a hint for nuclide identification of the sample.

---

6) NaI(Tl) crystals are sodium iodide mono-crystals doped with thallium.

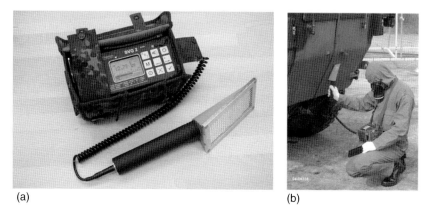

Figure 9.14 Contamination monitor with scintillation probe: (a) German armed forces SVG2 RADIAC[7] with the FHT-380SVG scintillation probe (ThermoFisher Scientific); (b) measurement of the remaining radioactive contamination on the surface after decontamination of the military vehicle with SVG2 equipment.

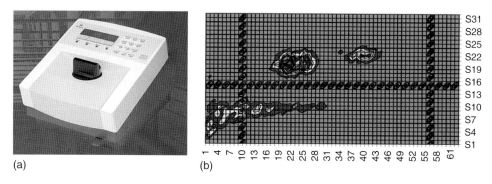

Figure 9.15 Liquid scintillation counter: (a) HIDEX Triathler LSC; (b) 2D-pulse height spectrum of radon (^{222}Rn) in a water sample.

9.4.5
Semiconductor Detectors

The main application for semiconductor detectors has been gamma and X-ray spectroscopy for more than 50 years. The use of semiconductor detectors in comparison to gas-filled detectors provides some advantages, especially concerning dose and dose-rate measurements in photon radiation fields. Because the density of semiconductors is about one thousand times higher than that of counting gases, the necessary detector volume at the same efficiency is much smaller for the solid-state detectors. Additionally, semiconductor detectors – mainly diodes with special characteristics – are radiologically very stable and mechanically robust. The

---

7) RADIAC stands for Radiation Detection Indication And Computation. The term is mainly used for military fitted detection and measuring devices for nuclear radiation.

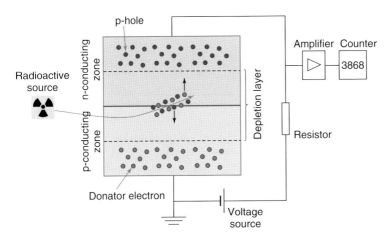

Figure 9.16 Schematic diagram of a semiconductor detector (Si-PIN diode).

standard semiconductor element for dose rate measurements is the silicon PIN[8] diode (Si-PIN diode). These PIN diodes are used as reverse-biasing photodiodes. An applied voltage causes an extensive depletion zone in the semiconductor structure. When a radiation particle interacts with the semiconductor material in the depletion zone, pairs of charge carriers (electron and defect electron) are separated by the applied voltage (Figure 9.16) [19].

For spectroscopic measurement of alpha radiation, silicon-based surface barrier detectors (e.g., PIPS[9] detector) are most common. They have a large active surface and are therefore able to cover a whole filter or swipe. These detectors consist of an n-conducting silicon substrate with a p-conducting oxide layer directly on top of the semiconductor. This barrier layer is covered with a thin gold layer that acts as electrical contact as well as radiation entrance window. The gold layer can be made in such a way that even alpha particles can penetrate it while, on the other hand, it is so robust that decontamination is possible. With an active detector area of 5 cm^2, barrier layer thicknesses of about 100 μm to 2 mm are possible; detectors with an active area of 50 cm^2 achieve barrier layer thicknesses up to 700 μm. Such planar detectors are used in aerosol monitors and activity measurement devices for both swipe and filter samples.

Photons have a much lower interaction probability with matter than alpha particles. Therefore, appropriate semiconductor detectors for gamma spectroscopy need a much larger depletion zone. This is achieved with highly purified semiconductors with very low impurity conductance. In practice extremely pure germanium mono-crystals are used, which are also known as *HPGe detectors*. During operation, such HPGe detectors must be cooled to very low temperatures to decrease the

---

8) A PIN diode is composed of a layer structure of positive doped semiconductor, an intrinsic layer and finally a negative doped semiconductor. The intrinsic layer is self-conducting and contains few free charge carriers and is therefore highly resistive.

9) PIPS detectors are silicon based with ion-implanted electrical contacts.

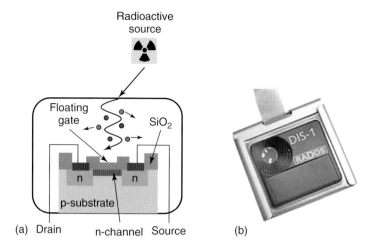

Figure 9.17 Direct ion storage detector: (a) schematic diagram; (b) DIS-1 dosimeter [Mirion Technologies (RADOS) GmbH].

intrinsic conduction of the germanium. Usually, the cooling is done with liquid nitrogen (−196 °C) but in recent years reliable electrical coolers have been developed, too. Large volume coaxial HPGe detectors can have a relative efficiency[10] up to more than 200% and an excellent energy resolution of about 0.15% (i.e., about 2 keV) at the 1332 keV peak of ^{60}Co. The most widely used detector types are the coaxial HPGe detector and the HPGe well-detector, which is designed for sensitive and accurate activity measurements because of its nearly 4π-measurement geometry. For standard applications concerning sample analysis, usually a p-type coaxial HPGe-detector is used. When energy resolution in the lower energy range is important (e.g., with identification of plutonium and uranium) an n-type detector would be preferred.

It is worth mentioning a relatively new detector on the radiation detector market that is known as a direct ion storage (DIS) detector and is used mainly for personal dosimeters (Figure 9.17). A DIS detector is a miniaturized ionization chamber based on a semiconductor element. Simply speaking, a modified electrical erasable programmable read only memory (EEPROM) cell is integrated in an also-miniaturized gas-filled chamber with electrically conducting walls. The dosimetric properties of the DIS detector can be adjusted by choice of the material of the chamber (wall) and the gas filling. The advantage of this technology is the ability to realize a small detector on an electrical basis that can also be used in pulsed radiation fields [20].

---

10) The relative efficiency of a HPGe detector is calculated from the quotient of the absolute efficiency of the detector to the absolute efficiency of a 3×3 inch NaI(Tl) scintillation detector.

9.4.6
Neutron Detectors

Because neutrons do not possess any electrical charge, they cannot be detected directly on the basis of electrical methods. Consequently, in an electrically working neutron detector the neutrons must interact with the detector material to produce charged particles in a first step, which are then electrically detected. Basically there are three methods for detecting neutrons:

1) In the nuclear reaction $^3_2\text{He} + ^1_0\text{n} \rightarrow ^3_1\text{H} + ^1_1\text{p}$ the neutron hits the helium-3 nucleus and a proton is ejected immediately. The remaining nucleus is hydrogen-3, which is also called *tritium*. Because tritium is radioactive, it decays again to helium-3 with a half-life of about 12 years.
2) In the nuclear reaction $^{10}_5\text{B} + ^1_0\text{n} \rightarrow ^7_3\text{Li} + ^4_2\alpha$ the neutron reacts with a boron-10 nucleus by emitting an alpha particle (Figure 9.18). The remaining nucleus is lithium-7. The alpha particle has a twofold positive charge and can be detected electrically.
3) In the nuclear reaction $^6_3\text{Li} + ^1_0\text{n} \rightarrow ^3_1\text{H} + ^4_2\alpha$ the neutron reacts with a lithium-6 nucleus by emitting an alpha particle. The remaining nucleus is tritium. The alpha particle has a twofold positive charge and can be detected electrically.

These three nuclear neutron reactions have, however, a reasonably high reaction probability (nuclear capture cross-section) only for thermal and epithermal neutron energies. To detect fast neutrons with reasonable sensitivity the fast neutrons have to be slowed down by collision reactions in a moderator (Figure 9.19).

There are also passive neutron detectors available such as, for example, neutron sensitive thermoluminescence elements, bubble detectors, and special semiconductors. The latter are sensitive also for fast neutrons and are widely used in modern military radiation detection and measurement equipment and have the capability to record the neutron part of the initial radiation of a nuclear weapon burst. The detectors are specially designed Si-PIN diodes wherein the neutron radiation causes permanent lattice damage, leading to a change of the current–voltage

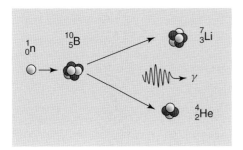

Figure 9.18 Schematic diagram of the nuclear reaction of neutrons with boron-10 to produce a charged particle for (indirect) neutron detection.

Figure 9.19 Examples of neutron detectors: (a) ^{3}He neutron probe including a $H*(10)$ dose equivalent moderator with SVG2 (ThermoFisher Scientific); (b) BF$_3$ neutron detection probe with SVG2 (ThermoFisher Scientific).

characteristic. The change of this characteristic is a measure of the received dose and is determined in a separate step using a special reader device [7, 15].

9.5
Gamma Dose Rate and Detection of Gamma Radiation

Measurement of the gamma dose rate in a radiation field is the most important measurement task in radiological and nuclear events. But often the term measurement is used imprudently and mixed up with the action that is better described by the term detection. Actually, the term "measurement" with respect to a dose rate of ionizing radiation describes the preferably accurate metrological determination of the ambient dose equivalent rate $dH^*(10)/dt$ according to the present legal physical dose quantity. When a dose rate (e.g., in μSv h^{-1}) is displayed on a dose-rate meter while searching for a radioactive source in theater on foot, on vehicle, or on helicopter it is generally not an accurate metrological determination of the real existing radiological situation – and in that sense no "real" measurement. In the latter case one should better talk of quantitative detection. The next sub-section explains the reasons for this differentiation in more detail, because it is fundamental in understanding how to make the right choice of appropriate dose rate detector equipment.

9.5.1
Metrological Dose Rate Measurements

The detector efficiency for commonly used gamma dose-rate meters – even in radiation protection applications – is in the range of a few counts per second (cps) per µSv h^{-1} dose rate. For example, the up-to-date RADIAC SVG2 of the German Armed Forces has an efficiency of about 0.5 cps for the low dose rate semiconductor detector. At a typical background radiation of 0.07 µSv h^{-1} one obtains a count rate of 0.035 cps. This count rate corresponds to an averaged time of 29 s between two consecutive detector pulses. To obtain a statistically reliable dose rate value the mean value has to be calculated by integrating over a certain time interval. To obtain a statistical accuracy of, for example, 20%, the integration interval would be 714 s, which is nearly 12 min. In this period of time the detector should be at the same place in a steady radiation field. At a dose rate of 10 µSv h^{-1}, the time interval reduces to 5 s for an accuracy of 20%. Most modern gamma dose-rate meters are designed with an automatic adaption of the integration time. Using that technique it is possible to detect a sharp rise in dose rate relatively quickly even with relatively insensitive gamma dose rate detectors, demanding a large integration time at low dose rates where the radiation field has to be steady and the detector and the radiation source are not allowed to move.

9.5.2
Energy Response of a Dose-Rate Detector

A gamma dose-rate meter for use in radiation protection scenarios has to measure the ambient dose equivalent rate $dH^*(10)/dt$, which describes the biologically assessed energy deposition of the existing gamma radiation field in human soft tissue. In practice, the detector is calibrated for a distinct photon energy. The 662 keV gamma radiation of ^{137}Cs is widely used in calibration for radiation protection purposes. The energy dependence of the detector material is often quite different from the energy dependence of the fluence-to-dose conversion factors for $H^*(10)$. This fact is mainly noticeable in the high over-response in the low energy regime. Therefore, radiation detectors for accurate dose rate measurements generally have to be energy filtered. For cylindrical radiation counting tubes this is usually done with well selected shielding elements made from metal foils. For semiconductor detector elements, metal filters in the form of spherical calottes are used. Such filtering elements are crucial because they determine the licensed energy range and angular range of the detector. In practice a relative detector efficiency of ±40% is allowed for dose-rate meters in legal applications. As can be clearly seen in Figure 9.20 the lower energy threshold is about 20 keV for the Geiger–Müller tube and about 50 keV for the semiconductor type of radiation detector on applying the above-mentioned ±40% deviation in efficiency. In this example a ^{137}Cs radioactive source with its characteristic gamma energy of 662 keV was used, as can be easily seen from the common relative efficiency value of "1" at the mentioned energy. Last but not least, the energy dependence for the high-energy regime is different between the two detector types. While the semiconductor shows a decreasing gradient, the Geiger–Müller tube shows an increasing gradient in the high-energy regime.

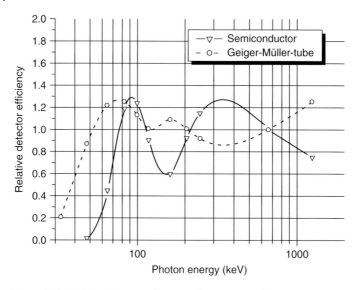

Figure 9.20 Relative detector efficiency of two gamma dose-rate meters, one based on a Si-PIN semiconductor detector and the other based on a GM-tube with appropriate energy filtering.

9.5.3
Quantitative Detection

Regarding the sensitive detection of single radioactive sources in theater during (fast) motion, the requirements placed on a radiation detector are quite different. The focus is not on a reliable metrological dose rate value with good statistics but on the rapid detection of a significant increase in the ambient gamma radiation field. For this purpose – as mentioned above – radiation detectors with extremely high efficiency and therefore with large detector volume are necessary. The integration time of the device also has to be set to a minimum value, so that changes of the dose rate can be displayed and processed immediately. For the common detection systems an integration time of 1 s has been established and large volume NaI(Tl) or plastic scintillators are used.

Compared to the detector efficiency of 0.5 cps per μSv h^{-1} mentioned for ordinary dose-rate meters such high efficiency detectors attain values of 2500 cps per μSv h^{-1} for a 0.75-l plastic scintillator used in the natural background rejection (NBR)[11] probe of the SVG2 kit or even about 20 000 cps per μSv h^{-1} for a 5-l plastic scintillator. Especially, the latter 5-l plastic scintillation detector is used

---

11) The NBR probe of the SVG2 kit is a high efficiency gamma radiation probe based on a 0.75-l plastic scintillator that can be used together with the German Armed Forces SVG2 basic device as an external probe. NBR (natural background rejection) allows the user to check if the energy distribution of the gamma radiation deviates from typical conditions of natural background radiation. This feature improves the ability to detect artificial radioactive sources.

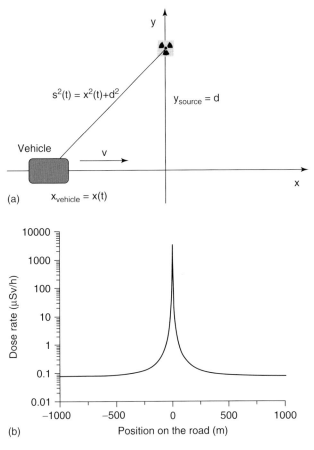

Figure 9.21 Vehicle-based detection of a point source: (a) geometrical arrangement of vehicle path and source location; (b) dose rate profile along the reconnaissance path caused by a 37 GBq ^{137}Cs radioactive source positioned 1 m from the road.

in stationary portal monitor systems or in special mobile detection systems that are intended to be used in high-performance radiological detection vehicles or helicopters [21].

Figure 9.21a shows a scenario with a radioactive source positioned at a distance d from the road. Imagine a reconnaissance vehicle driving along the road in positive x-direction with a constant velocity v. Figure 9.21b shows the resulting dose rate produced by a 37 GBq ^{137}Cs source positioned 1 m from the road ($d = 1$ m) at coordinate $x = 0$. The source produces a relatively sharp dose rate peak in the vicinity of the source. About 200 m away from the source (on the road at $x = -200$ m, $y = 0$) the dose rate is not significantly higher than the natural background radiation of about 0.1 µSv h^{-1}. The dose rate at $x = 0$ directly beside the source at 1 m distance gives a relatively high dose rate of about 3200 µSv h^{-1}. If you wanted

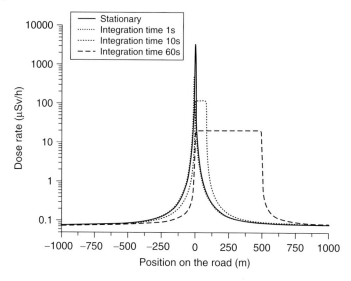

Figure 9.22 Displayed dose rate on a dose-rate meter with fixed integration times (1, 10, and 60 s) while passing the radioactive ^{137}Cs source at 30 km h^{-1}.

to measure the exact dose rate with an ordinary dose-rate meter with standard efficiency you would have to wait at every measuring point until the detector is in equilibrium, which means until the dose rate dependant integration time is over. This may be up to a few minutes depending on the type of dose-rate meter and the actual dose rate value. Assuming a fixed integration time for a dose-rate meter, Figure 9.22 shows the effect of passing a point source with a velocity of 30 km h^{-1} in a vehicle. The continuous line shows again the real dose rate along the path. Even with an extremely short integration time of 1 s, the measured dose rate peak value is only 560 µSv h^{-1} and for the higher integration times of 10 and 60 s only 120 and 20 µSv h^{-1}, respectively. What can also be seen is the fact that with large integration times the response to a peak value in terms of the real dose rate is very bad. The dose-rate meter with an integration time of 60 s would display a dose rate of about 20 µSv h^{-1}, even at a distance of 500 m away from the source, where the real dose rate is in the range of natural background radiation again.

This example scenario shows the need for short integration times for efficient detection purposes for gamma (and neutron) radiation fields. The velocity of the vehicle was only 30 km h^{-1}. The mentioned effects become even more important when the relative velocity between radioactive source and detector is higher, as, for example, in airborne detection with a helicopter.

To get a sufficient number of pulses during integration time, in particular at low dose rates, high efficiency, and therefore large volume, detectors have to be chosen. But the larger the detector volume the more difficult it is to fulfill the requirements for energy compensation, dose rate, and angular dependence. The dose rate value displayed by devices with those high efficiency probes generally have a lower quality and are therefore often not licensed for radiation protection issues. Therefore, it is recommended to use the term "quantitative detection" rather than "dose rate measurement."

We now give a short overview of the applications that drive radiation detector research at present and in the future.

9.6
Conclusions and Outlook

It has been shown that qualified detection and measurement of ionizing radiation demand a profound knowledge of the fundamental concepts of radiation detection technology and health physics. The simplest task is certainly the measurement of the dose rate in a present gamma radiation field because appropriate radiation meters have already been built and calibrated to measure the ambient dose equivalent rate. But even in that case the choice of appropriate device concerning energy range, efficiency, and dose rate range has to correspond with the actual measuring task. If other types of radiation are involved, additional probes are necessary to obtain a comprehensive overview of the real radiological situation. A further more complicated analysis is to identify single radionuclides and to determine the quantity, that is, activity, of the different radionuclides representing the actual radioactive source. Also for this task, a variety of analysis equipment is on the market. While some years ago the latter equipment could only be found in stationary laboratories, systems have been increasingly simplified and miniaturized and are therefore able to be fielded for use in the military or civil defense theater.

In the case of a large-scale radiological or nuclear event it is imperative to quickly identify exposed or contaminated individuals for the purpose of medical intervention and contamination control and further to identify first responders who must be restricted from further exposure. In the absence of a personal physical dosimeter, in high dose scenarios individual exposure to radiation can be estimated using damage to the chromosomes of white blood cells. This method of dose estimation is called *biological dosimetry* [22]. However, we can define several areas that influence detector technology. Beside nuclear power industry and nuclear safeguard activities, homeland security is the newest application driver for improved and new ionizing radiation detection technology [16]. The detection of clandestine nuclear materials leads to the problem of distinguishing between the relatively weak emissions of the material being sought from the background radioactivity. To underpin the difficulty we should be aware that industrial, medical, and natural sources can obscure the sought-after material. Ongoing research is needed to improve and enhance X-ray scanners to produce radiographic 2D density maps of the content of containers [23]. It seems that the threat of radiological or nuclear terrorism will not be eliminated soon and, therefore, research into new detector systems will be ongoing.

References

1. Laughlin, J.P. (2005) The natural radiation environment, *Seventh International Symposium on Natural Radiation Environment (NRE-VII)*, vol. 7, Elsevier Science. ISBN: 978-0-0804-4137-5.
2. Shleien, B. (1998) *Handbook of Health Physics and Radiological Health*, 3rd edn, Lippincott Williams and Wilkins. ISBN: 978-0-6831-8334-4.
3. L'Annunziata, M.F. (2003) *Handbook of Radioactivity Analysis*, 2nd edn, Academic Press. ISBN: 978-0-1243-6603-9.
4. Planel, H. (2004) *Space and Life: An Introduction to Space Biology and Medicine*, 1st edn, CRC Press. ISBN: 978-0-4153-1759-7.
5. Hajek, M. (2010) Radiation exposure of space and aircrew. *Proceedings of Third European IPRA Congress, Helsinki, June 14–16, 2010*, STUK, Radiation and Nuclear Safety Authority, Finland. ISBN: 978-952-478-551-8. Available at http://www.irpa2010europe.com (accessed on 20 April 2012)
6. Lantos, P. (1993) *Radiat. Prot. Dosim.*, **48** (1), 27–32.
7. Knoll, G.F. (2010) *Radiation Detection and Measurement*, 4th edn, John Wiley & Sons, Inc., Hoboken. ISBN: 978-0-4701-3148-0.
8. Cember, H. and Johnson, T.A. (2008) *Introduction to Health Physics*, 4th edn, McGraw-Hill. ISBN: 978-0-0714-2308-3.
9. Greening, J.R. (1985) *Fundamentals of Radiation Dosimetry*, 2nd edn, Taylor & Francis. ISBN: 978-0-8527-4789-6.
10. Stabin, M.G. (2010) *Radiation Protection and Dosimetry: An Introduction to Health Physics*, 1st edn, Springer. ISBN: 978-1-4419-2391-2.
11. NATO Standardization Agency (NSA) (2004) Commander's Guide to Radiation Exposures in Non-Article 5 Crisis Response Operations, 2nd edn, STANAG 2473, NATO/PfP Unclassified.
12. Gilmore, G. (2008) *Practical Gamma-Ray Spectroscopy*, 2nd edn, John Wiley & Sons, Ltd, Chichester. ISBN: 978-0-470-86196-7.
13. Ross, H. (1991) *Liquid Scintillation Counting and Organic Scintillators*, 1st edn, CRC Press. ISBN: 978-0873712460.
14. Maiello, M. (2010) *Radioactive Air Sampling Methods*, 1st edn, CRC Press. ISBN: 978-0-8493-9717-2.
15. Attix, F.H. (1986) *Introduction to Radiological Physics and Radiation Dosimetry*, 1st edn, Wiley-VCH Verlag GmbH, Weinheim. ISBN: 978-0-4710-1146-0.
16. Wehe, D.K. (2006) *Nucl. Eng. Technol.*, **38** (4), 311–318.
17. Lecoq, P. et al. (2010) *Inorganic Scintillators for Detector Systems: Physical Principles and Crystal Engineering*, Springer. ISBN: 978-3-6420-6615-3.
18. Moses, W.W. (2002) *Nucl. Instrum. Methods Phys. Res., Sect. A*, **487**, 123–128.
19. Lutz, G. (2007) *Semiconductor Radiation Detectors: Device Physics*, Springer. ISBN: 978-3-5406-4859-8.
20. Kahilainen, J. (1996) *Radiat. Prot. Dosim.*, **66** (1–4), 459–462.
21. Philliou, T. (2006) Mobile radiation detection for military and civil defense. Presentation by THERMO Electron Corporation, http://gis.esri.com/library/userconf/proc04/docs/pap2225.pdf (accessed 20 April 2012.)
22. Blakely, W.F. (2008) Early biodosimetry response: recommendations for mass-casualty radiation accidents and terrorism. Presented at Proceedings of the 12th International Congress of the International Radiation Protection Association (IRPA 12), Buenos Aires, October 19–24, 2008.
23. Pruet, J. et al. (2005) *J. Appl. Phys.*, **97** (9), 094908–094908-10.

Part IV
Technologies for Physical Protection

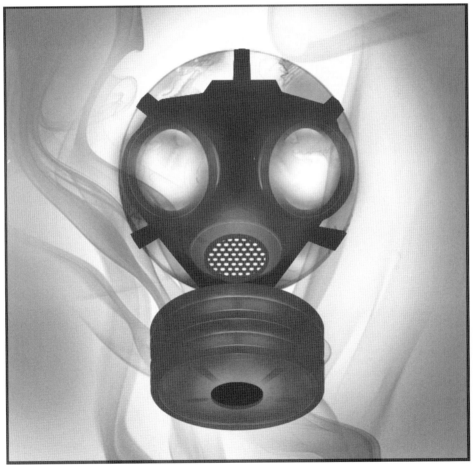

copyright by Jörg Pippirs, http://www.artesartwork.de

CBRN Protection: Managing the Threat of Chemical, Biological, Radioactive and Nuclear Weapons,
First Edition. Edited by A. Richardt, B. Hülseweh, B. Niemeyer, and F. Sabath.
© 2013 Wiley-VCH Verlag GmbH & Co. KGaA. Published 2013 by Wiley-VCH Verlag GmbH & Co. KGaA.

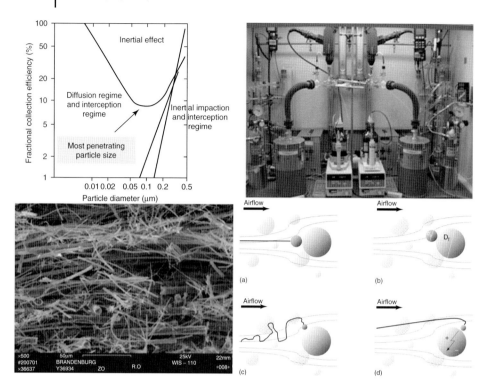

10
Filter Technology – Clean Air is Required
Andre Richardt and Thomas Dawert

The earliest recorded proposals and concepts for effective removal of toxic compounds can be dated in the fifteenth century, where Leonardo da Vinci envisioned a wet fine cloth for defense against a sulfide of arsenic and verdigris powder. However, World War I with the massive use of chemical warfare agents saw the beginning of intensive research to protect soldiers from chemical warfare agents, such as mustard gas and phosgene. Chemical impregnates that aid activated carbon to remove high-volatility vapors and non-polar contaminants were discovered and developed over subsequent decades to provide soldiers and first-time responders with state of the art masks and filter technology for protection not only from chemicals but also biological agents and nuclear particles. Nowadays we increasingly understand how effective filters can be designed against air released chemical, biological, and nuclear agents and how they can be implemented in an overall concept for physical protection.

10.1
Filters – Needed Technology Equipment for Collective and Individual Protection

We can accentuate that physical protection against a chemical, biological, radiological, and nuclear (CBRN) threat is nothing less than a part of employment protection and that filter technology is the key technology for effective individual and collective protection (COLPRO). The whole filtration process is a critical bottleneck in a strategy to design effective physical protection against CBRN threats. Therefore, we need to look at some aspects of air-filtration and air-cleaning systems before we discuss collective protection (COLPRO) and individual protection (IP) as parts of physical protection further. Properly designed, installed, and maintained air-filtration and air-cleaning systems can reduce the effects of a CBRN agent release, either outside or within a building, shelter, vehicle, or individual protection (IP) suit, by removing the contaminants from the air supply without having too much impact on platforms, buildings, or individuals (Figure 10.1).

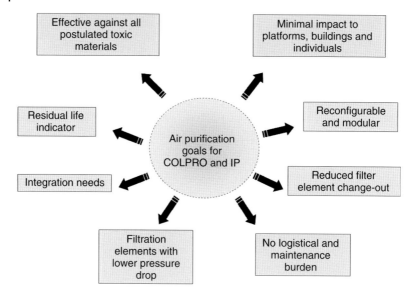

Figure 10.1 Air purification goals for collective protection (COLPRO) and individual protection (IP).

For platforms the avoidance of a CBRN contamination by entering a protected unit is necessary to avoid crosscontamination. To choose the right filter technology for effective physical protection from chemical, biological, or radiological attacks, we have to answer the following questions:

- What are the principles of air purification and what types of air-filtration and air-cleaning systems are effective for various CBRN agents?
- Are we able to implement these systems in existing ventilation systems?
- Can we incorporate these systems into buildings, vehicles, and so on?
- Do we know how to properly maintain the air-filtration and air-cleaning systems installed in our systems [1]?

Based on these questions, this chapter aims to provide an overview of air-filtration and air-cleaning principles. We will also cover some physical and physicochemical basics and how a filter can be evaluated. For further details we refer to additional literature and standard text books. Based on this information, we will try to answer the question of how to select the right filter technology.

10.2
General Considerations

We can assume that proper air filtration and air cleaning, combined with other protective measures, is able to reduce the risk and mitigate the consequences of a CBRN attack. It is common knowledge, that air filtration and air cleaning systems are able to remove various toxic contaminants from unit's airborne environment and that the filter design or air-cleaning media will depend upon the nature of the contaminant [23]. According to the literature, air filtration means the removal of toxic aerosol contaminants from the air, and air cleaning means the removal of gases or vapors from the air [2, 25]. As a general rule, filter based adsorbents collect only toxic gases and vapors, particulate filters remove aerosols. To choose the best adsorbent for the removal of a toxic contaminant depends on the knowledge of the characteristics of the toxic gas or vapor (see Chapter 3). Also, the size of the particulate filter in combination with the type of the filter has to be considered. Although larger sized aerosols can be collected on lower efficiency filters, it is of high importance that the effective removal of smaller sized and highly harmful aerosol requires high efficiency filters [23]. In addition to proper filter and adsorbent selection, several issues must be considered before installing or upgrading filtration systems:

1) **Technical aspects.** To purify a high volume of air it should be possible to use a pre-filtration unit to separate large particles such as fine sand and large inorganic debris. With this simple method it is possible to increase the life-time of the downstream filters. Furthermore, filter bypass avoids air filtration systems. Filter bypass occurs when air flows around the filter module, decreasing both collection and separation efficiency, and thus defeating the intended purpose of the filtration system. Poorly fitting filters, poor sealing of filters in their framing systems, missing filter panels, or leaks and openings in the air-handling unit between the filter bank and blower are reasons why filter bypass occurs (Figure 10.2).

Without addressing filter bypass we will provide little if any benefit by simply improving filter efficiency [3]. However, we have to point out that filter bypasses have a smaller relative influence on the filter loading rate for a lower than a particulate filter of a higher efficiency. This is important especially for scenarios

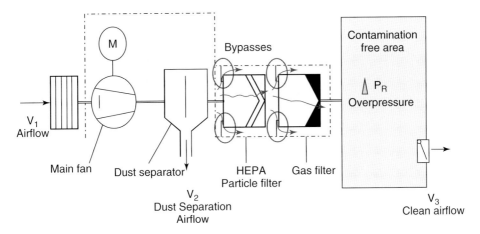

Figure 10.2 Schematic view of possible filter bypass.

in which higher concentrations of large diameter particles contact the filter. The reasons for this differing influence are:
a. if the filter is poor at removing particles, it matters less if some air bypasses the filter;
b. high efficiency filters are likely to have a larger pressure drop across them when compared with lower efficiency filters.

The increased pressure drop drives a larger volumetric flow rate of air through the same size bypass gap [3, 4].

2) The envelope of the protected unit matters. Filtration and air cleaning affect only the air that passes through the filtration and air-cleaning device, whether it is outdoor air, re-circulated air, or a mixture of the two. Therefore, system integration, the interfaces between the different sub-units, the overpressure, and the configuration of the locks are further important points, which will be discussed in Chapter 12.

3) Economic aspects are another issue that affects all filtration systems. Lifecycle cost should be considered (initial installation, replacement, operating, maintenance, etc.). High efficiency filters and adsorption filters are more expensive than the commonly used system filters. Also, fan units may need to be changed to handle the increased pressure drop associated with upgraded filtration systems.

We cannot expect filtration alone to protect, for example, critical infrastructure from an outdoor CBRN release. This is particularly so for systems in which no make-up air or inadequate overpressure is present. Instead, we must consider air filtration in combination with other steps, such as unit pressurization and envelope air tightness, to increase the likelihood that the air entering the unit actually passes through the filtration and air-cleaning systems [1]. For further details see Chapter 12.

CBRN agents may be transported in the air as an aerosol or a gas. Chemical warfare agents with relatively high vapor pressure are gaseous, while many other chemical warfare agents could potentially exist in both states. Biological and radiological agents are largely aerosols (Figure 10.3). The fact that chemical agents may exist in either state of aggregation requires a certain order of the filter elements – particulate filters must be located upstream of the vapor filter so that the off-gassing contaminants can be removed by the downstream gas-phase filter.

10.3
What are the Principles for Filtration and Air-Cleaning?

We can simply state that filtration and air cleaning remove unwanted material from an air stream. The collection mechanisms for particulate filtration and air-cleaning systems are very different. With the following description of the principles governing filtration and air cleaning we will briefly provide an understanding of the

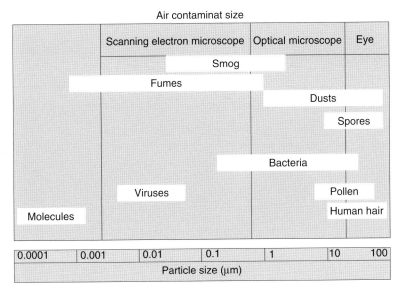

Figure 10.3 Common air contaminants and their sizes [5].

most important factors we should consider when selecting or enhancing filtration systems [6–8].

10.3.1
Particulate Filtration

For the purification of contaminated air different filter technologies have been developed. However, the most important filter types for CBRN protection are non-woven micro-fibrous filter mats made of a (poly-disperse) fiber glass mixture. We will therefore focus on this filter type. The fibers range in size from less than 1 μm to greater than 50 μm in diameter and the filter packing density ranges from 1% to 30%. High efficiency fibrous filters can be found worldwide today and make up a large percentage of the filters used. Under CBRN aspects we find them in applications like:

- respiratory protection (CBRN individual protection),
- clean rooms (CBRN collective protection),
- building ventilation systems (CBRN collective protection).

We can divide the most common filter technologies and media into:

- cyclone filters, depending on the size of the particles,
- fibrous filters (Figure 10.4),
- porous membrane filters (e.g., capillary porous membrane filters),
- fabric filters,
- electrical precipitator.

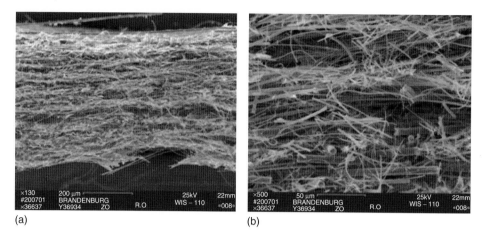

Figure 10.4 Scanning electron microscope (SEM) image of a cross-sectioned high-efficiency particulate air (HEPA) media, showing randomly oriented fibers: (a) 130-times magnified and (b) 500-times magnified.

It is a popular misunderstanding that fibrous filters behave like a sieve where particles above a certain size are trapped and smaller particles pass through. Porous membrane filters in liquids do function this way, but fibrous air filters work by actually trapping smaller and larger particles more effectively than mid-sized particles in a deep bed filtration manner. In general, a fibrous filter consists of a large number of randomly oriented fibers. These fibers form a dense material or mat that captures and retains particles throughout its depth or thickness (Figure 10.4).

The fiber material, filter media thickness, fiber diameter, and density of the mat enable fibrous filters to function. We can express filter performance in terms of penetration rate and efficiency rate:

$$\text{Penetration rate } (P)\colon P = \frac{\text{Particle downstream concentration}}{\text{Particle upstream concentration}} \times 100\% \quad (10.1)$$

$$\text{Efficiency rate } (E)\colon E = 100\% - P \quad (10.2)$$

It is important to understand that the efficiency of a fibrous filter varies for different particle sizes and flow rates. However, to specify the efficiency of a fibrous filter without also stating the pertinent particle size and size distribution, respectively, and flow makes no sense. The 'worst-case' particle size is called the most penetrating particle size (MPPS) – because these particles are more likely to get though the filter than any other size (Figure 10.5). Particles that are either smaller or larger than the MPPS have a lower tendency to penetrate. The MPPS is a function of the:

- structure of the filter media,
- velocity of the airflow through the filter,
- chemical and physical nature of the particles and filter media.

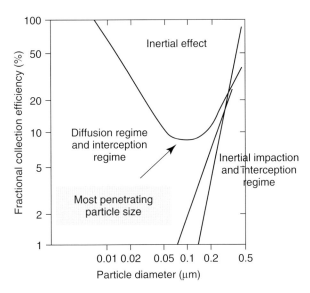

Figure 10.5 Schematic diagram: physical mechanisms of a collection efficiency curve. Fractional collection efficiency versus particle diameter for a mechanically operating filter with physical interactions on the surface. The minimum filter efficiency will shift, based upon the type of filter and flow velocity. Note: The dip for the most penetrating particle size (MPPS) and dominant collection mechanisms is based upon particle size [9].

Most penetrating particle size (MPPS): the size of the particles that achieve maximum penetration through the filter medium.

Various physical mechanisms contribute to a high efficiency fibrous filter's effectiveness in capturing particles. The most predominant are:

- inertial impaction,
- interception,
- diffusion,
- electrostatic attraction (Figure 10.6) [1, 10].

The main collection mechanisms of deep-bed mechanical filtration are impaction (Figure 10.6a), which occurs when a heavy particle traveling in the air stream and passing around a fiber deviates from the air stream due to mass inertia and collides with and adheres to the fiber.

Interception occurs when a large particle, because of its size, collides with a fiber in the filter that the air stream is passing through (Figure 10.6b).

Diffusion occurs when the random (Brownian) motion of a very small particle causes that particle to contact a fiber by chance (Figure 10.6c).

Currently, electrostatic attraction, the fourth mechanism, plays a minor role in mechanical CBRN filtration [1, 25]. Weak electrostatic forces are responsible for the attraction and smaller particles are retained on the fibers by this force [25] (Figure 10.6d). The fiber contact is usually made at the stream averted side of the

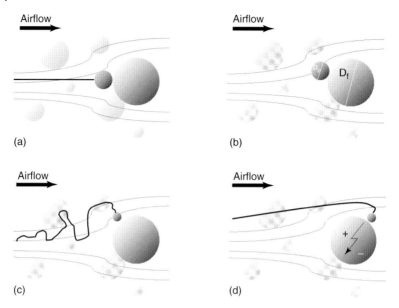

Figure 10.6 Four primary filter collection mechanisms: (a) inertial impaction, (b) interception, (c) diffusion, and (d) electrostatic attraction.

fabric. Smaller particles are retained on the fibers by a weak electrostatic force. However, the permanent electric currents can not be maintained under operational conditions for a longer period of time. Therefore it plays a minor role for CBRN filtration. By solving this issue, this could be an effective mechanism for particles the range of high micrometer to low micrometer scale.

The first three mechanisms discussed apply mainly to mechanical features. All the effects (Figure 10.6a–d) are specifically influenced by certain particle size ranges, and mass. Large particles above 0.4 µm diameter will be captured due to both the impaction and interception mechanisms. Particles in the size range of 0.1–0.4 µm are generally considered as the most penetrating and are captured by both the diffusion and interception filtration mechanisms. Small particles, below 0.1 µm in diameter, are captured by the diffusion mechanism. The various filtration mechanisms result in an efficiency versus particle size curve (the well known collection efficiency curve), which has an upside down bell shape (Figure 10.5). A fibrous filter is for many CBRN HEPA filter applications least effective at removing particles in the 0.1–0.4 µm particle diameter range. To test a filter for worst case situations, it should be tested with an aerosol at the most penetrating particle size region. As an example we will show the required filtration efficiency for high efficiency particulate air (HEPA) and ultra-low penetration air (ULPA) filters at the most penetrating particle size in accordance to the European Norm EN 1822-1 (Table 10.1).

At present electrostatic filters play a minor role for CBRN filter applications. They contain electrostatic charged fibers, which actually attract the particles to the fibers

Table 10.1 Filter class depending on filtration efficiency [11].

Filter class	Total		Local	
	Efficiency (%)	Penetration (%)	Efficiency (%)	Penetration (%)
E10	85	15	–	–
E11	95	5	–	–
E12	99.5	0.5	–	–
H13	99.95	0.05	99.75	0.25
H14	99.995	0.005	99.975	0.025
U15	99.9995	0.0005	99.9975	0.0025
U16	99.99995	0.00005	99.99975	0.00025
U17	99.999995	0.000005	99.9999	0.0001

and collect them by the electrostatic forces, in addition to retaining them by simple collision and subsequent adhesion. Electrostatic filters rely on charged fibers to dramatically increase collection efficiency for a given pressure drop across the filter [1].

10.3.2
Gas-Phase Air Cleaning

Most CBRN systems may be equipped with adsorbent filters, designed to remove pollutant gases and vapors from the environment. Adsorptive separation consists of the following mechanisms (Figure 10.7) [12]:

1) physical adsorption (physisorption).
2) chemical adsorption (chemisorption).

Physical adsorption results, for example, from the weaker van der Waals or hydrogen binding between a molecule of gas or vapor phase and the adsorbent's surface, and only the physical adsorption capacity will be regenerated. In addition chemical adsorption is characterized by strong interactions between a molecule to be adsorbed (adsorptive) and the solid surface, resulting from electrostatic interactions or hydrogen bonding (Figure 10.7). Solid adsorbents, varying from activated carbon, silica gel, activated alumina, zeolites, porous clay minerals, to molecular sieves, are useful because of their large internal surface area, stability, and low cost. The inorganic adsorbents, preferably functionalized silica supports can be regenerated safely by application of heat, dillution by inert gases, such as air stream, or other effects or other processes. Both capture mechanisms remove specific types of gas-phase contaminants from air. Understanding the precise removal mechanism for gases and vapors is often difficult due to the nature of the adsorbent and the processes involved. The knowledge of adsorption equilibrium (maximum capacity for adsorptive binding onto the adsorbent's surface) helps for understanding air treatment basically. Adsorbent performance additionally depend on such properties such as mass transfer, as well as adsorption rates. We can define some of the most important parameters of gas-phase air cleaning.

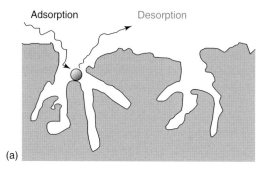

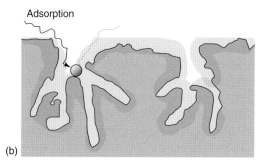

Figure 10.7 Two mechanisms for capturing and controlling gas-phase air contaminants [12]: (a) physical adsorption with possible desorption; (b) chemical adsorption with very low possibility of desorption.

Important parameters of gas-phase air cleaning [1]:

- **Breakthrough concentration:** The downstream contaminant concentration above which the sorbent is considered to be performing inadequately. Breakthrough concentration indicates the agent has broken through the sorbent, which is no longer giving the intended protection. This parameter is a function of the kind of adsorbent, its surface functionalization, kind of adsorptive, to be separated, the loading history, the actual relative humidity and temperature of the gas stream treated, air flow rate, quality of the fixed-bed arrangement, etc.
- **Breakthrough time:** the elapsed time between the initial contact of the toxic agent at a reported concentration on the upstream surface of the adsorbent bed and the time, when given breakthrough concentration on the downstream side of the adsorbent's bed is reached.
- **Challenge concentration:** the airborne concentration of the hazardous agent entering the adsorptive medium.
- **Residence time:** the length of time the hazardous agent spends in contact with the sorbent. This term is generally used in the context of superficial velocity, which is calculated by dividing the volumetric airflow rate by the filter surface area.

- **Mass transfer zone or critical bed depth:** Interchangeably used terms, which refer to the adsorbent bed depth required to reduce the chemical vapor challenge to the breakthrough concentration. Mass transfer describes the mechanisms (convection, different types of diffusion) transferring the vapor compounds from the bulk phase towards the adsortion sites onto the solid surface, including its inner surface, like pores. Mass transfer zone usually refers to the zone, where compounds are transported towards the adsorption sites, that means, where the adsorption-desorption equilibrium is not yet reached.

Chemical impregnates aid activated carbon [13] to remove high volatile vapors, which usually possess a low or medium molecular weight, and non or low polar contaminants to decompose them (Table 10.3 below). High-molecular chemicals – such as O-isopropyl methylphosphonofluoridate (GB), which is a nerve gas (sarin), and bis-(2-chloroethyl) sulfide (HD), which is a vesicant (sulfur mustard) and beyond that a low vapor-pressure chemical – are effectively removed by physical adsorption (Table 10.2). Reactive chemicals have been successfully impregnated into activated carbon to decompose chemically high-vapor pressure agents such as the blood agents cyanogen chloride (CK) and hydrogen cyanide (AC). One type of impregnated activated carbon, ASZM-triethylenediamine (TEDA) carbon, has been used in U.S. military nuclear, biological, and chemical (NBC) filters since 1993. This material is a coal-based activated carbon that has been impregnated with copper, zinc, silver, and molybdenum compounds, in addition to TEDA. ASZM-TEDA carbon provides a high level of protection against a wide range of chemical warfare agents. Table 10.3 provides a list of chemical impregnates and the air contaminants against which they are effective. For activated carbon, the possible mechanism of degradation for hazards is shown in Table 10.2.

The agent adsorption and decomposition by activated carbon can be enhanced by using chemical impregnates (Table 10.3). These impregnating agents are very useful in removing lightweight and highly volatile vapors and non-polar contaminants [1, 14].

Table 10.2 Mechanisms of agent vapor filtration by activated carbon [1].

Agent	Filtration mechanism
Nerve agents (e.g., sarin)	Strong physical adsorption, followed by slow hydrolysis
Blister agents (e.g., sulfur mustard)	Strong physical adsorption, followed by slow hydrolysis
Choking agents (e.g., phosgene)	Weak physical adsorption, followed by agent decomposition
Blood agents (e.g., arsine)	Weak physical adsorption, followed by agent decomposition

Table 10.3 Chemical contaminants and useful chemical impregnates [14, 15].

Chemical contaminant	Impregnate
Acid gases, carbon disulfide	Potassium carbonate
Aldehydes	Manganese(IV) oxide
Ammonia	Phosphoric acid
Arsine, phosphine	Silver
Hydrogen cyanide	Zinc oxide
Hydrogen mercaptans	Iron oxide
Hydrogen sulfide	Iron oxide, potassium permanganate
Mercury	Sulfur
Phosgene, chlorine, and arsine	Copper/silver salts
Radioactive methyl iodide	Zinc oxide

In chemical adsorption the gas or vapor molecules react with the adsorbent material or with reactive agents impregnated into the adsorber matter, which undergoes chemical conversion with the contaminant or converts it into more benign chemical compounds. Potassium permanganate is a common chemisorbent, impregnated into an alumina or silica substrate and used to oxidize formaldehyde into water and carbon dioxide. Other more complex reactions bind the contaminants to the adsorbent substrate, where they are chemically altered. Chemical adsorption is usually slower than physical adsorption and is normally irreversible. Several very toxic vapors (e.g., AC) are not retained on activated carbon by physical adsorption due to their high volatility and weak bonding forces to the carbons surface. The traditional approach to provide protection against such materials is to impregnate the adsorbent material with a reactive component to decompose the vapor (Table 10.3). Usually, the vapor is converted into an acid gas byproduct, which must also be removed by reaction with adsorbent impregnation, which may lose reactivity over time. Weathering of the impregnate is a particular concern for blood agents, such as AC and CK.

10.4
Test Methods

Several methods for the testing of particle and gas filter are available. However, we will not describe the different test methods in deep detail for particle and gas filter testing. Before we discuss some filter testing standards, we have to refine the different applications for filters into coarse dust, fine dust, and suspended particles filters (Table 10.4).

Table 10.4 Applications for different filter types from different norms.

Applications for coarse dust filters

	G1–G3, MERV 1–6	G3–F5, MERV 6–12	F5, MERV 12
General	Less effective against smoke	Limited separation of pollen. Very low effects against soot and smoke	Separation of pollen. Low effects against soot and smoke
Special application	Low requirements for purification of air. Pre filtration	Pre filter located upstream of fine dust filter	Pre filter located upstream of fine dust filter and protection buildings

Applications for fine dust filters

	F5–F7, MERV 12–13	F7–F9, MERV 13–15	E10, MERV 16
General	Separation of pollen. Low effects against soot and oil dust	Effective against all kinds of dust. In parts effective against tobacco smoke	Very effective for particles, for example, soot and oil dust. Effective against germs
Special application	Additional air for groceries	Air conditioning systems for laboratories	Additional air for radiological laboratories. Air conditioning systems for laboratories

Applications for suspended particle filters

	E10–E12	E12–H14	H14–U17
General	Very effective against germs, radioactive dust, and all kind of smoke and aerosols	Very effective against germs, radioactive dust, and all kind of smoke and aerosols	For special cases
Special application	Additional air for nuclear power stations. Special laboratories	Exhaust air for nuclear technical stations. Separation of oil dust	Sterile bacteriological laboratories. Nuclear power stations. Filter for suspended particles in gas defense facilities

10.4.1
Particle filter testing methods

To fulfill the different applications, several filter testing standards have been developed and it has become very difficult and confusing to try to work through this maze of different standards.

1) EN 779,
2) DIN EN 1822,
3) Eurovent Class 3928,
4) ASHRAE 52.1 and 52.2 (American Society of Heating, Refrigerating, and Air Conditioning Engineers).

Table 10.5 gives an overview of the most important particulate filtration test standards.

Table 10.5 Some important test standards for particulate filtration tests.

Type	ASHRAE 52.1	ASHRAE 52.2	DIN-EN 779	DIN-EN 1822	Eurovent class
Coarse dust filter	<65% arrestance (a)	MERV 1	G1	–	EU1
	65–70% a	MERV 2	G2		EU2
	70–75% a	MERV 3			
	75–80% a	MERV 4			
	80–85% a	MERV 5	G3		EU3
	85–90% a	MERV 6			
	>90% a	MERV 7	G4		EU4
		MERV 8			
Fine dust filter	>95% a	MERV 9	F5		EU5
		MERV 10			
	>98% a	MERV 11	F6		EU6
		MERV 12			
	>98% a	MERV 13	F7		EU7
		MERV 14	F8		EU8
		MERV 15	F9		EU9
High efficiency particulate air filter (HEPA)	85% DOP	MERV 16	–	E10	EU10
	95% DOP	–	–	E11	EU11
	99.5% DOP	–	–	E12	EU12
	99.95% DOP	–	–	H13	EU13
	99.995% DOP	–	–	H14	EU14
Ultra-low penetration air filter (ULPA)	99.9995% DOP	–	–	U15	EU15
	99.99995% DOP	–	–	U16	EU16
	99.999995% DOP	–	–	U17	EU17

Table 10.6 Test substances for filter testing.

Test substance	Particle size (μm)
Paraffin	<1 (61%)
NaCl	0.3–0.6
Di-Octyl Phthalate (DOP)	0.3
Diethylhexyl sebacate (DEHS)	0.1–0.3
Silica dust	0.2 (90%)

The evaluation of filter performance is an important aspect for their safe application and operation. In order to proof the filters, different substances can be used (Table 10.6). However, the particle size (Dp) depends on different parameters such as viscosity, concentration, or temperature. Therefore, is it necessary to test the system under defined and reproducible conditions.

In Europe, DIN EN 779 is applied for testing and classification of coarse and fine dust filter elements and for the HEPA and ULPA grade filters DIN EN 1822 is probably the best available technology. However, notably, several other test standards are available.

10.4.2
Gasfilter tests

Filter replacement schedules have been developed, based on measurements of cyanogen chloride (CK) and hydrogen cyanide (AK) breakthrough time as a function of environmental conditions, including the most unfavorable (hot and humid conditions). Figure 10.8 depicts a typical breakthrough curve for CK

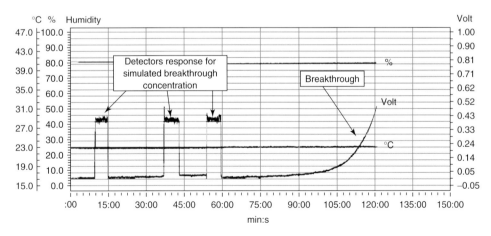

Figure 10.8 Calibration of the breakthrough concentration during cyanogen chloride (CK) agent challenge for subsequent automatic breakthrough identification.

with a pre-humidified filter type at 23 °C and 80% RH (relative humidity). In this example a gasfilter element has been challenged with a constant concentration cyanogen chloride (CK) until breakthrough was detected. The detector, responsible for measuring the breakthrough, was calibrated during the challenge. The results, in terms of breakthrough time, has been compared to the filter specifications.

10.5
Selection Process for CBRN Filters

Based on knowledge of the filter principles and the results of the filter test methods we are able to think about which CBRN filter is the best one in a given scenario [16]. Importantly, before selecting a filtration and air-cleaning strategy that includes a potential upgrade in response to perceived types of threats, an understanding of our system to be protected and its CBRN system has to be developed. Initially, we have to answer several questions (Figure 10.9):

1) Which types of air contaminants have to be separated from the air?
2) How long must the CBRN filter hold on?
3) Which amount and concentration of CBRN particles can be expected?
4) Which resistance of the filter system subject to the respective application is acceptable?
5) Where must the filter work and what climate conditions have to be considered?
6) Are the needed energy resources available (COLPRO)?
7) How clean does the air need to be for the occupants, and how much money and efforts can be spent to achieve that desired level of air cleanliness? What are the total costs and benefits associated with the various levels of filtration?
8) What are the minimum airflow needs for the unit? [17–19] (Figure 10.9).

The answers to these questions could be a guide in making appropriate decisions about what types of filters and/or adsorbents should be installed in the needed CBRN system, how efficient those filters and/or adsorbents must be, and what procedures we should develop to maintain the system. Because of the wide range of systems to be protected and CBRN system designs, no single off-the-shelf system can be installed in all units to protect against all CBRN agents. Some system components could possibly be used in a large number of units; however, these systems should be designed on a case-by-case basis for each unit and application.

It is important to recognize that improving physical protection is not an all or nothing question. Because many CBRN agents are extremely toxic, high contaminant removal efficiencies are needed; however, many complex factors can influence the human impact of a CBRN release (i.e., agent toxicity, physical and chemical properties, concentration, wind conditions, means of delivery, release location, etc.). Incremental improvements of the removal efficiency of a filtration or

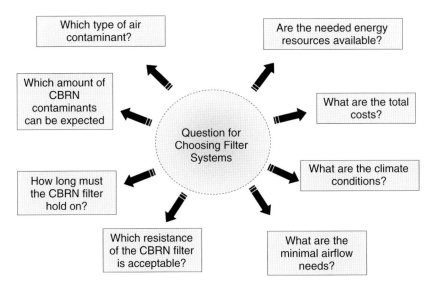

Figure 10.9 Important questions for choosing the right filter system for collective protection (COLPRO) and for individual protection (IP).

air-cleaning system are likely to reduce the impact of a CBRN attack to a protected area and its occupants while generally improving indoor air quality.

If we are able to answer these questions properly we should be able to select the best filter for the given situation. We have to emphasize that CBRN filters are critical system components.

Filter trains (see Chapter 12) often consist of two or more sets of filters; therefore, it is important to consider how the entire filtration system will perform and not just a single filter. The outermost filters are coarse, low-efficiency filters (pre-filters), which remove large particles and debris while protecting the blowers and downstream higher class filters from clogging, and potentially protect other mechanical components of the ventilation system against damage. These relatively inexpensive pre-filters are not effective for removing sub-micrometer particles. Therefore, the performance of the additional downstream filters is critical. These may consist of single or multiple filters to remove sub-micrometer particles. Different filters can be compared for collection efficiency and particle size [20, 21]. In the example shown in Figure 10.10, particles in the 0.1 to 0.3 µm size range are the most difficult to remove from the air stream and require high-efficiency filters. For typical CBRN aerosol dispersions (particulates) in the 1–10 µm range, high-class HEPA filters provide high efficiencies greater than 99.9999% in that particle size range, assuming there is no leakage around the filter and no damage to the fragile pleated media. This high level of filtration efficiency provides protection against most aerosol threats. Chemical aerosols removed by particulate filters include tear gases and low volatility nerve agents, such as VX; however, a vapor

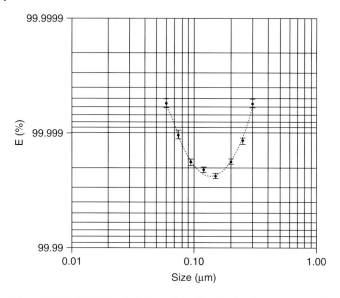

Figure 10.10 HEPA detailed view of the fractional collection curve of a CBRN-COLPRO filter in the particle size range 60–300 nm (test aerosol: paraffin, flow velocity through filter medium 2.5 cm s^{-1}).

component of these agents could still exist. Biological agents and radioactive particulates are also efficiently removed by HEPA filters.

10.6
Conclusions and Outlook

Filtration and air-cleaning systems may protect a unit and its occupants from the effects of a CBRN attack. Although it is impossible to completely eliminate the risk from an attack, for COLPRO the filtration and air-cleaning systems are important components of a comprehensive plan to reduce the consequences. CBRN agents can be removed effectively by properly designed, installed, and well-maintained filtration and air-cleaning systems. These systems have other benefits besides reducing clean-up costs and delays, should a CBRN event occur. These benefits include improving unit cleanliness, improving CBRN system efficiency, potentially preventing cases of respiratory infection, reducing exacerbations of asthma and allergies, and generally improving unit indoor air quality. Poor indoor air quality has also been associated with eye, nose, and throat irritation, headaches, dizziness, difficulty concentrating, and fatigue [22]. Initially, we must fully understand the design and operation of the critical infrastructure and CBRN system. Backed with that knowledge, along with an assessment of the current threat and the desired level of protection we want from our system, we can make an informed decision regarding our unit filtration and air-cleaning needs. In some situations, the existing system may be adequate, while in others major changes or improvements may be merited. To optimize effectiveness, we should minimize air infiltration and eliminate filter bypass. Maintenance plans and operations should ensure that the system works as intended for long periods. Life-cycle analysis will ensure that filtration and air-cleaning options satisfy our safety needs while providing protection to the protection system occupants.

For individual protection (IP), approaches are needed to reduce breathing resistance and to improve comfort of respiratory equipment. This includes improved aerosol filtration media and so-called low resistance respiratory valves. Furthermore, carbon-based filtration media, inorganic adsorbents, immobilized onto non-woven matter and improved sealing and blown air systems can increase the time where individual protection (IP) equipment can be worn.

Several new technologies are under investigation and/or have been developed to enhance or augment CBRN filtration systems. Many of these technologies have taken novel approaches to removing contaminants from the unit air stream, e.g. applying selective ligands onto functionalized inorganic adsorbents [24]. New catalytic filtration systems, pressure swing adsorption, and pressure–temperature swing are on the brink of coming into service in the very near future. While some of these new systems may be highly effective, many are unproven. Therefore, before we commit to one of these new technologies for the protection of units and its occupants, these systems have to be tested thoroughly.

References

1. NIOSH (2003) Guidance for Filtration and Air-Cleaning Systems to Protect Building Environments from Airborne Chemical, Biological, or Radiological Attacks, Department of Health and Human Services Centers for Disease Control and Prevention, NIOSH, - National Institute for Occupational Safety and Health, DHHS (NIOSH), Publication No. 2003-136.
2. Dickenson, T.C. (1997) *Filters and Filtration Handbook*, 4th edn, Elsevier Advanced Technology, Oxford. ISBN: 1-85617-322-4.
3. Waring, M.S. and Siegel, J.A. (2008) *Indoor Air*, **18** (3), 209–224.
4. Ward, M. and Siegel, J.A. (2005) *ASHRAE Trans.*, **111**, 1091–1100.
5. Hinds W.C. (1982) *Aerosol Technology*, John Wiley & Sons, Inc., New York, pp. 172–182. ISBN-0 471 08726.
6. NAFA (2001) *Guide to Air Filtration*, National Air Filtration Association, Washington, DC.
7. ASHRAE (2000) *ASHARE Handbook: HVAC Systems and Equipment*, American Society of Heating, Refrigerating, and Air-Conditioning Engineers, Inc., Atlanta, GA.
8. Gail, L. and Hortig, H.P. (2004) *Reinraumtechnik*, Springer. ISBN: 3-540-20542-X.
9. Lee, K.W. and Liu, B.Y.H. (1982) *Aerosol Sci. Technol.*, **1**, 147–161.
10. Hutten, I.M. (2007) *Handbook of Nonwoven Filter Media*, Elsevier Science & Technology. ISBN-10:1-85617-441-7.
11. DIN (2011) EN 1822-1:2009. *High Efficiency Air Filters (EPA, HEPA and ULPA) – Part 1: Classification, Performance Testing, Marking*, Austrian Standards Institute, Vienna.
12. Karge, H.G. and Weitkamp, J. (2008) *Adsorption and Diffusion. Molecular Sieves*, Science and Technology, vol. 7, Springer. ISBN-10:3-540-73965-3.
13. Lodewyckx, P. (2006) Adsorption of chemical warfare agents in activated

carbon surfaces in environmental remediation, in *Interface Science and Technology* (ed. T.J. Bandosz), Elsevier, Oxford, ch. 12.

14. CBIAC (2001) Development and evaluation of CARC with Triosyn® additive, May Newsletter.

15. Bandosz, T.J. (2006) *Carbon Surfaces in Environmental Remediation*, Academic Press Inc. ISBN-10:0-12-370536-3.

16. Bollinger, N. (2004) NIOSH Respirator Selection Logic, DHHS (NIOSH) Publication No. 2005-100. Available at *http://www.cdc.gov/niosh/docs/2005-100/pdfs/05-100.pdf* (accessed 7 May 2012).

17. Kowalski, W.J. and Bahnfleth, W.P. (2002) MERV filter models for aerobiological applications. *Air Media* (Summer issue), 13–17.

18. Kowalski, W.J. and Bahnfleth, W.P. (2003) Immune building technology and bioterrorism defense. *HPAC Eng.*, **75** (1), 57–62.

19. Hitchcock, P.J. (2006) *Biosecur Bioterror: Biodefense Strategy, Pract., Sci.*, **4** (1), 1–15.

20. Hanley, J.T., Ensor, D.S., Smith, D.D., and Sparks, L.E. (1994) *Indoor Air*, **4**, 169–178.

21. Viner, A.S., Ramanathan, K., Hanley, J.T., Smith, D.D., Ensor, D.S., and Sparks, L.E. (1991) Air cleaners for indoor air pollution control, in *Indoor Air Pollution; Radon, Bioaerosols, and VOCs* (eds J.G. Kay, G.E. Keller, and J.F. Miller), Lewis Publishers.

22. Spengler, J.D. and Chen, Q. (2000) *Annu. Rev. Energy Environ.*, **25**, 567–601.

23. Preventative systems, Global Security, Alexandrai, Virginia, US, *http://www.globalsecurity.org/security/systems/prevention.htm* (accessed 10. May 2012).

24. Krogman, K.C., Lyon, K.F. and Hammond, P.T. (2008) *J Phys Chem B*, **112** (46), 14453–14460.

25. Gustin, J.F., (2005) *Bioterrorism: A Guide for Facility Manager*, The Fairmont Press, Inc, ch. 3, pp. 35–48, ISBN-0-88173-468-3.

11
Individual Protective Equipment – Do You Know What to Wear?
Karola Hagner and Friedrich Hesse

Immediately after the first use of chemical weapons during World War I efforts to cope with the threat of weapons of mass destruction were increased by all military organizations. The first step in the development of defensive equipment against chemical weapons were gas masks and – later, after the development of blister agents – impermeable clothing [1, 2]. The first gas masks were crude, heavy, and easy to break and more a sign to the troops that some protection during a gas attack was available. With the further development of chemical warfare agents, more sophisticated masks with a better protection were developed and introduced. After World Wars I and II research into new, more advanced technologies for individual protection equipment was intensified. Modern protective equipment provides high levels of protection and even considers egonomic requirements and is therefore better to wear than the heavy protective equipment known from World War I.

11.1
Basics of Individual Protection

Individual protection is the key to human survival in a contaminated environment. Its purpose is to create a safety zone around the human that cannot easily be penetrated by hazardous substances. Without the fundamental feature of individual protection, any mission imposing chemical, biological, radiological, and/or nuclear (CBRN) hazards cannot be fulfilled devoid of health impairment or death. The purpose of individual protective equipment (IPE) that we will address in this chapter is to protect human life against external hazards spread by accident, means of weapons of terrorists, or weapons of mass destruction (WMD). As toxic industrial materials (TIMs) coincide to a certain degree with chemical warfare agents protection against them is also partly provided [3]. Consequently, the items described are relevant not only for military personal but also for firefighters and homeland security and civil defence personal, as all of them face similar threats [4]. It is important to understand that universal endless protection cannot be expected from any protective equipment [5].

11.2
Which Challenges for Individual Protection Equipment (IPE) Can Be Identified?

CBRN hazards have to be kept away from the respiratory tract and the skin; therefore, generally an all-over protective shell is required.

CBRN IPE allows individuals to continue their tasks and to survive in contaminated environments or to safely handle contagious or toxic subjects. It seems logical that IPE needs to be adapted to the threat. However, threat analysis is a complex subject and, thus, so also is the identification of the IPE needed. The following questions and answers can give guidelines:

1) **In what area is the IPE employed?** The first discrimination is the area of usage. Indoor protective equipment may be integrated in a laboratory's infrastructure, especially to energy and clean air supply. The level of protection might be less severe if only small amounts of known contaminants are handled and if the contaminants can be worked on in flues. Field missions on the other hand require in most cases high-level stand-alone solutions, especially if the threat is unidentified and mission times are long. Protective equipment for

decontamination purposes has to protect against splashing water and the decontaminants as well.

2) **Is the hazard known?** A known hazard allows adaptation of the protective equipment to the toxic threat, which provides the highest effectiveness. If the hazard is unidentified, the chosen protective equipment should be able to protect against a broad variety of contaminants. Detection must be carried out.

3) **For how long does IPE protect?** The protection or retention time of the equipment depends on the materials used, the kind of contaminants, and the level and time of exposure. The protection time can vary from a few minutes to even days.

4) **How do I know IPE works?** Detection for the surveillance of the contaminated environment is essential. At the same time the analysis of the protection capability of the employed IPE is required. Detailed knowledge of the application and the limits of IPE must be learnt thoroughly by the operator in advance. Then the user can rely on the performance of the material as well as on their abilities to employ it properly. The user must be aware of the danger of degradation of the protection level of IPE during the appliance. IPE must not be used without this training as the protective abilities must be known to ensure safety during missions (Figure 11.1).

In the following we describe how to design, evaluate, and use IPE for CBRN protection. Descriptions of the most important items give details of the requirements, effects, and limits of use. The important issues of quality assurance and work place safety complete this review on IPE.

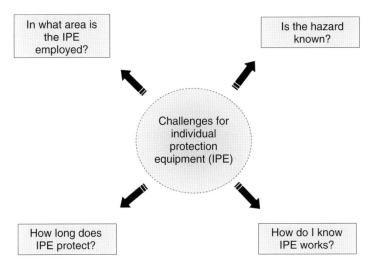

Figure 11.1 Challenges for individual protective equipment.

11.3
The Way to Design Individual Protective Equipment

Considering protective equipment, several different items are in co-action (Figure 11.2). The person inside the protective system is covered permanently with barrier layers to prevent any contact with contamination. The materials for the barrier layers are chosen according to the hazards. The way a protective system is assembled is commonly similar, regardless of the materials used. All items of a protective system should offer the same level of protection, as the weakest part governs the overall protection level. As a number of different items belongs to a protective system, their compatibility is crucial.

Going through the different contributing objects we will start with the face seal of the respirator. It must provide a tight interface to a broad range of face forms. It must also remain tight regardless of the movements or actions performed. This applies to all items: during CBRN missions, for example, in military use as well as fire brigades, the exposure of skin or any openings in the protective system are absolutely forbidden because of potential toxic skin effects. Hence resilient materials are used to prevent tears; extremely strained areas should be reinforced by additional material layers. To achieve a reliably enclosed system, protective suits are usually designed with generous fit and wide overlaps to cover areas such as trouser legs and boot legs or sleeves and gloves. To prevent the shifting of sleeves and legs, fixations such as pull cords, elastics, or Velcro® can be employed.

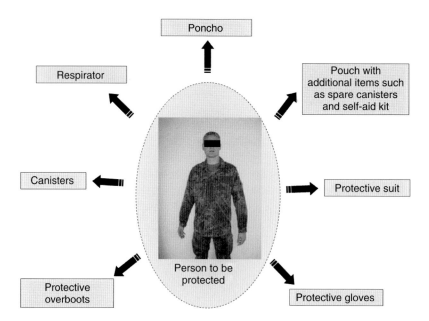

Figure 11.2 Overview of items needed for a modern permeable protective system.

Protective suits can be designed either as coveralls or as garments consisting of two pieces; three piece garments can also be found that have separate hoods. The advantage of a coverall is the reduced number of openings, as each opening is a possible entrance area. Two and three piece garments need a wide overlap of the single components (such as trousers and jacket) to provide permanent coverage, and supporting pull cords and/or elastics keep the garments in place. All openings need to be closed thoroughly. A protective jacket can be designed without a hood if head protection is realized separately.

If zippers are applied to close a protective suit, either the zipper should have the same protective features as the suit material or the zipper area needs to be covered with protective material to prevent hazardous substances from penetrating through the zipper into the suit. Depending on the suit type, impermeable gastight zippers can be utilized.

Permanent fixations are available to link boots and trouser legs or sleeves and gloves; this can often be found for impermeable suits. The additional items can be sewn and/or glued into the suit. The pre-assembled systems have disadvantages: damaged or non-fitting suit items cannot be exchanged easily. To achieve the right fit for each user, a large number of suit varieties is needed. Alternatively, non-permanent connections can be realized by circlips or clamping collars. Hoods must be designed to fit with the respirators used. If the hoods are not integrated into the suit or jacket, a separate head cover could be realized by using an additional hood with a shroud. To ensure the tight connection between hood and respirator, hoods can have either integrated seals made from rubber or pull cord or/and elastic closures. To ensure the compatibility of all items and to verify the suit design, visual controls are the first step to disclosing leakages. Gaps or openings should not be visible if extreme movements are performed with donned protective systems. Impartial evaluation methods of leak tightness control and design efficiency of the protective system are helpful. System tests can be carried out with either robotics or volunteers.

11.4 Function

The safety zone that IPE creates around the human body to prevent entry of hazardous substances is realized by various barrier layers; their composition depends on the required function. Barrier layers used for protective equipment can be permeable, selectively permeable, or impermeable. A combination of different barrier types is possible. Because chemical agents tend to penetrate most materials if the contact time is long enough, knowledge of the function of the different barrier layers is vital:

- **Permeable barriers (Figure 11.3a)**: these are textile based filtration laminates, made of air-permeable materials with integrated adsorptive layers that today usually consist of activated carbon. These materials are especially designed for protection against chemical warfare agents by binding them onto their surface.

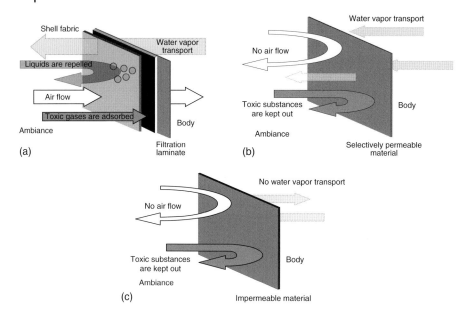

Figure 11.3 Different barrier layers: (a) permeable barrier layer; (b) selectively permeable barrier layer; (c) impermeable barrier layer.

Permeable barriers are used for different types of protective equipment such as overgarments, undergarments, combat suits, gloves, and/or socks.
- **Selectively permeable barriers (Figure 11.3b)**: These barriers contain membranes that are tight to air and toxic substances in all forms but allow water vapor molecules to pass through for sweat evaporation. Thus the membranes separate the toxic substances. Selectively permeable barriers are used for suits, ponchos, and/or pouches.
- **Impermeable barriers (Figure 11.3c)**: Materials completely tight to both air and toxic substances in all forms. Note also that the reverse penetration of sweat and body moisture does not occur. Impermeable barriers are used for respirators, boots, gloves, suits that protect against splashing waters, suits that protect against high-level contamination, ponchos, and/or pouches.

Depending on the barrier type used, CBRN agents can be adsorbed onto a specific adsorbent surface or mechanically separated. Therefore, after exposure the contamination remains either on the outside or on the adsorptive layers of the protective equipment. A substance dissolved in liquid or gaseous phases binds to the (inner) surface of a solid substance (called adsorbent) because of certain physicochemical interactions (see Figure 10.6). The adsorbent materials mostly have a large inner surface; selective adsorbents are also being developed (see Table 10.3).

The retention time of the protective equipment depends on the type and doses of the contamination, the level and time of exposure, the material composition, and the thickness of the barrier layer. Protective systems need to be chosen according to the mission requirements [5]. The retention time can vary from only a few minutes up to days or even weeks. Resulting from this, especially if in contact with an unknown hazard, is the need to be able to detect and identify the threat. Note that on continuous contact with CBRN agents the protective barrier layers might degrade or be penetrated, depending on the type and amount of the attacking substance. Especially, some chemical warfare agents exhibit corrosive features, and provide sufficient diffusivity if high contact time and amount are given [6].

Radiological and nuclear particles may stay on outside the barrier but ionizing radiation – especially β- and γ-radiation – penetrates the barrier layers. Effective mechanical separation can be achieved for α-radiation and biological hazards. However, protection can neither be expected to work universally nor last eternally but degrades with the degree and time of exposure of contamination.

11.5
Ergonomics – a Key Element for Individual Protection Equipment

Ergonomic aspects play a central role for IPE, to ensure its effective and safe use considering the limits of human performance (Figure 11.4).

> - "*Ergonomics* (or human factors) is the scientific discipline concerned with the understanding of interactions among humans and other elements of a system, and the profession that applies theory, principles, data, and methods to design in order to optimize human well-being and overall system performance." International Ergonomics Association [7].

A first ergonomic aspect to consider is the *sizing* of the different items. Protective equipment must be neither too small nor too big as it is meant to support survival in contaminated environments and not to hinder the wearer: wearing a well-fitting boot prevents the danger of stumbling and permits long missions, well-fitting (and non-slippery) gloves maintain tactility and allow operating devices, suits adapted to the body size reduce bulkiness and material weight and allow the reduction of insulating air-layers, and only a well-fitting respirator contributes to survival. Logistically, too many items, with each in various sizes, are not welcome: a compromise has to be found that contributes to the necessity of individually adapted equipment and the limits of logistic possibilities.

A second ergonomic aspect to consider is the *physiological burden*. The burden of using IPE originates from the breathing resistance of the canister, the additional weight of the protective equipment that needs to be carried, and the heat stress due to insulation of the protective barrier layers that surround the complete body.

One of the main challenges while wearing protective equipment is coping with heat stress. Sweat evaporation plays a central role in the body's own cooling

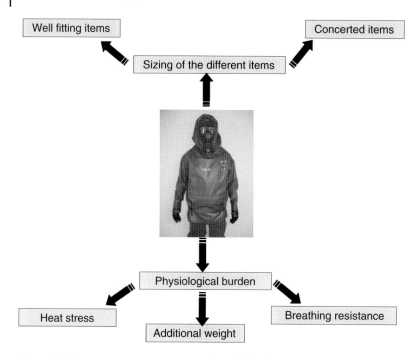

Figure 11.4 Important ergonomic aspects for individual protection equipment.

system, but while wearing IPE it is often limited or might even be completely impossible. The reduced or non-existent evaporation of sweat leads to an increase in the body's core temperature, which in the worst case can lead to a deadly heat stroke. Impermeable systems cause the highest heat stress; some limited relief can be expected from cooling systems such as blower units and cooling vests. Even permeable suits cannot be worn eternally as they interfere with ventilation and sweat evaporation, too. During a CBRN mission, the climate situation (hot as well as moderate temperatures) and the workload influence the sustainability and the health of the user due to the necessity to permanently cover the whole body with insulating protective materials [5, pp. 556–560]. Consequently, the management of work and rest time must be established [8]. Liquid intake is essential as the loss of fluid might be extreme under hot climate conditions and dehydration leads to severe health problems [9]. Military equipment is usually designed to meet these requirements; to prevent dehydration feeding valves are, increasingly, integrated into military respirators. Protective equipment needs to be adapted for the mission: anticipated threats need to be identified, at least roughly in type and concentration; otherwise it has to cover universal hazards. Overprotection leads to unnecessary heat stress and weight load.

> *Note:* The basic principle has to be "as much protection as necessary, the least burden possible [10]."

11.5 Ergonomics – a Key Element for Individual Protection Equipment | 303

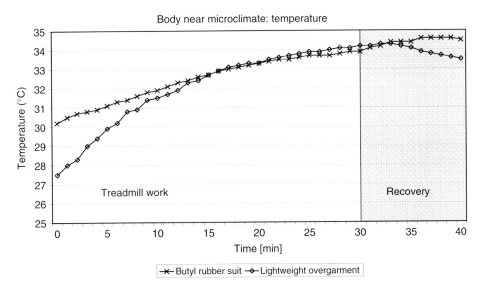

Figure 11.5 Body near microclimate investigations: Near skin temperature versus time. The whole body was covered with the indicated wear; heat transfer from inside the suit to the ambiance could only pass through the material (see Figure 11.6).

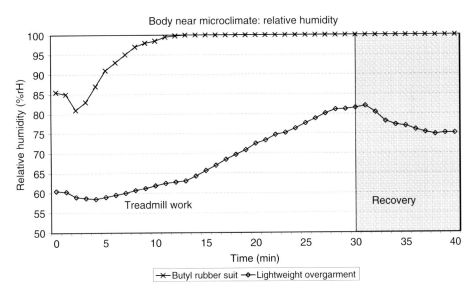

Figure 11.6 Body near microclimate investigations: Near skin relative humidity versus time. Inside the impermeable butyl rubber suit saturation was reached after about 12 min, while the permeable suit material allowed a humidity exchange with the dryer ambiance (see Figures 11.5–11.7) [13].

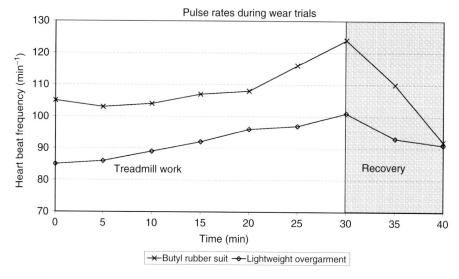

Figure 11.7 Body near microclimate investigations: Pulse rate during the wear trials versus time. The higher pulse rate at the start might result from the more complicated and time consuming donning procedure of an impermeable suit (see Figure 11.6).

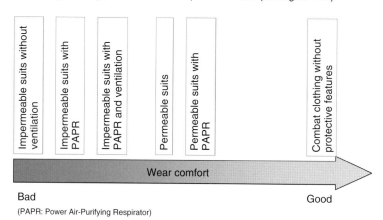

(PAPR: Power Air-Purifying Respirator)

Figure 11.8 General overview of wear comfort of protective clothing compared to non-protective clothing (diagram not to scale).

For the development of optimal protective wear, investigations can be conducted in a controlled atmosphere, representing a defined (here: a moderate[1]) climatic burden. The above diagramed wear trials included a 30 min work period and a 10 min recovery time. To simulate a medium-heavy workload, the volunteers walked without burden on a treadmill without slope at a velocity of 4 km h^{-1}.

---

1) The temperature was 25 °C and the relative humidity 50% rH. Radiant heat as emitted from the sun was not simulated but would additionally increase the burden when working outside.

Throughout, the temperature and relative humidity of the body near microclimate and the pulse rate were measured (Figures 11.5–11.7).

To interpret the results one must know that the average adult's heart rate at rest is about 70 beats per min; on average the maximum heart rate is 220 beats per min minus the person's age [11, 12].

A burden that should not be underestimated is the effect of breathing while wearing a respirator [5]. Breathing is more difficult as ambient air can only be inhaled by overcoming the canister's resistance, which means additional work; it contributes to the work load. Relief can be gained by using a blower support for the respirator [powered air-purifying respirator (PAPR)] (Figure 11.8).

11.6
Donning and Doffing – Training Is Required

The donning and doffing of IPE require training, as the time for donning must not be neglected as some situations might require the donning of IPE within a few minutes only. The right fit and accurate wearing are crucial for the tightness of system and safety during mission. Training in the use and knowledge of IPE are essential to ensure the user's confidence in the equipment, to prevent leakage, and to assure a fast pace, especially if the contact with contamination is unexpected. We want to emphasize that users of IPE need to be aware of the danger of leakage – potential leakage areas need special control.

Some military protective suits such as overgarments are designed especially to be donned by a single person only. Impermeable protective suits on the other hand very often need a second person to be donned correctly. During missions, the leak tightness of a CBRN suit can be controlled visually by using the four-eyes principle. One characteristic of the overgarment is the possibility of donning it in several steps adapted to the hazard until the full level of protection is reached. The purpose of the so-called "mission oriented protective postures" (MOPPs) is to keep the physiological burden for the utmost time as low as possible but to have the protective measurements at hand if required (Table 11.1).

The contamination hazard does not stop after a mission in a CBRN environment ends. Unless wearing contaminated IPE, the hazard remains and is even carried into clean areas. Therefore, special procedures for doffing contaminated IPE need to be established; it cannot be doffed like normal clothing. The doffing must follow a special procedure that needs to be validated and trained for as the risk of contamination during the removal of the protective equipment is high. The destruction and cutting off of contaminated suits should be considered as a fast and safe procedure: suits can be cut open and fall on the floor. Doffing must take place in a toxic-free area only but persons who help to remove contaminated IPE need to wear adequate protective equipment as well; at minimum a respirator with an adequate canister, gloves, and an apron.

Decontamination (and disinfection) might be time consuming: the decontamination time has to be considered carefully while planing the mission, especially for protective equipment with wear time restrictions. If blower units or self-contained

Table 11.1 Mission oriented protective posture (MOPP) levels and military equipment worn at the different levels [14].

MOPP level	Meaning	Equipment
MOPP 0	Knowledge that opponent has CBRN capabilities	Combat clothing (IPE accessible)
MOPP 1	Attack suspected	+ Overgarment (detection equipment, other protective items kept ready)
MOPP 2	Attack possible	+ Overboots (other protective items kept ready)
MOPP 3	Attack probable	+ Mask/hood (other protective items kept ready)
MOPP 4	Attack imminent	+ Gloves = Full level of protection

breathing apparatus (SCBA) are used, their period of application need to be considered as both the energy and the clean air supply are limited. The protective equipment's barrier must protect against the decontaminants as skin exposure to decontaminants might lead to health problems. It is important to remember that contaminated IPE is hazardous and has either to be decontaminated or disposed.

11.7
Overview of IPE Items – They Have to Act in Concert

IPE must be regarded as a system: several different items are employed together and they have to be compatible to ensure tightness and optimal protection. The requirement that a single CBRN IPE should offer universal protection to all substances cannot be fulfilled [5]. A threat analysis before choosing IPE is crucial. CBRN IPE is not suitable for all other toxic threats and is generally worn by military only or in exceptional civil CBRN situations by fire-fighters or home security. Military requirements for IPE take into account military threats and mission requirements such as high robustness, durability, ease of decontamination, possible reusability, and longer mission times than IPE designed for non-military use.

The following subsections introduce the most important items. A recommendation for special items cannot be given as it would depend on the mission and actual conditions; all pictures shown are meant to give examples for clarification. A multitude of items is available on the market and changes frequently as ongoing optimization is in progress. Manufacturers of IPE usually provide data sheets with the features of their items according to national and international civil IPE standards. Military standards, however, may contain classified information and are

Table 11.2 Levels of protection of civilian protective equipment according to NIOSH [15]. Each level has to be regarded as a system of respiratory and body protection [1].

Level	Degree of protection
A	Highest level of full protection: a self-contained breathing apparatus (SCBA) and a completely encapsulating chemical protective suit
B	Highest level of inhalation protection with lower level of body protection: SCBA and non-encapsulating chemical resistant garments, gloves, and boots, which guard against chemical splash exposures
C	Lower level of respiratory protection provided by an air-purifying respirator (APR), same level of body protection as level B: the hazard is known and the equipment offers sufficient protection. APR and non-encapsulated chemical-resistant clothing, gloves, and boots
D	No respiratory and only minimal skin protection: standard work garment without respirator

therefore not always available in the public domain. Different categories can be found to classify IPE as a system; commonly used are the NIOSH categories level A to D (Table 11.2), the European classification uses the categories 1a, 1b, 1c and 2 [10].

11.7.1
Respiratory Protection

Though in a CBRN scenario full body protection is necessary, respiratory protection is the most important safety measurement as the inhalation tract is extremely vulnerable. This fact originates from the huge surface of the lung, which is "25–50 times larger than the entire surface of the skin," [5, p. 551] and the fact that the lung tissue is due to its composition more sensitive to agent exposure than the normal skin – agent permeation in lung tissue is much faster than through skin [16]. Respiratory protection has to be chosen according to the hazard [17]. Modern respirators use canisters with different layers of impregnated activated carbon and particle filters to protect against a wide range of hazards. Respirators can only be used in atmospheres containing more than 17–19% of oxygen and if the dosage of the contaminant does not exceed the canister's capacity. Otherwise a SCBA must be used [10]. Respirators are manufactured in various styles. For CBRN use full-facepiece respirators, while for special tasks helmets are common.

- **Full-facepiece respirators** are adequate for general use as high protection factors (PFs) can be achieved. They offer a broad range of protection and can ideally be used with different canister types and a SCBA. They are easy to use and to maintain and do not require a special infrastructure.
- **CBRN helmets** are applied for special tasks only, such as jet pilots. Helmets are usually heavier, bulkier, and more complex than respirators, as very often

electronic devices for communication and protection are integrated into it, which require an energy supply. The clean air for breathing is, for example, provided via an oxygen mask, while eye protection is achieved separately by ventilation. If the regular operation mode fails, additional protective measures come into place. Helmets are highly specific and complex to employ.
- **Loose fitting facepieces or visors** need to be combined with a ventilation unit, otherwise the achieved protection factor (APF) is very low as the tightness of the system is not realized. Ventilation requires a blower unit with a reliable energy supply. The limited energy supply of batteries might lead to mission restrictions.
- **Half-face piece masks** are also not appropriate for missions under CBRN conditions as they leave parts of the face and the eyes unprotected. Simple *hoods* are also inappropriate as they are not adapted to their users faces but are designed to be "one-fits-all;" tightness can, for example, be achieved by a neck seal but they have limited operational capabilities. Special hoods are available for the integration of military respirators; they are required if the protective suits do not have integrated hoods – their function though is additional head protection against contaminated dust, insects, and particles but not the inhalation protection that is provided by a tight fitting full-facepiece respirator.

The level of protection achievable can be described with the PF. The higher this dimensionless number is the better the protection. It is frequently utilized for respirators but applies for protective clothing/systems as well. The PF required for a hazardous scenario depends on the contaminant. Manufacturers of protective items usually provide datasheets with information on the retention capabilities. If protection against a special or uncommon substance is required, extra tests should be carried out for reliable mission predictions. Referring to respirator protection, the PF indicates the proper fit of the face seal as well as the tightness of all other elements such as exhalation valve and canister seal. A high PF can only be achieved if the person wearing the respirator has a smooth and even skin, facial hair in the area of the face seal leads to leakages.

Minimum required PF: The minimum required PF differs for each contaminant. If a concentration of chlorine of 22.5 mg m^{-3} has been measured and the workplace or occupational exposure limit (OEL) is 1.5 mg m^{-3}, the minimum required PF is 15, resulting from the following example:

$$\text{Minimum required PF} = \frac{\text{Concentration at workplace}}{\text{Exposure limit}} = \frac{22.5 \frac{mg}{m}}{1.5 \frac{mg}{m}} = 15$$

OELs are set by competent national authorities or other relevant national institutions as limits for concentrations of hazardous compounds in workplace air.

APF: The APF can be evaluated by measuring the particle concentration in the ambient atmosphere outside ($n_{ambiance}$) the protective equipment divided by the respective figure inside the protective area (n_{inside}), for instance of a respirator.

A minimum particle concentration inside the protective equipment must be achieved to gain a result. If the value is measured in a static situation without movement, the result is likely to be better than a value measured in a dynamic environment, where movements might lead to leakages, for example, in a respirator's face seal. As an example:

$$APF = \frac{n_{ambiance}}{n_{inside}} = \frac{1250}{50} = 25$$

For the correct respirator protection, it is as important to select the right filter class to purify the ambient air and remove the toxic substances (see Chapter 10). The concentration of a substance a person is exposed to must always be regarded in combination with the exposure time as it plays a significant role concerning the effects a hazardous substance might have [18, 19]. The product of multiplying the concentration and exposure time is called the dosage.

Example: Chlorine has been released due to leakage in a pipe. A detector mounted in a distance of 1 m indicates a chlorine concentration of 20 mg m^{-3}. Repair of the leakage requires 10 min.

Questions:

1) How must the person repairing the pipe been protected?
   The dosage a person facing repairing the leakage will be:

   $$\text{Dosage} = 20 \frac{mg}{m^3} \times 10\,\text{min} = 200 \frac{mg \times min}{m^3}.$$

2) Which APF is required for the person repairing the pipe?

   The person must wear a proper fitted full-facepiece respirator with an adequate canister as chlorine has an immediate pulmonary effect and causes eye irritation or damage. According to the Regulations for Respiratory Protection of the United States Department of Labor a minimum APF of 50 can be achieved with the equipment (Table 11.3) and a canister appropriate for chlorine such as canister type B3 [20].
   With this equipment, the achieved dosage reduction is:

   $$\text{Dosage reduction} = \frac{200 \frac{mg \times min}{m^3}}{50} = 4 \frac{mg \times min}{m^3}$$

   The resulting concentration can be calculated by dividing the reduced dosage by the time:

   $$\text{Concentration} = \frac{4 \frac{mg \times min}{m^3}}{10\,\text{min}} = 0.4 \frac{mg}{m^3}$$

Table 11.3 Examples for protection factors of different full-facepiece respirators [20].

Respirator type	Assigned protection factor
Air-purifying full-facepiece respirator	50
Power air-purifying full-facepiece respirator	1000
Self-contained breathing apparatus (open/closed circuit) in positive pressure mode with full-facepiece respirator	10 000

Table 11.4 Acute exposure guideline levels (AEGLs) for chlorine [21] [see Chapter 6 (Table 6.1) for AEGLs].

	Chlorine^a exposure (mg m^{-3})				
	10 min	30 min	60 min	4 h	8 h
AEGL-1	1.45	1.45	1.45	1.45	1.45
AEGL-2	8.12	8.12	5.8	2.9	2.059
AEGL-3	145	81.2	58	29	20.59

^{a}Chemical abstract service (CAS) number 7782-50-5.

This concentration is below the NIOSH recommended exposure limit of 1.45 mg m^{-3} (for a 15 min ceiling) (Table 11.4) [21]. If the resulting concentration would have exceeded the recommended exposure limit, a higher PF must have been chosen: depending on the dosage expected either a PAPR or a SCBA would have to be applied.

11.7.2
Respirator Design

Respirator designs vary [5, 22, p. 551–556]. As the functionality is affected considerably by the design, some of the following statements should be considered carefully (Figure 11.9):

- **Position of the canister**: Respirators are available with different canister positions: (i) centric canister with centric connection, (ii) canister with eccentric connection, and (iii) centric side canister. A centric and low canister position is common but might lead to problems if the close approach of the face toward equipment is necessary. The canister might prevent this because it often increases the distance between face and equipment. The canister connection is usually beyond the field of view, colliding with an object holds the danger of dislocating the respirator, which could lead to leakages. A solution that allows further approach toward

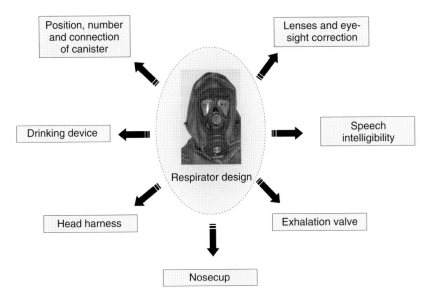

Figure 11.9 Parameters for respirator design.

equipment is an eccentric canister position that allows wearing of the canister either on the left- or right-hand side.

- **Number of canisters**: Some respirators provide the utilization of two canisters at the same time. Increasing the number of canisters reduces the breathing resistance but also increases the weight.
- **Exhalation valve**: It leads the exhaled air to the contaminated outside. As sweat and saliva usually accumulate inside the respirator during long extensive missions, the exhalation valve should allow them to be drained off. This is only possible if its position is very low on the facepiece.
- **Separated lenses or panoramic lens**: Military respirators often have separated lenses. The reason is their higher robustness and, additionally, they allow the respirator to be folded and stored in a pouch without destroying it. Separate lenses might suffer from the prejudice of offering a smaller effective field of view than a panoramic lens. Investigations have proven, though, that the field of view depends merely on the distance between the eyes and the lenses and on the integration of the lenses into the facepiece.
- **Eye-sight correction**: Glasses cannot be worn with a full-facepiece respirator as their bars would lead to leakages in the face seal. Special glasses with individual optical lenses need to be integrated inside the respirator (Figure 11.10).
- **Canister connection**: Canisters are often attached via a thread connection to the respirator. The widespread standard thread size for canisters is M40. Standardization allows the use of different canister types with the same respirator facepiece and exchange with filters from other sources, which might be applicable if the supply of canisters is lacking in an emergency situation. Non-standard

Figure 11.10 Respirator with integrated glasses.

connections exist such as bayonet joints with different diameters and depths.
- **Head harness**: The construction of the harness is important not only for the protection but also for the wear comfort of a respirator. The harness straps with adjustable buckles are attached to the facepiece; their purpose is to keep the respirator in position when worn. Individual adjustment should take place during a respirator fit test to verify the PF. The use of harnesses with long rubber straps is common but often leads to punctual pressure on the head that might over long periods of work provoke headaches. Rubber harnesses are also well-known for painfully tearing out hair. Alternatives are textile harnesses with large contact areas to the head, comparable to a skull cap.
- **Drinking device**: Modern military respirators provide incorporated feeding valves for fluid and nutrition intake under CBRN conditions. In the absence of a feeding valve, feeding is impossible without opening the protective barriers. The physiological burden caused by the CBRN protective equipment under all climatic conditions requires a high fluid intake to prevent dehydration or other heat related conditions [8].
- **Speech intelligibility**: Communication is important during missions, either by direct contact with persons or via radio. To guarantee high quality speech intelligibility, either passive speech membranes or active communication systems need to be integrated inside the respirator. Active systems require a reliable energy supply.
- **Nosecup**: The nosecup is a very important component that nowadays can be found in almost every respirator. It is an inner mask inside the respirator facepiece that covers mouth and nose. Its purpose is to reduce the volume inside the respirator and hence the volume to be filtered; thereby, the ventilation efficiency increases. The nosecup is designed in such a way that during inhalation the air stream is also guided over the lenses inside the respirator to prevent the misting of the lenses; additionally, it directs the warm and moist exhaled air directly to the exhalation valve. Missing or misfitting of nosecups leads not only to misting of the respirator, which impairs the view, but often leaves an unfiltered dead space volume inside the respirator, where the carbon dioxide produced during exhalation accumulates. Over time, the carbon dioxide accumulation leads to an

oxygen reduction, causing negative physiological effects such as performance impairment, headaches, or dizziness.

11.7.2.1 Air-Purifying Escape Respirator (APER) with CBRN Protection

Escape hoods should provide enough protection to persons to leave a hazardous situation within a short period of time (15–60 min). As they are designed to be employed by untrained personal, their usage needs to be easy. It is recommended to mark the escape hoods with labels indicating the approved hazards and the protection time [23].

The hood usually covers the head, neck, and in some cases even parts of the upper chest and is designed as one-size-fits-all item for adults. Special models of CBRN escape hoods are available for babies and children. The lung power of infants cannot be expected to be strong enough to overcome the breathing resistance of the canister; hence blower support is necessary for non-adult escape hoods. Escape hoods need to be pulled over the head only; some models require the additional adjustment of a nose cup and some fastenings. Problems with non-ventilated simple hoods are the danger of carbon dioxide accumulation inside the hood caused by misfitting or missing nosecups or the misting of the visor.

Misting inside the escape hood and the stress of breathing through a filter can be avoided if a forced airflow and overpressure is created by blowing filtered air inside the hood (Figure 11.11). Usually this leads to a higher PF as airborne particles cannot penetrate inside. Ventilation requires an integrated blower unit and a reliable energy source and leads to a rather bulky system, the use of which might be more complicated and the price of which is considerably higher than for non-ventilated systems.

Figure 11.11 Model of a ventilated escape hood. Reprinted with permission from Dräger Safety AG & Co. KGaA, Lübeck, Germany.

11.7.3
Air-Purifying Respirators (APRs) with Canisters for Ambient Air

Full-facepiece respirators with canisters for ambient air are most common for inhalation protection. In international armed forces usually every soldier is equipped with an air-purifying respirator (APR) that is used with military combination filters that provide a broad range of protection against CBRN hazards (Figure 11.12). Canisters always need to be chosen according to the expected hazard [2]. APR is adequate as long as the contamination does not exceed the canister's capacity and as long as enough oxygen is available in the ambient atmosphere.

Respirators are manufactured in different sizes and always require special adjustment to the individual's face and head to provide the ideal protection level.

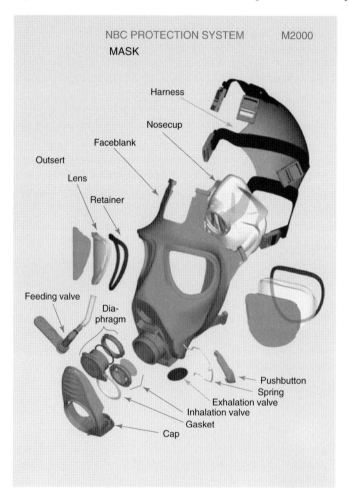

Figure 11.12 Components of military respirator M2000. Reprinted with permission from MSA Auer GmbH, Berlin.

Before a mission, each respirator should be tested to ensure the right fit: the PF must be measured. APR has several advantages for common use:

1) reliable,
2) low level of maintenance required,
3) easy to store and transport,
4) non-bulky,
5) robust,
6) lightweight.

The respirator facepiece material must possess the same protective capabilities as the rest of the protective equipment; especially, chemical warfare agents must not penetrate the material easily or blind the lenses. The retention time of the respirator depends on the materials used and on thickness of the materials. The user needs to be aware that chemical warfare agents will eventually penetrate the protective material. Typically, respirators are reusable; hence even without being contaminated, hygienic cleaning is necessary after utilization to remove sweat and saliva. All cleaning and decontamination procedures that are performed on the respirator need to safely remove contaminants without damaging it. A considerable number of CBRN APR is available on the market. Standardization requirements are published for APR in Reference [25]. Breathing while wearing a non-powered respirator is exhausting as the canister represents a resistance to the airflow; thus inhalation has to overcome a breathing resistance. This additional workload can be reduced by using PAPR because the filtered ambient air is provided from the ventilation unit, which will be discussed in the next section.

> *Note:* an unnecessarily high breathing resistance is a physiological burden and leads to exhaustion of the mask carrier.

11.7.3.1 Respirators with Blower Support (Powered Air-Purifying Respirator, PAPR)

A PAPR additionally provides a powered blower unit over APR. Each person carries their own ventilation unit, linked via a hose with the respirator, attached with a hip belt or as a rucksack. The number of canisters required varies with the different systems that are available on the market. Overpressure generated from the blown-in air reduces the intrusion of external contamination and increases the PF. A PAPR requires a high level of maintenance. Batteries deliver the energy supply of the blower units. An alarm should indicate critical battery states. Several models allow the user to independently breathe filtered ambient air even in the case of a blower failure though the breathing resistance increases significantly. Manufacturers have promoted mission times up to several hours for PAPR. They vary depending on the system chosen, as does the air flow. Standards have been established for CBRN PAPR from NIOSH [26]. PAPR is adequate as long as the contamination does not exceed the canister's capacity and as long as enough oxygen is available in the ambient atmosphere. The blower units are in most cases encapsulated systems, both CBRN hardened and liquid-tight to allow decontamination. Notably, the ventilation unit generates a noise level that might be disturbing, especially during a military mission.

11.7.3.2 Respirators with Self-contained Breathing Apparatus (SCBA)

A SCBA provides the highest degree of respiratory protection as the air for the user is provided from a compressed air bottle that needs to be carried around and not from the ambiance [27] [21N], but even a SCBA cannot be expected to offer universal protection as substances may permeate through the material used for the system; hence even a SCBA has to be chosen according to the hazard. Standards for CBRN SCBA have been established [28]. A CBRN mission with SCBA means that the person to be protected wears a full-facepiece respirator connected to a supply of compressed air that can either be delivered from an open circuit or closed circuit system. The most commonly employed type is the open circuit positive pressure SCBA [29]:

- **Open-circuit system**: Compressed air cylinders contain a limited supply of clean dry air and therefore possess a limited time of use, which depends mainly on the breathing rate and the breathing rate in turn depends on both the physical and physiological state of the person. Considerable variation concerning the mission time must be taken into account: a compressed air supply of 1600 l may be sufficient for a time period of approximately 20–50 min; hence the application of open-circuit systems is limited to nearby areas and short-time missions. Notably, the weight, depending on the type, might be up to 18 kg [10].
- **Closed-circuit systems**: A closed-circuit system contains an oxygen supply. The exhalation air is not released into the ambiance via an exhalation valve but is regenerated inside the system. The carbon dioxide of the exhalation air is bonded inside a regeneration cartridge, while oxygen from the systems' supply substitutes the consumed oxygen. Heat is generated due to chemical reactions inside the regeneration cartridge, which leads to high temperatures on the surface of the regeneration cartridge: the temperature of the inhalation air might increase up to $45\,°C$. Because of the heat generated, the close-circuit system can only be worn inside a protective suit if a very efficient heat removal exists, otherwise the temperature inside the suit would exceed the dangerous threshold for the user. Depending on the system utilized, mission times can reach up to 4 h [5]. Again, it should not be neglected that the weight, depending on type, might be up to 16 kg [10].

Both SCBA types require monitoring during missions to control the remaining time of use, to report any technical problems, and to initiate alerts. As the air supply is limited, missions need to be planned carefully as at the end of the air supply the equipment has to be doffed. After a mission, trained personal service the SCBA equipment to ensure its safe re-use. The application of SCBA requires a high level of maintenance from trained personal and can only be utilized by physically fit persons who, depending on national regulations, might need to acquire certain medical certificates. Sophisticated knowledge of the equipment and theoretical and practical training are required. Thus, SCBA teams are only small, highly specialized units with a clear task. National regulations determine the number of

SCBA missions per person per day. It must be taken into account that the extremely dry air provided by a SCBA contributes to dehydration [5].

11.7.4
Canisters

Different types of canisters are available. It is important to select the right filter to purify the ambient air and remove the toxic substances (refer to Chapter 10) [30]. The necessity of choosing a canister according to the hazard is not only founded on the basic separation characteristic of the filtration technology. The resulting increase in weight, filter resistance, and a bulky form if more and more capabilities need to be integrated into a single canister have to be considered. The negative consequences of a large canister might be a higher vulnerability to fouling and its intrusion in the field of view [31].

The canister weight is a neglected aspect. Depending on the sources used, the weight limit for a CBRN canister mounted on a respirator is up to 500 g. A high canister weight can lead to neck strain and further physiological stress. To prevent this, the canister can be connected to a hose and then attached on the back or on the load carrying system, though this might lead to problems with a dead space volume. Still, the weight has to be carried, though the location might be less interfering. Canisters need to be changed before depletion. Especially if dealing with CBRN hazards, a preventive canister exchange should be performed. The given time is based on the specific threat scenario so as to exclude any possibility of CBRN agent breaking through the protective barrier. Soldiers usually carry a spare canister in a pouch with them so as to be prepared to exchange the canister if necessary. Canister depletion can be recognized if:

1) smells and/or taste can be experienced while wearing a respirator with a gas filter;
2) the breathing resistance increases significantly while using a particle filter;
3) smells and/or taste can be experienced or the breathing resistance increases significantly while using a combination filter.

So far we have discussed the protection of the respiratory tract. The following subsection outlines skin protection.

11.7.5
Body Protection

The skin is the body's largest organ. Thus an attack on it has to be effectively countermeasured. Mustard gas (HD) was the first warfare agent developed especially for skin attack; for further chemicals refer to Chapter 3. The protective equipment therefore needs to be extended to cover the whole body, with special boots, gloves, coats, and suits that were first made of butyl rubber to achieve sufficient protection.

Especially when considering body protection, it has to be kept in mind that wearing personal protective equipment means creating a closed shell, comparable to an individual microcosm with its own climatic conditions, noise, and smells [5]. The closed "universe" prevents or slows down the penetration of toxic substances but leads to problems for the encapsulated individual because of the non-existent exchange with the environment: the missing heat and humidity exchange lead to a severe physical burden. Different categories (Table 11.5) and types can be found to classify suits, for example, according to the EU directive 89/686/EEC

Table 11.5 Types of category III protective clothing (such as chemical protective and contamination suits) and their meaning [32, 33].

Type	Meaning
1	Gas-tight clothing
2	Non gas-tight clothing
3	Liquid-tight connections
4	Spray-tight connections
5	Protection against airborne solid particulates
6	Limited protective performance against liquid chemicals

Some general information on body protection must be considered before any mission can start, such as:

1) What kind of protection is appropriate for the hazard? *A threat analysis needs to be made.*
2) Depending on wear time restrictions and the number of users estimated: *How many suits, boots, gloves are needed?*
3) Is the protective equipment re-usable? *The time needed for the decontamination process must be considered in planning the number of equipment needed.*
4) Does the contaminated equipment need to be disposed of somewhere? *Methods for safe storage and disposal must be planned.*

11.7.6
Protective Suits

Protective suits vary in material and functionality. Suits that are appropriate for CBRN missions can be impermeable, selectively permeable, or permeable (Section 11.3). Valid for all CBRN suits is the need to cover – together with gloves and boots and respiratory protection – the whole body. Protective suits can be disposable or re-usable. Reusability includes launderability after mission for hygienic reasons but can also mean decontamination from CBRN agents. The applied procedures have to be investigated and established. Permeable protective suits are typical for long-term military missions and can be worn for several hours, some types even for days. They provide protection especially against chemical agents. Impermeable suits for military use provide significantly higher levels of protection but can be worn only for a short time period as wear time restrictions need to be followed because of health risks originating from the impermeable shell

[34]. The utilization of both permeable and impermeable suits includes a threat analysis and detailed mission planning; logistic requirements must be considered. If an encounter with an unknown hazard or high concentrations is likely, a main requirement for the protective equipment is a wide range of protection against different types and concentrations of toxic substances. The use of impermeable barrier layers for skin protection and a higher degree of respiratory protection provided from a SCBA might be advisable.

In the following we explain further the most important items for CBRN body protection.

11.7.6.1 Permeable Protective Suits

Permeable suits provide breathability, thus allowing ventilation and water vapor transport. Sweat evaporation is possible to a degree. They therefore provide a comparatively good wear comfort, especially with impermeable suits in mind. Their main purpose is to allow long-time mission in CBRN contaminated environments; their wear time varies from several hours up to a few weeks. They are made from air-permeable durable resilient fabrics to allow them to stay undamaged even under heavy mechanical stress during long missions. These suits can be made as coveralls or in two or even more pieces if they, for example, have separate hoods; they are usually combined with protective gloves and boots. These are the items of protection that – together with the respirator – are provided to every soldier in the field in the case of a CBRN hazard, they are typical military equipment. Materials used for permeable protective suits have an integrated air-permeable filtration layer that typically contains activated carbon as it is an approved adsorbent with a high and rather durable adsorptive capacity especially for chemical warfare agents. The activated carbon can have different forms such as powder, beads, or cloth. It should be kept in mind that the amount of activated carbon in a filtration layer adds to the total weight of the garment and that the thickness and the density of the carbon layer influence the heat insulation of the suit.

Permeable protective suits can be designed either as undergarments, overgarments, or as combat clothing with integrated CBRN protection. The chosen design has consequences on the use during missions: a CBRN undergarment is similar to functional underwear with additional protective features. It needs to be donned before a mission starts as it is worn underneath the combat clothing. An overgarment is only donned if necessary and otherwise carried in a pouch. Combat clothing with integrated CBRN protection can be employed like normal combat clothing but degrades faster as at present the protective layers are not as durable as the non-protective fabrics. The advantage of both overgarment and CBRN combat clothing is flexibility to adapt the level of protection according to the actual hazard (see Table 11.1 MOPP level).

Manufacturers provide information on launderability and wear time forecasts for their materials. Degradation of the protective layer during use caused by mechanical stress leads to a loss in protection and is not always noticeable as, for example, tears in the fabric. To ensure the required level of protection during missions, the monitoring of a permeable protective suit's life cycle is required. Protection is not only provided from the adsorbent but also incorporate additional features – either into the material or as added extra finishes – such as oil and water repellency, flame retardency, and antistatic behavior.

A common and high quality treatment for oil and water repellency is the application of a fluorocarbon finish on the outside of the outer fabric as it provides chemical warfare agent repellency as well because drops would stay on the outer side of the protective suit. The fluorocarbon finish though does not prevent the negative effects of contact with all petroleum, oil, and lubricants (POLs). Even if oil might be repelled, contact with the other substances must be prevented as they are capable of resolving the adsorbed chemical warfare agents from the adsorbent and reactivating them, even letting them pass through the protective shield onto the skin. If contact with POL cannot be prevented, additional protection such as an apron must be worn [5]. Antistatic behavior and/or flame retardent features can either be incorporated in the fibers used to construct the material or added as extra finishes. If permeable suits are meant to be decontaminated and re-used after CBRN exposure, all suit materials and components must be chosen to resist the degradation of the decontamination process. Permeable suits can often be laundered several times by validated processes.

11.7.6.2 Impermeable Protective Suits

The application of impermeable suits includes a threat analysis and detailed mission planning, logistic requirements must be considered, and adequate detection equipment is advisable. These suits are worn if permeable suits do not provide sufficient protection. Owing to the threat scenario expected, they might need to be worn with a higher degree of inhalation protection. Impermeable protective suits are required if:

1) high concentrations of contamination are expected;
2) the type of contamination is unknown;
3) decontamination missions are planned where additional protection against splashing water and aggressive decontaminants is needed;
4) aerosol protection is necessary.

Several materials can be utilized for impermeable protective suits, differing in either reusable or disposable barriers. The latter need to be available in large quantities if long-time missions are expected as they can only be employed once. Strategies for their safe disposal need to be established. Reusable suits must be decontaminated, cleaned, and inspected after each mission with special procedures to ensure safe and hygienic re-use.

Disposable materials can be extremely lightweight compared to reusable ones, so a weight reduction of up to 75% and even more can be achieved. They can, for example, be made of non-woven fleece layers with polymer coatings. Depending on the coatings used, excellent long-time protection against chemical warfare agents can be expected and, as the materials are impermeable, particle tightness is given. Disadvantages are the commonly absent stability against heat and flame exposure, and sometimes the rustling noise of the material, which might be undesirable for

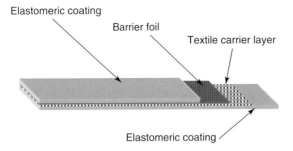

Figure 11.13 Common construction of a multilayer laminate for chemical protective suits.

military missions. Products with simple closures, such as cuffs, zippers, and elastic ribbon insertions, are critical to prevent leakages during missions. They promote the dangerous bellows effect [35]: the user generates changes in the air pressure inside the suit when moving, comparable to an air-pump if openings are present. Contamination from the environment can literally be sucked into the protective shell as long as leakages exist. The bellows effect can be prevented by tight sealing. To close the suit openings, the use of duct tape is common but tests prove that even the extensive use of duct tape does not necessarily lead to complete tightness.

The composition of reusable materials is more complex compared with disposable ones, hence their price is significantly higher. Reusable materials can be laminates of different materials, each contributing special protective features (Figure 11.13). The mechanical features usually are significantly higher than for disposable materials. The disadvantages of the multilayer laminates might be their stiffness, weight, and price.

Multilayer laminates are used for general purpose chemical protective suits, as they offer a wide range of protection, sometimes including limited heat and flame protection. For CBRN protection only, butyl coated fabrics are common. They are usually smoother and more light-weight though their protection time depends on the material and its thickness. It has to be considered that contact with POL products degrades the protection of butyl rubber materials drastically, from several hours down to a few minutes only [5, p. 571]. However, gas-tightness requires a special suit construction. Not all protective suits made from impermeable materials are necessarily gas-tight; this requirement is achieved by a sophisticated construction and the use of gas-tight components only. Gas-tight connections for gloves and boots can be achieved with circlips or by gluing the items in. Some impermeable suits have integrated boots or boot replacements; suits with integrated impermeable socks require additional foot protection such as boots or overboots. Gas-tight zippers are required.

Because of the missing exchange with the ambiance, wear time restrictions apply for impermeable protective clothing, and mission times can be limited to 30 min only so as not to endanger the wearer's health [34]. To increase the wear comfort, the mission time, and the PF of impermeable suits, the ventilation of the suits is possible. Commonly, the inhalation protection is also connected to the ventilation unit as it supports the overcoming of breathing resistance and hence leads to a reduction in workload and physical burden. Ventilation units can either be worn individually or can be static installations that supply the users via a hose connection with air. The use of non-self-sufficient systems is applicable if the limited radius of movements is acceptable. The additional weight of a ventilation unit to be carried means an extra workload from an ergonomic point of view. In particular, the batteries used today are often very heavy.

The air transported into an impermeable suit from a ventilation unit leads to an overpressure inside the protective system; integrated overpressure valves allow a controlled exhaust of the additional air. The airflow generated inside the suit can contribute to the body cooling as it supports the sweat evaporation and transports the humid air out of the suit as long as the air blown inside the suit is capable of absorbing humidity. Ideally, moderately chilled, dry air is used for the suit ventilation [36]. For effective body cooling, a minimum flow rate of about $500 \, l \, min^{-1}$ should be achieved [8].

11.7.7
Protective Gloves

The main function of gloves is the protection of the hands but at the same time the flexibility, the sense of touch, and dexterity should be maintained with the least limitations possible. The design and the material of protective gloves cannot simply be derived from the suits as the latter materials often are too stiff, too bulky, and not stretchable enough to be applicable for gloves.

Impermeable gloves are required if unknown contaminants are handled, contaminated surfaces are likely to be touched, and/or the contact with liquids is expected. Impermeable gloves usually mean bad wear comfort, especially if worn for a long time. The skin macerates due to the permanently humid or even wet ambiance of the hands, which leads to a degradation of the natural skin barrier ("Washerwomen's hands") and hence to a higher permeability for toxic substances. To improve the wear comfort and to reduce skin maceration, usually thin permeable gloves made from bibulous materials are worn underneath the impermeable gloves.

Impermeable gloves can, for example, be made from butyl rubber, polychloroprene, or fluoroelastomer. These materials provide a wide range of protection; the thickness of the materials' effects the retention time. Butyl rubber usually offers an excellent CBRN protection but would degrade dramatically in direct contact with oil and lubricant (POL). For POL protection, fluoroelastomer gloves can be used [5] but they might not offer long-time protection against sarin (GB). Polychloroprene gloves might not provide sufficient long-time chemical agent protection [37].

Another option to achieve better wear comfort for the hands is to use permeable or partly permeable gloves if impermeable gloves are not necessarily required. Partly permeable gloves can have integrated impermeable areas, for example, at the palm of hand, where protection against direct liquid contact is required.

Protection against biological threats requires impermeable gloves. It can be realized by wearing thin nitrile or latex gloves. For security reasons two impermeable pairs should be worn one over the other. The dexterity remains good as these types of gloves generally allow a smooth and tight fit. Both latex and nitrile gloves usually provide protection against chemical agents only for a short time [37]. Ideally, the outer glove should provide a high mechanical resistance to prevent tears or holes. A simple method to detect holes in the outer glove is to wear a colored near-to-skin inner glove with a transparent outer glove, if liquids penetrate through a hole in the outer glove, staining can be observed. If the inner glove offers a level of protection, it allows the change of a damaged or contaminated outer glove without opening the protective system [38].

All protective gloves must also provide sufficient protection against the decontaminants used. Ambidextrous gloves facilitate logistics.

11.7.8
Protective Footwear

Foot protection has to be adequate for the situation. For large amounts of splashing waters, decontaminants, and contaminants, rubber boots can be worn. Because of the bootleg, a tight connection to an impermeable protective suit is possible. Protective footwear, especially rubber boots, should provide a proper fit and a good sizing to contribute to workplace safety. A bad fit might cause stumbling and leads to an unnecessary increase in the physical burden as the user experiences problems with traction [5]. The construction of the sole should be non-skid and allow easy cleaning to avoid the spread of contamination. The same materials used for protective gloves are appropriate for protective boots [39]. A higher retention time of the boots can be expected as boots are made of thicker material than gloves. For missions requiring the features of combat boots, especially when an extended wear time is expected, additional foot protection can be realized as protective overboots or protective socks (instead of regular ones). Protective socks need to be donned precautionarily. Protective overboots are donned only if needed and should fit smoothly over the combat boot to prevent stumbling and to not hinder walking or driving a vehicle. Donning over the combat boot should be fast and easy; overboots are usually closed with fast closure systems such as draw strings or elastics. If not used, they are folded together and stored in a pouch. To facilitate logistics, some manufacturers offer ambidextrous overboots.

11.7.9
Pouches

All IPE already mentioned in this chapter must be designated to their individual user because their right fit and sizing are essential for their function and therefore for security during the mission. All items belonging together have to be stored in the same place to ensure rapid availability of the equipment. To keep the items together, it is helpful to have them stored in pouches, which can be transported

Figure 11.14 Respirator pouch for protected storage of respirator and canisters.

easily and assigned to their owner again if marked properly. It is common to have sets of permeable or impermeable suits stored in this manner. During missions, items not needed or spare parts can be stored within reach or carried around in a pouch.

A typical pouch, carried by each soldier during a mission is the respirator pouch (Figure 11.14). It contains not only the respirator but also spare canisters, a self aid kit, a poncho, and so on. The pouch shown can either be worn on an adjustable carrying strap over the shoulder or attached to a load carrying system. Pouches can be made of protective materials if their function is also to protect their content, for this purpose the use of rubber coated fabrics is common.

11.7.10
Ponchos

CBRN ponchos are additional protective equipment that is used only if needed. A poncho is usually worn on top of a permeable protective suit to provide protection against heavy rain, and high amounts of CBRN warfare agents. Ponchos can be made of any impermeable material that offers the needed level of protection against the expected liquid contamination as well as rain. Depending on the purpose, their design can be rather simple, such as a plastic layer with strings or belts for fixation comparable to a tarpaulin, or more sophisticated, comparable to clothing with an integrated hood and closures (Figure 11.15). They can be either disposable or reusable.

11.7.11
Self-Aid Kit

A self-aid or first aid kit is usually carried by each soldier. It contains items for both common injuries and CBRN decontamination. The items of the kit usually

Figure 11.15 Bundeswehr CBRN poncho.

allow cleaning, disinfection, wound treatment, and personal decontamination; additionally, special means for the treatment of nerve agent toxication, such as atropin auto-injectors, can be part of the kit [40].

11.7.12
Casualty Protection

A CBRN emergency very likely includes the necessity to deal with casualty handling, either contaminated or not (Figure 11.16). In a contaminated environment, the CBRN protection and transportation of casualties might not be possible with an ambulance to the hospital for further medical treatment immediately. The ability of the casualty to breathe with a respirator might be impaired.

Transport and protection for casualties can be provided by casualty hoods and bags that are offered in different varieties. Integrated windows allow surveillance of the casualty, filtered ambient air is provided from a ventilation unit. Casualty bags can be constructed like sheltered stretchers or can be similar to sleeping bags with hoods and a filtered air supply. Very sophisticated versions allow basic medical treatment via integrated gloves and medical equipment. If transport is not required, casualty hoods might be sufficient. The risks and effects of lethal contamination must be taken into account to ensure safety for the survivors and the unconcerned population: special bags for contaminated bodies are available.

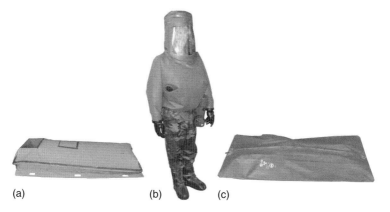

Figure 11.16 Casualty protection: (a) casualty bag, (b) casualty hood, and (c) body bag. Images, copyright by Blaschke Wehrtechnik, Austria.

11.8
Quality Assurance

As the health and life of users rely on the flawless functionality of IPE, severe legal restraints exist for marketing IPE. In Europe, the EU directive 89/686/EEC includes the superior guideline for IPE – among other things it defines basic safety requirements. The EU directive needs to be transposed into national law for each member state; national transpositions might differ. For employment protection reasons it is important that IPE without national quality certificates should never be used. Information has to be supplied from the manufacturer for each IPE item such as detailed instructions for the handling, protective performance, limits of use, and more. Expressly not covered under this directive is IPE designed and manufactured specifically for use by the armed forces. The standards for civilian IPE as well as their test procedures might both not be sufficient in all cases for military IPE. Manufacturers provide lists of substances their equipment protects against and information on which test was employed for the evaluation. If a threat analysis unveils missing information on hazardous substances, additional tests in accredited laboratories could close this gap. IPE should be tested to gain information on its performance when challenged with the expected level of contaminants and all other substances it could come in contact with such as decontaminants, disinfectants, and so on [e.g., petroleum, oil and lubricant (POL)].

During IPE procurement, a quality management has to be installed to ensure a constant quality for the whole tender; it might be necessary to carry out crucial performance tests along the production process. Methods to evaluate the protective features can, for example, be found in DIN EN ISO 6529 for the "determination of [the] resistance of protective clothing materials to permeation by liquids and gases" [41] and in DIN EN ISO 6530 for the "Test method for resistance of materials to penetration by liquids." [42]

11.9
Workplace Safety

Workplace safety is essential for the positive outcome of a mission; its basis is regular training and detailed knowledge of the protective equipment and of the requirements on how to act in a CBRN emergency. It includes knowledge of the protective level to be chosen for a hazard, training in the use of the protective equipment, knowledge of the limits of the equipment provided, and of the human physical properties and limitations. National regulations are established for workplace safety concerning IPE but do not always apply for emergency situations. For missions in hot climate conditions, work-break regulations need to be established to ensure the health of the user wearing IPE that imposes thermal insulation and an additional work load. The fluid supply is extremely important [8].

Workplace safety is not limited to how to safely utilize IPE but must include detailed training of complex emergency situations to impart the knowledge. International studies and practical experience show that psychosocial knowledge and action also support the operational performance and provide, hence, a contribution to workplace safety as stress levels decrease and personnel feel more competent and secure. This training prepares for how the population will probably react in the case of a CBRN emergency and how to deal with these additional effects. To handle these problems contributes to the coordination and safety of missions [43].

11.10
Future Prospects

Future CBRN scenarios will probably be very different from those of cold war times. Focusing on an increase in asymmetric threat, localized events appear especially likely. But still, as CBRN agents are meant to incapacitate the human being as well as to poison the opponent's assets, protection of the individual will also in the future be of significant importance. More emphasis might be placed on the potential attacker's substances – additionally to the classic CBRN warfare agents, a widespread collection of hazardous compounds with strongly varying physical-chemical as well as toxicological characteristics is given with TIMs.

Future CBRN protection will likely be integrated in combat clothing and include a method to observe the degree of usage. The results of today's research and development work would consequently lead to:

1) self cleaning surfaces
2) "intelligent" textiles
3) the use of robots.

Coatings made from nanostructures applied to IPE's surfaces, nanotubes containing reactive substances incorporated into the protective material for self-decontamination during and after the mission (1), and integrated, miniaturized washable electronic devices for control of the soldier's health status as well as the detection for an improved guidance (2) might allow tasks that are not possible today on a user's scale. For some missions, the employment of robots might be an option if lives are in imminent danger or if the workload is too demanding for human beings (3).

References

1. Croddy, E.A., Wirtz, J.J., and Larsen, J.A. (2005) *Weapons of Mass Destruction: An Encyclopedia of Worldwide Policy, Technology, and History*, Vol. I: Chemical and Biological Weapons and Volume II, ABC-Clio Inc., pp. 226–227, ISBN-13: 978-1851094905.
2. Marrs, T.T., Maynard, R.L., and Sidell, F. (2007) *Chemical Warfare Agents: Toxicology and Treatment*, 2nd edn, John Wiley & Sons, Inc., pp. 159–170, ISBN: 978-0-470-06002-5.
3. Dishovsky, C. (2006) in *Medical Treatment of Intoxications and Decontamination of Chemical Agents in the Area of Terrorist Attacks*, NATO Security Science Series, vol. 1 (eds C. Dishovsky, A. Pivovarov, and H. Benschop), Springer, pp. 3–11. ISBN: 978-1-4020-4169-3.
4. NATO Civil Emergency Planning Civil Protection Committee, Project on Minimum Standards and Non-Binding Guidelines for First Responders Regarding, Planning, Training, Procedure and Equipment for Chemical, Biological, Radiological and Nuclear (CBRN) Incidents, Guidelines For First Response To a CBRN Incident, http://www.nato.int/docu/cep/cep-cbrn-response-e.pdf (accessed 01 January 2011).
5. Romano, J.A., Lukey, B.J., and Salem, H. (2007) *Chemical Warfare Agents: Chemistry, Pharmacology, Toxicology, and Therapeutics*, Informa Healthcare. ISBN-13: 978-1420046618.
6. James, A.D. (2006) *Science and Technology Policies for the Anti-Terrorism Era*, NATO Science Series, Series V – Science and Technology Policy, vol. 51, IOS Press. ISBN: 1-58603-646-7.
7. International Ergonomics Association, http://www.iea.cc/01_what/What%20is%20Ergonomics.html (accessed 09 May 2012).
8. Glitz, K.J., Seibel, U., and Leyk, D. (2005) *Wehrmedizinische Monatszeitschrift*, **49** (1), 16–20.
9. McLellan, T.M., Cheung, S.S., Latzka, W.A., Sawka, M.N., Pandolf, K.B., Millard, C.E., and Whitey, W.R. (1999) *Can. J. Appl. Physiol.*, **24** (4), 349–361.
10. BGR/GUV-R190 (2011) Deutsche Gesetzliche Unfallversicherung e.V. (DGUV), Berlin, 1–174, http://www.publikationen.dguv.de/dguv/pdf/10002/r-190.pdf (accessed 31 May 2012).
11. Fox, S.M. and Haskell, W.L. (1970) in *Cardiology: Current Topics and Progress*, 6th edn (eds M. Eliakim and H.N. Neufeld), Academic Press, New York, pp. 149–154.
12. Tanaka, H., Monahan, K.D., and Seals, D.R. (2001) *J. Am. Coll. Cardiol.*, **37**, 153–156.
13. Jürgens, H.W. Christian-Albrechts-Universität zu Kiel, (2001 – 2008) Forschungsgruppe fuer Industrieanthropologie, Controlled Wear Trials for the Evaluation of the Wear Comfort of NBC Protective Clothing, Reports for the WIS, Munster.
14. Nadeau, J. (2003) Just-in-time and continuous physical protection for military personnel in a chemical or biological environment, in *A Summit for People and Technology, Human Factors and Ergonomics Society 47th Annual Meeting*, Human Factors and Ergonomics Society, pp. 1948–1952.
15. CDC NIOSH Publications http://www.cdc.gov/niosh/docs/2009-132/ (accessed 14 May 2012).
16. Seifert, S.A., VonEssen, S., Jacobitz, K., Crouch, R., and Lintner, C.P. (2003) *Clin. Toxicol.*, **41** (2), 185–193.
17. Rengassy, A., Zhuang, Z., and BerryAnn, R. (2004) *Am. J. Infect. Control*, **32** (6), 345–354.

18. Hartmann, H.M. (2002) *Regul. Toxicol. Pharmacol.*, **35** (3), 347–356.
19. Watson, A., Opresko, D., Young, R., and Hauschild, V. (2006) *J. Toxicol. Environ. Health Part B: Crit. Rev.*, **9** (3), 173–263.
20. Assigned Protection Factors for the Revised Respiratory Protection Standard, Occupational Safety and Health Administration (OSHA), U.S. Department of Labor, OSHA 3352-02 (2009). Available at *http://www.osha.gov/Publications/3352-APF-respirators.pdf* (accessed 14 May 2012).
21. U.S. Environmental Protection Agency (2011) Acute Exposure Guideline Levels, Chlorine Results. Available at *http://www.epa.gov/opptintr/aegl/pubs/results56.htm* (accessed 01 January 2011).
22. Wetherell, A. and Mathers, G. (2007) Respiratory protection, in *Chemical Warfare Agents: Toxicology and Treatment*, 2nd edn (eds T.C. Marrs, R.L. Maynard, and F.R. Sidell), John Wiley & Sons, Ltd, Chichester. ISBN-978-0470013595.
23. NIOSH Statement of Standard for Chemical, Biological, Radiological and Nuclear (CRBN) Air-Purifying Escape Respirator (2003) Atlanta, National Institute for Occupational Safety and Health, U.S., *http://www.cdc.gov/niosh/npptl/standardsdev/cbrn/escape/standard/pdfs/cbrn-esc-att-a.pdf* (accessed 31 May 2012).
24. Anthony, T.R., Joggert, P., Janues, L. *et al.* (2007) *Ann. Occup. Hyg.*, **51**, 703–716.
25. NIOSH Statement of Standard for Chemical, Biological, Radiological, and Nuclear (CRBN) Full Facepiece Air Purifying Respirator (APR), *http://www.cdc.gov/niosh/npptl/standardsdev/cbrn/apr/standard/aprstd-a.html* (accessed 15 May 2012).
26. NIOSH *Respirator Standards*, National Institute for Occupational Safety and Health, Atlanta, GA, *http://www.cdc.gov/niosh/npptl/respstds.html*, (accessed 15 May 2012).
27. Occupational Safety and Health Administration (OSHA) and U.S. Department of Labor, (2009) Assigned Protection Factors for the Revised Respiratory Protection Standard, *http://www.osha.gov/publications/3352-APF-respirators.pdf*, (accessed 01 June 2012).
28. NIOSH Approved Respirator Standards, Self Contained Breathing Apparatus (SCBA) with CBRN Protection, *http://www.cdc.gov/niosh/npptl/RespStandards/ApprovedStandards/scba_cbrn.html* (accessed 15 May 2012).
29. Schnepp, R. (2009) Hazardous Materials Awareness and Operations, International Association of Fire Chiefs, National Fire Protection Association, Jones & Bartlett Pub (Ma); First Edition, ISBN-13: 978-0763738723.
30. Currie, J.,, Caseman, D.,, and Anthony, T.R. (2009) *Ann. Occup. Hyg.*, Vol. 53, (5), 523–538, *http://annhyg.oxfordjournals.org/content/53/5/523.full.pdf* (accessed 16 May 2012).
31. Wetherell, A. (2003) The UK General Service Respirator, Defence Scientific and Technical Laboratory Porton Down, Salisbury SP4 0JQ, United Kingdom, *http://www.dtic.mil/cgi-bin/GetTRDoc?AD=ADA452235* (accessed 01 June 2012).
32. PPE Guidelines (2010) Guidelines on the Application of Council Directive 89/686/EEC of 21 December 1989 on the Approximation of the Laws of the Member States relating to Personal Protective Equipment (Version 12. April 2010).
33. DIN EN 14126 (2004-01). *Protective Clothing – Performance Requirements and Test Methods for Protective Clothing Against Infective Agents.*
34. BGR/GUV-R 189 (2007) Deutsche Gesetzliche Unfallversicherung e.V. (DGUV), Berlin, *http://www.publikationen.dguv.de/dguv/de/dguv/pdf/10002/r-189.pdf*, (accessed 31 May 2012).
35. Public Health Response to Biological and Chemical Weapons: WHO Guidance, (2004) Annex 4: Principles of Protection, *http://www.who.int/csr/delibepidemics/annex4.pdf*, (accessed 31 May 2012).
36. Glitz, K.J., and Restorff, W.von, Reduction of Heat Stress in Chemical Protective Suits, Central Institute of

the Federal Armed Forces Medical Service Koblenz, Koblenz, Germany, http://www.lboro.ac.uk/microsites/lds/EEC/ICEE/textsearch/98articles/Glitz1998.pdf, (accessed 31 May 2012).
37. Lindsay, R.S. (2001) Swatch Test Results of Commercial Protective Gloves to Challenge by Chemical Warfare Agents: Executive Summary, Research and Technology Directorate, DIR, ECBC, ATTN: AMSSB-RRT, APG, MD 21010-5424, http://www.dtic.mil/cgi-bin/GeTRDoc?AD=ADA440404, (accessed 31 May 2012).
38. Biologische Gefahren I, Handbuch zum Bevölkerungsschutz (2007) *Bundesamt für Bevölkerungsschutz und Katastrophenhilfe*, 3rd edn, pp. 568–569, ISBN-13: 978-3-939347-06-4.
39. Occupational Safety and Health Administration (OSHA), U.S. Department of Labor, (2005) Best Practices for Hospital-Based First Receivers of Victims from Mass Casualty Incidents Involving the Release of Hazardous Substances, *http://www.osha.gov/Publications/osha3249.pdf*, (accessed 31 May 2012).
40. Müller, M., Schmiechen, K. (2009) Forschung Im Bevölkerungsschutz, *Bundesamt für Bevölkerungsschutz und Katastrophenhilfe*, Band 10, Hrsg. ISBN-13: 978-3-939347-22-4.
41. DIN EN ISO 6529, Schutzkleidung – Schutz gegen Chemikalien – Bestimmung des Widerstands von Schutzkleidungsmaterialien gegen die Permeation von Flüssigkeiten und Gasen (ISO/DIS 6529:2011); Deutsche Fassung prEN ISO 6529:2011.
42. DIN EN ISO 6530, Schutzkleidung – Schutz gegen flüssige Chemikalien – Prüfverfahren zur Bestimmung des Widerstands von Materialien gegen die Durchdringung von Flüssigkeiten (ISO 6530:2005); Deutsche Fassung EN ISO 6530:2005.
43. Psychosoziales Krisenmanagement in CBRN-Lagen (2011), Bundesamt für Bevölkerungsschutz und Katastrophenhilfe (BBK), ISBN: 978-3-939347-34-7.

12
Collective Protection – A Secure Area in a Toxic Environment
Andre Richardt and Bernd Niemeyer

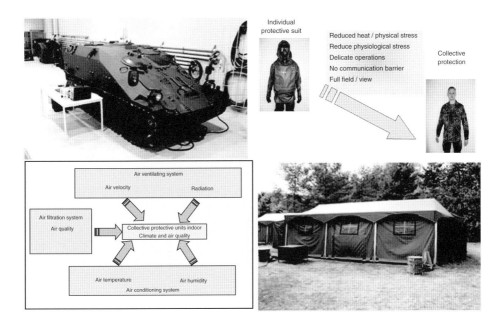

Protection of personnel during a chemical, biological, or radiological/nuclear (CBRN) threat from a wartime or terrorist attack, or from an industrial incident, is one of the most important issues in a technical concept for CBRN protection. In the hazard zone individual protective equipment (IPE) and collective protection (COLPRO) sheltering are the most important ways to protect persons from the impact of hazardous substances. Even for the evacuation operation, collective protection and individual protection equipment are needed. Because, after a CBRN incident, both the soldiers and first time responders need a place to rest, research into better collective protection strategies was intensified after World War I. Over the years, different strategies have been developed to protect soldiers, first time responders, and the population from the effects of CBRN incidents. During the cold war, protection against a nuclear threat was especially under investigation.

CBRN Protection: Managing the Threat of Chemical, Biological, Radioactive and Nuclear Weapons,
First Edition. Edited by A. Richardt, B. Hülseweh, B. Niemeyer, and F. Sabath.
© 2013 Wiley-VCH Verlag GmbH & Co. KGaA. Published 2013 by Wiley-VCH Verlag GmbH & Co. KGaA.

Better understanding of airflow and the design of collective protection units has led to improved protection concepts. Nowadays, a diversity of advanced filtration materials, improved tent materials, and innovative designs has led to systems that are transportable and easy to maintain.

12.1
Why Is Collective Protection of Interest?

Collective protection is on a par with the other components of chemical, biological, or radiological/nuclear defence. If collective protection fails we have a vulnerability that could be exploited. Collective protection has a simple concept: to establish and maintain a toxic-free area (TFA) and to provide the persons and all equipment inside this zone with uncontaminated conditioned air (Figure 12.1). However, the technical realization of this approach is challenging. Collective protection serves personnel in several different and important ways. Often, collective protection is only considered to be a room pressurized with clean air. This means that conditioned, purified, ventilated air has to be supplied at a pressure above the normal atmospheric pressure. With this overpressure CBRN substances can be prevented from infiltrating. However, an effective collective protection unit is far more complex. One important question that we want to answer in this chapter is: How should we design units for collective protection units to resist chemical, biological, radiological, and nuclear agents? Owing to the different national regulations for CBRN scenarios and applications different nations use

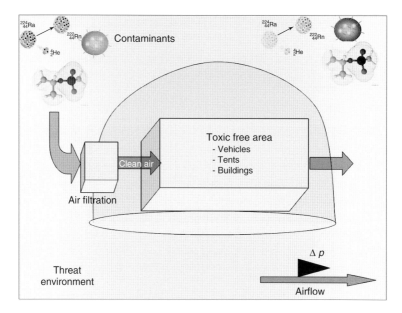

Figure 12.1 Concept of collective protection (COLPRO).

different approaches and standards for collective protection. Therefore this chapter aims, based on some considerations about indoor air quality (IAQ), to show the general idea and design features of a collective protection system that can be transferred to and adapted for special applications. For these special applications we should have in mind the major challenges pertaining to stand-alone collective protective systems, as:

1) availability,
2) portability,
3) functionality,
4) integrity of the barrier.

Functionality includes ingress and egress that minimize or eliminate the possibility of internal contamination. Further challenges, for example, for protecting crews in vehicles include integrating collective protection measures into vehicle designs. Improved protection in any environment depends on "air filtration" (Figure 12.1), which consists of particle filtration and adsorbent technologies, as well as the availability of protective equipment [1].

- **CBRN collective protection**: protection provided to a group of individuals in a chemical, biological, radiological, and nuclear environment, which permits a CBRN-threat free area and thus relaxation of individual CBRN protection.
- **Indoor air quality (IAQ)**: refers to the nature of conditioned air that circulates throughout the space/area where we work and live.

Collective Protection: a Secure Area in a Toxic Environment
The concept of collective protection is to establish and maintain a toxic free area and to provide the persons inside this zone with uncontaminated conditioned air (Figure 12.1). In this area relief, difficult operations, and free communication without the burden of individual protection equipment is possible.

However, before we can look at collective protection design details we have to discuss some general questions that have to be considered if we want to display a collective protection system. How can collective protection improve indoor climate and air quality? Here we can provide an idea about the complexity of achieving comportable indoor air quality. However, this topic is in the focus of several research groups, and even an international society of air quality and climate (ISIAQ) was founded (*www.isiaq.org*, accessed 3 April 2012). We breathe conditioned air for a significant part of our lives and it is important to understand that indoor air quality refers not only to comfort, which is affected by temperature, humidity, and odors, but also to harmful contaminants within the conditioned space. In our daily life it is normal for us to open windows if the room feels stuffy and unaired. However, in the case of a CBRN incident or an accident with harmful chemicals it could be deadly if we open windows or doors. Therefore, the question of achieving an acceptable level of indoor air quality in a collective protective unit is essential to survive in an environment where the outer air is contaminated. This question often competes with other challenges such as, mobile firepower, mine protection, and speed for armored vehicles as well as required space and equipment for other tasks and saving of energy in collective protective units.

Figure 12.2 summarizes some major parameters affecting high level indoor air quality (IAQ). We know that air temperature concerns the convective heat loss from the body and by expired air [2]. Heat losses have to be paid back by external energy input arising from both heating and humidifying the respired air. Air temperature is only one parameter influencing the thermal state of the human body. Activity and clothing are further important parameters to be considered in assessing thermal comfort (Table 12.1).

As well as air temperature, air velocity (v) is another important ambient parameter for the convective heat exchange between the human body and the ambient air [3]. The influence of v on the heat transfer coefficient is an almost quadratic function function. This means that at very low v small velocity changes have a larger effect on the convective heat transfer than similarly small velocity variations at high v (Figure 12.3) [calculated from the human heat balance model munich energy balance model for individuals (MEMI)] [4, 5].

Air humidity, another point we have to consider, is certainly the most complex parameter in assessing the climate in collective protection indoors. The humidity of ambient air influences three mechanisms of water loss from the human body:

1) diffusion of water vapor through the skin and its evaporation;
2) evaporation of sweat from the skin surface;
3) humidification of the respired air.

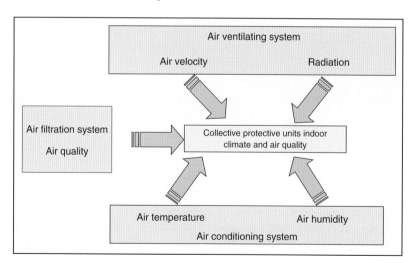

Figure 12.2 Most important parameters affecting the indoor climate and air quality of collective protective units [3].

12.1 Why Is Collective Protection of Interest?

Table 12.1 Comfort air temperature (T_{comf}) at different levels of activity and with different types of clothing with air velocity (v) = 0.1 m s^{-1}; air vapor pressure = 10 hPa and mean radiant temperature (T_{mrt}) = air temperature (T_a) [3].

Activity	Work metabolism (W)	Clothing (clo)[a]	T_{comf} (°C)
Resting	0	0.5	31
	0	1.0	29
Standing	43	0.5	27
	43	1.0	23
Light work[b]	100	0.5	22
	100	1.0	16
Heavy work[c]	200	0.5	12

[a] clo 1: corresponds to a person wearing a typical suit; 1 clo = 0.155 m^2 K W^{-1}.
[b] Light work: office work, walking, and light machine work.
[c] Hard work: heavy machine work, exercise.

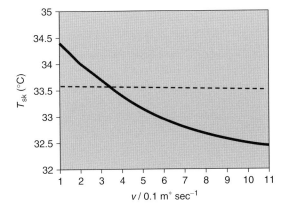

Figure 12.3 Relationship between air velocity (v) and mean skin temperature (T_{sk}, °C) of a man [35 years old, sitting, 1.0 clo (clothing), 60 W work metabolism]. Dotted line (– – –) represents comfort skin temperature [3].

Figure 12.4 gives typical values of these mechanisms, demonstrating that under the assumed conditions sweat evaporation and respiratory humidification consume similar energies of −12, and −11 W, respectively. Another important ambient parameter for thermal comfort is the thermal radiation of indoor surfaces (walls, floor, and ceiling). For typical indoor conditions (with no air-conditioning), the largest heat flow from the body occurs by radiation (−59 W), followed by convection (−46 W) (Figure 12.4). A lack of understanding and neglecting the influence of thermal radiation for thermal comfort can lead to thermal discomfort in collective protection units. Normally, air quality is the point that we should have first in mind when we think about providing persons in collective protection units with

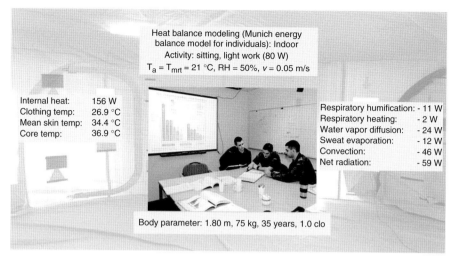

Figure 12.4 Heat fluxes and body temperatures under typical indoor conditions calculated with the energy balance model MEMI (munich energy balance model for individuals) [5, 8].

uncontaminated conditioned air. As seen, avoiding contamination of air from outdoor CBRN substances is only one aspect of air quality. Another important point is sources of pollutants like formaldehyde, carbon monoxide, or nitrogen oxides produced inside the collective protective unit [6]. Over time out-gassing of furniture and electrical equipment or other phenomena can lead to higher indoor concentrations of these harmful substances [7]. However, the last important point we have to discuss is the respiratory quotient (RQ), which is calculated from the ratio:

$$RQ = P_{CO_2 \text{ removed}} / P_{O_2 \text{ consumed}} \text{ from the body}$$

In general, there always has to be enough oxygen (O_2) in the indoor air of the collective protective unit. During respiration the ratio of the transformed O_2 molecules to the carbon dioxide (CO_2) molecules generated (calculated either as the molar or the partial pressure quotient as RQ) is close to one (Figure 12.5). Generally, this means that the concentration of O_2 decreases by the same amount as the concentration of CO_2 increases. An illustrative example of the importance of this statement is given when a concentration of CO_2 of 0.5% is reached. This value of 5000 ppm is the threshold concentration for work places in many countries, while at this level the O_2 concentration is still 20.5%. In other words, long before there will be a risk of having too little O_2 in the indoor air in the collective protective unit there will be a level of CO_2, which has to be prevented [9, 10]. Therefore, beside the filtration of air the main function of artificial ventilation in collective protection rooms with no natural ventilation or with many occupants is not to provide enough O_2 (as commonly assumed) but to remove CO_2 and other indoor pollutants.

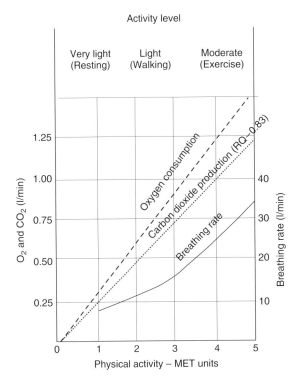

Figure 12.5 Dependence of oxygen consumption and carbon dioxide production on activity level, MET = metabolic equivalent, which reflects the energy costs of activity.

12.2
Collective Protection Systems – Required for Different Scenarios

Now have the basics concerning the complexity of what is required we aim to find the best fitting conditions for different collective protective scenarios:

1) **Fixed collective protection**: collective protection systems integral to static facilities; these may be hardened, semi hardened, or unhardened.
2) **Transportable collective protection**: Stand-alone collective protection systems capable of being deployed in the area of operations. They will usually be unhardened but may be capable of erection within buildings or other enclosures.
3) **Mobile collective protection**: collective protection systems integral to land, sea, or air platforms.
4) **Hybrid collective protection**: Platform installed systems that supply breathing air direct to individual occupants. These may sometimes be used in association with mobile systems [1].

When we think about these different applications, the question of where collective protection is typically required should be answered first. Typically, collective protection is required to provide a safe environment for:

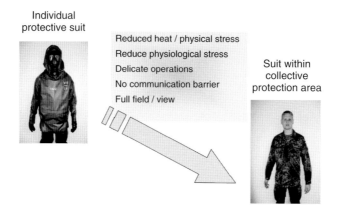

Figure 12.6 Advantages of collective protection compared to individual protection equipment.

- command/control/communication areas,
- medical areas,
- rest/relief areas.

In all these cases there should be no need to wear cumbersome CBRN protective clothing and masks and no need to cope with the burdens associated with individual protective equipment (Figure 12.6).

To achieve these advantages, the minimal requirements for basic collective protection is to provide a group of people with uncontaminated (preferably conditioned) air and protection from skin contact with biological, chemical, or radiological agents. Therefore, CBRN collective protection is just a special form of employment protection. Another point we can highlight is that perfect collective protection does not affect the physical, and mental condition of the users. It also depends less on the level of training than on the equipment available. Furthermore, as a main difference between collective protection and individual protection, the former systems should provide a higher protection level over long time periods. Collective protection can therefore be seen as preferred to individual protection. Based on these basic requirements and on the definition for collective protection we can divide the different collective protection applications further into:

- **mobile collective protection**: vehicles, ships, and aircraft;
- **transportable collective protection**: portable shelters, tents, vans, and containers;
- **fixed site collective protection**: critical infrastructure, permanent shelters.

Generally, collective protection applications can be permanent or temporary. There are three basic methods for providing collective protection equipment:

(i) overpressure
(ii) hybrid, and
(iii) passive systems (no overpressure) systems.

They deliver different advantages and disadvantages, as seen in Figure 12.7.

12.2 Collective Protection Systems – Required for Different Scenarios | 339

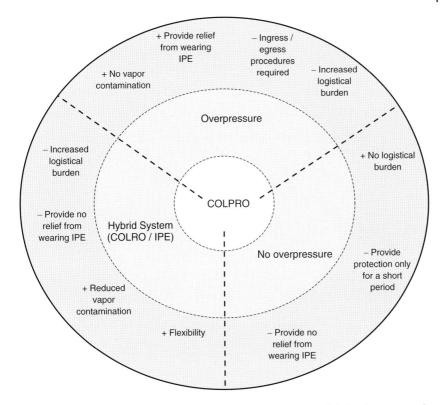

Figure 12.7 Types of collective protection (COLPRO) equipment and their advantages and disadvantages; IPE: individual protective equipment. Adapted from Reference [11].

All the properties of collective protection systems can be covered by properly considering the overpressure system. Therefore, although a hybrid system is a possible solution for special applications (armored vehicles) we will not consider such a system here Systems with filtration can be divided further:

1) **Filtration with pressurization.** This is the highest class of protection and used against wartime military threats. Here we can define the need to protect personnel for large-scale release of agents over an extended period of time. An effective CBRN filtration and overpressure system has to be provided by the system to resist a continuous large-scale threat to a 40-km h^{-1} wind. A contamination control area (CCA) and an ingress and egress airlock are required. It is possible that the filtration and pressurization system can be operated continuously or maintained in a standby mode. This means that it is energized only when there is a CBRN alert.

2) **Filtration with *little* or no pressurization.** This class of protection provides a lower level of protection than the former system and is applicable to a CBRN terrorist attack with little or no warning as these attacks normally produce a short duration small-scale release of agents. Nevertheless, the continuously

operating CBRN filtration units must be able to filter the incoming air effectively. The CBRN filtration system will be required to provide a low overpressure. The aim of the elevated pressure is to prevent the penetration of agents into through the toxic free area envelope at wind speeds of 12 km h^{-1}.
3) **Passive protection**. This application is mainly useful for an industrial incident with a short duration of release. Closing building openings and turning off ventilation systems are helpful measures (as discussed in Reference [12]).

Only integrating an overpressure system into the whole device the system has the chance of fulfilling the prerequisites that there should be no need to wear cumbersome protective clothing and masks and no need to cope with the difficulties associated with individual protective equipment (IPE).

Therefore, we focus on the *filtration with pressurization* as the main important feature for collective protection. If we want to go into more in detail for the design of collective protection, we face the question of what further associated equipment beside treatment for purified air is needed. We can identify alarm and air conditioning systems and protective entrances as necessary equipment to improve collective protection systems. If we employ automatic detectors, this equipment may automatically initiate protective actions such as shutdown of ventilation systems, closing outside air intakes, or turning on filtration systems. Unfortunately, as we have learned in Chapter 9, only the detection of radiological agents can be achieved in time and the detection of radiological agents is easier than the detection of biological or chemical agents. Current biological detection technology requires a minimum analysis time of approximately 15 min to detect the presence of possible biological agents. However, even then the rate of false positive alarms is significantly high. We can state nearly the same case for chemical detection. We see shortcomings in response time and false alarms. For all three types of sensor technologies the maintenance requirements, cost, and the quantity of sensors needed at air intake locations have also be taken into account. Beside the detector limitations we have to distinguish between wartime threats and terrorist threats. We can assume that during wartime audible and visual indicators in combination with detector technologies are available to identify a CBRN incident. In contrast, it seems that for terrorist threats only the incapacitation of people is an indicator of the use of CBRN agents. Based on these assumptions, the utilization of detectors for terrorist threats should be limited to [12]:

- first entry determination by first time responders,
- monitoring casualties before and during medical treatment,
- determining the extent of the hazard,
- determining when protective measures are no longer needed.

In addition to filtration and pressurization and the detection system, cooling systems for the incoming air are also of interest in modern collective protection set-ups. Air conditioning/cooling can reduce heat stress when operating under extremely hot and/or humid conditions. It is well known that mission oriented protective posture (MOPP) gear significantly increases the potential for heat stress, especially under heavy working conditions (Figure 12.5).

12.3
Basic Design

Based on the introductory considerations for indoor air quality and features of collective protection systems requirements, we can identify several functional items necessary for collective protective systems as demonstrated in Figure 12.8.

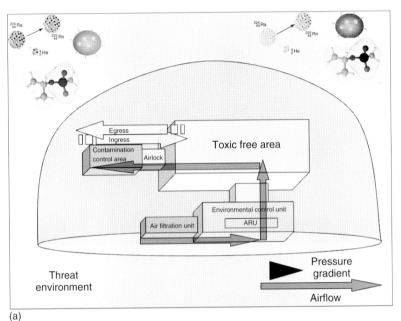

(a)

(b)

Figure 12.8 (a) Basic elements of collective protection; (b) example of a collective protective tent.

1) air filtration unit (AFU) and auxiliary equipment;
2) environmental control unit (ECU);
3) contamination control area (CCA);
4) airlock;
5) air regeneration unit (ARU);
6) toxic free area (TFA) (Figure 12.8).

Based on this general list for collective protective items we will discuss each item in more detail.

12.3.1
Air Filtration Unit (AFU) and Auxiliary Equipment

The air filtration system is the most important part of the air filtration overpressure system. The first design parameter is the airflow capacity of the collective protection overpressure filtration system. The question is "How much new conditioned clean air from the outside is required?" It can be answered as the sum of the following four components:

1) multiply the number of persons by the required fresh clean air (in cubic meter per second, $(m^3 \cdot s)$);
2) the toxic free area envelope air leakage rate at the designed differential pressure;
3) the ventilation air intake rate that meets exhaust requirements;
4) the airlock airflow necessary to achieve the required purge rate and the reestablishment of the needed overpressure.

Auxiliary equipment is critical to a totally integrated process and in addition to offering paths for contamination, and along with all other contamination control area/airlock (A)/toxic free area equipment, it must be optimized for weight, size, and cost considerations. The main aim of the air filtration system is to provide

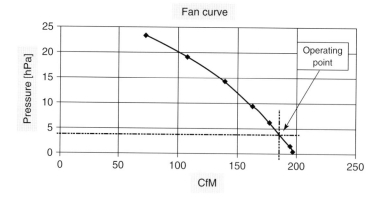

Figure 12.9 Example of the determination of the operating point of an air filtration unit. The operating point is the point of intersection of fan curve and system flow resistance (dotted points).

enough uncontaminated air (in cfm) to the toxic free area. Therefore, it is necessary to know the fan operating curve of the air filtration system [13] in order to determine the operating point for the system (e.g., see Figure 12.9) and to minimize pressure fluctuations due to loaded filters, overpressure changes, and so on.

When we design a filtration system based on knowledge of the technical principles of filter technologies (Chapter 10) we should consider the following (seen in Figure 12.10):

1) **Bag-in/bag-out housings:** Any system that filters dangerous contaminants should incorporate bag-in/bag-out housings. These housings are necessary to

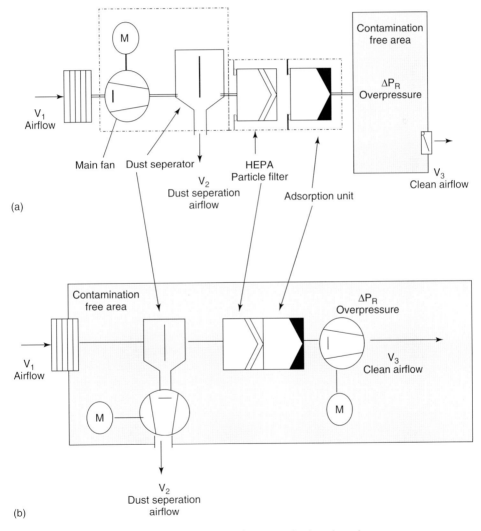

Figure 12.10 Air filtration systems: (a) blow-through system; (b) draw-through system.

contain the contaminated filters and protect maintenance personnel during filter changes.
2) **Particulate filtration**: Must be provided at the gas entrance into the system in order to remove coarse particles. Fabric or cyclone filtration is suitable.
3) **High efficient particle air (HEPA) filters as a backup system:** The removal of fine particles, for example, dust or small liquid droplets from aerosols, requires high-performance filtration. This filter is located next to the "entrance particle filter" and upstream of the following adsorption system.
4) **High efficient gas adsorber (HEGA):** This is an adsorption separation to eliminate gaseous hazardous substances that are not recovered by the prior filtration systems as well as outgassed compounds from the particle filters used to deal with aerosols.
5) **Filter trains:** Any combination of pre-filters (coarse and fine dust filter) HEPA- and HEGA-filters, which have to be constructed for easy and safe maintenance under hazardous conditions.

We can find several effective filtration systems from both military and commercial suppliers. The military filtration units are typically provided as Government-furnished equipment (GFE), with the advantage of being pre-approved for use on Government installations. Commercial equipment, available of the shelf, requires additional Government quality testing. The filtration systems are of modular sectional design and will require filter sections:

- coarse filter,
- pre-filter,
- HEPA filter,
- one or two stages of adsorbers.

We should now be able to arrange filter systems for different applications (Figure 12.11).

For particulate filtration the arrangement shown in Figure 12.11a is sufficient. Mainly gaseous substance elimination with removal of potential coarse particles (e.g., aerosols) is provided by the set-up in Figure 12.11b. To get rid of coarse and small particles as well as of gas compounds the device shown in Figure 12.11c is suggested. The greatest effort has to be provided for the separation process shown in Figure 12.11d, which additionally sets a HEGA filter downstream of the HEPA filter. This additional HEGA filter avoids the danger that particles removed by the HEPA filter can be outgassed again. With this basic knowledge we are able to understand how specialists can arrange different filter systems for various applications. The actual type and number of filters required will depend upon what has to be filtered and how much conditioned air is needed.

12.3.2
Environmental Control Unit (ECU)

The ECU has to provide different temperature and humidity requirements, and gas flow for various activities. The gas flow is related to the overpressure of a toxic

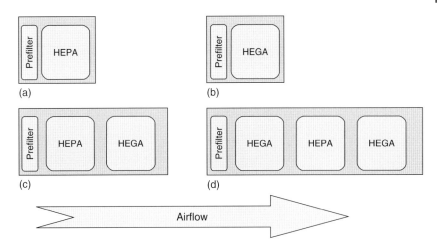

Figure 12.11 Examples of several types of filtration applications with high efficiency gas adsorber (HEGA) and high efficiency particulate air (HEPA) filters: (a) streams containing particles only; (b) streams with vapor phase contaminants only; (c) recirculating applications where particulates and vapor phase contaminants have to be removed; (d) recirculation applications with an additional HEGA filter to remove outgassing contaminants from the particles removed by the HEPA filter.

free area as well as to the leakage rate of the total shelter. The minimum toxic free area overpressure for stationary collective protection systems should be 2 hPa. The maximum toxic free area overpressure should not exceed 12 hPa. In terms of envelope leaks, it is nearly impossible to create a totally impermeable toxic free area and so, therefore, it is necessary to determine the envelope air leakage rate and to pinpoint the air leakage, which is in the range of a low percentage of the whole gas flow through. This can be achieved during pressurization testing [11].

12.3.3
Contamination Control Area (CCA)

- **Contamination control**: procedures to avoid, reduce, remove, or render harmful (temporarily or permanently) nuclear, biological, and chemical contamination for the purpose of maintaining or enhancing the efficient conduct of military and civil operations.

In general, the objective of the contamination control area is simply to minimize the spread of contamination. To achieve this goal, rigid and established operating procedures must be followed. These procedures consist of:

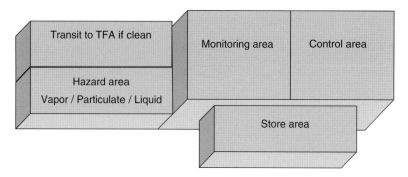

Figure 12.12 Parts of a contamination control area (CCA).

1) **Monitoring**: initial monitoring upon arrival to determine the preliminary site as well as personnel contamination.
2) **Decontamination**: decontamination procedures to minimize the spread of contamination.
3) **Control lines**: strict contamination control line procedures to control contamination spread during response/recovery/restoration operations.

To avoid any cross contamination, personnel must be monitored through the whole ingress procedure (Figure 12.12). Trained personnel are needed to decontaminate potentially contaminated people with established decontamination procedures (Chapter 13). What about CBRN contaminated people who are injured? Here we need special procedures [CBRN triage (Chapter 16)] to avoid the contamination of medical facilities and personnel by contaminated injured people.

12.3.4
Airlock – the Bottleneck for Ingress and Egress

A critical point during the whole process of ingress and egress of persons is the carryover of hazardous substances into the toxic free area zone.

> - **Airlock**: In collective radiological, biological, and chemical protection an airlock is a compartment with two doors between the toxic free area and the contamination control area or source of nuclear, biological, and chemical hazard, which is purged by clean air to allow personnel to pass from one area to another without contaminating the clean one [1].

To avoid this carryover, airlocks can be used where personnel or material have to enter the toxic free area during a CBRN incident, and where the conditions outside the protected area, the threat environment, can contaminate people and supplies. Therefore, obviously, personnel who want to enter the toxic free area have to pass through an airlock. Based on these essential considerations it is necessary that personnel and material are properly decontaminated before entering the airlock. The requirements for airlocks are to (i) retain a constant overpressure from the toxic free area toward the threat environment, (ii) restrict the migration of airborne contaminants into the toxic free area, and (iii) purge contaminants from personnel before entering the toxic free area. In general, the airlock pressure is lower than in the protective area and higher than in the unprotected area (Figure 12.8a). Uncontaminated air from the toxic free area will enter the airlock near the ceiling and exhaust near the floor (Figure 12.13). Airlocks can be single stage or two stages. Another aspect of airlock requirements is the calculation of the number of airlocks, which depends on the number of personnel/material that ingress and egress during a given time period. An airlock is also required for resupplying food, bottled water, emergency evacuation items, and mission-critical material. Furthermore, we can subdivide between stand-alone airlocks and integral airlocks. However, the airflow purge rate through the single stage airlock can provide a 1000-fold reduction in contaminants.

To get an idea of the design of an airlock system we describe the most common features of a single-airlock system. Owing to overpressure, air flows from the protective area or filter blower cascades through the airlock, thereby continuously purging contaminants from the airlock (Figure 12.13). Airflow from the top of the chamber to the bottom enhances protection by reducing contaminated air in the breathing zone (Figure 12.13). In most cases, vans and shelters modified for collective protection and buoyed in non-contaminated areas use a single-compartment protective entrance. Before entering the airlock from a contaminated area, personnel must remove their mission oriented protective posture (MOPP) gear except gloves and mask. Minor exposure to chemical agent vapor is possible between overgarment

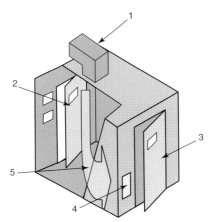

Figure 12.13 Single-stage airlock diagram with the most important features: (1) supply air duct from toxic-free area to perforated diffuser in airlock ceiling with airflow measuring station, back-draft damper, and airflow control damper; (2) inner door; (3) outer door; (4) exhaust air louver with overpressure valve, provides static pressure drop; (5) direction of airflow inside the airlock from the supply air duct to the exhaust air louver.

removal and entrance into the airlock. Clothing tends to adsorb any chemical agent vapor in the atmosphere during this brief exposure. The amount of agent adsorbed depends on the agent concentration in the atmosphere, length of exposure, type of agent, type of clothing exposed, and climatic conditions. These data cannot be calculated exactly. The air purge in the airlock flushes out the contaminated air brought in. It also reduces the amount of adsorbed agent on clothing before the soldier enters the protective shelter. After a soldier and piece of equipment enter the protective shelter, monitoring ensures that hazardous levels of agent were not carried inside.

Integrating a contamination control area to a single airlock system creates a two-stage airlock. Entering soldiers remove mission oriented protective posture (MOPP) gear in the contamination control area. This airlock design provides better control of the liquid and vapor hazards of entry and exit [14].

Finally, some thoughts about the air regeneration unit (ARU). The aim of the air regeneration unit is to regenerate the indoor air to a comfortable level for the inhabitants. As mentioned in the introductory remarks to the most important parameters affecting collective protective unit indoor climate and air quality, it is necessary to calculate the dimensions of the air regeneration unit in such a manner that the highest possible protective and comfortable level can be achieved. We must always have in mind that in a CBRN incident it is nearly impossible to open doors or windows to get relief from indoor stuffy air.

12.3.5
Toxic-Free Area (TFA)

At the beginning of this chapter we explained/defined the term indoor air quality. The hazardous effects of poor indoor air quality on the health of individuals beside the effects of CBRN substances are obvious. Health related symptoms like headache, dizziness, fatigue memory loss, and so on are symptoms of growing concern, especially in the medical area [15]. To reinforce the term we repeat that the terminology "indoor air quality" refers to the nature of the conditioned purified air throughout the space of the collective protection unit. This means that we have specific requirements for ventilation and filtration. The filtration system not only has to remove the CBRN particles from the outside coming air but also to dilute/remove contamination in the form of air-borne microorganisms, outgassing chemicals, and displeasing odors. For the toxic free area we can calculate the total required space from the number of people sheltered and the required floor area per person.

12.4
Conclusions and Outlook

We have learned in this chapter that collective protection (COLPRO) refers more or less to indoor air quality. Our first questions were: How to protect pople in a room? What are the most important parameters for an acceptable indoor air quality? Based on this knowledge we discussed the most important technical features for the design of a general collective protective system. However, we must keep in mind that for different applications we can find numerous variations and that a process of design sophistication is needed if we want to address the users need in the best possible way. However, for the future we can expect progress in the following areas:

1) standardization of equipment;
2) fielding of technology improvements;
3) full integration of COLPRO into standard shelter systems;
4) fielding of novel filtration technologies.

Especially, the the research in new catalytic filtration systems for pollution abatement [16] and the fielding of novel catalytic filtration technologies [17] could lead to increased filter capacity and performance, reduced logistics burden, reduced cost, weight, and power and expanded protection capability against toxic industrial chemicals (TICs) and toxic industrial materials (TIMs).

References

1. NATO (2006) NATO AAP-21(B): NATO Glossary of Chemical, Biological, Radiological and Nuclear Terms and Definitions, North Atlantic Treaty Organization.
2. Höppe, P. (1981) *Int. J. Biometeorol.*, **25**, 172–132.
3. Höppe, P. and Martinac, I. (1997) *J. Biometeorol.*, **42**, 1–7.
4. Höppe, P. (1984) Die Energiebilanz des Menschen. Thesis, Wissenschaftlicher Mitteilung Meteorological Institute, University of Munich, Nr. 49.
5. Höppe, P. (1993) *Experientia*, **49**, 741–746.
6. Wolkoff, P. and Nielsen, G.D. (2001) *Atmos. Environ.*, **35**, 4407–4417.
7. Moschandreas, D.J., Relwani, S.M., O'Neill, H.J., Cole, L.T., and Macriss, R.A. (1985) Characterization of emission rates from indoor combustion sources. GRI Report No. 85/0075, Gas Research Institute, Chicago, IL.
8. Matzarakis, A., Rutz, F., and Mayer, H. (2007) *Int. J. Biometeorol.*, **51**, 323–334.
9. Rigos, E. (1981) *Umschau*, **81**, 172–174.
10. Huber, G. and Wanner, H.U. (1983) *Environ. Int.*, **9** (2), 153–156.
11. US Marine Corps (1992) Collective Protection, FM 3-4 NBC-Protection, Chapter 6, Headquarters Department of the Army, US Marine Corps, Washington DC.
12. US Army Corps of Engineers (1999) Design of Collective Protection Shelters to Resist Chemical, Biological and Radiological Agents (CBR) Agents, ETL 1110-3-498, Department of the Army, US Army Corps of Engineers, Washington, DC.
13. Grondzik, W. (2007) *Air-Conditioning System Design Manual*, 2nd edn, ASHRAE Special Publications, Butterworth-Heinemann. ISBN: 978-1-933742-13-7.
14. National Institute of Building Sciences (2008) Unified Facilities Criteria: Security Engineering: Procedures for Designing Airborne Chemical, Biological, and Radiological Protection for Buildings, Unified Facilities Criteria, UFC 4-024-01, United States of America.
15. Kaushal, V., Saini, P.S., and Gupta, A.K. (2004) *JK Sci. J. Med. Educ. Res.*, **6** (4), 229–232.
16. Heck, R.M., Farrauto, R.J., and Gulati, S.T. (2009) *Catalytic Air Pollution Control: Commercial Technology*, John Wiley & Sons. ISBN 0470275030.
17. Zhao, J., and Yang, X. (2003) *Building and Environment*, **38** (5), 645–654.

Part V
Cleanup after a CBRN Event

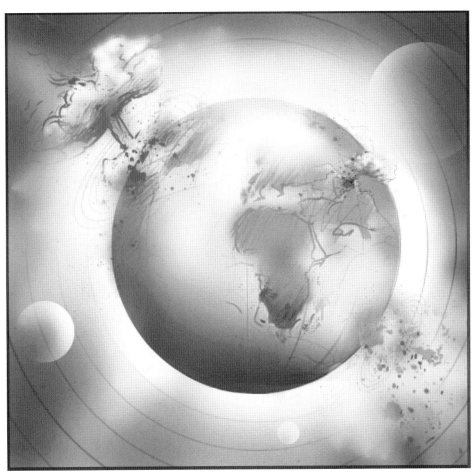

copyright by Jörg Pippirs, http://www.artesartwork.de

CBRN Protection: Managing the Threat of Chemical, Biological, Radioactive and Nuclear Weapons,
First Edition. Edited by A. Richardt, B. Hülseweh, B. Niemeyer, and F. Sabath.
© 2013 Wiley-VCH Verlag GmbH & Co. KGaA. Published 2013 by Wiley-VCH Verlag GmbH & Co. KGaA.

352 | *Cleanup after a CBRN Event*

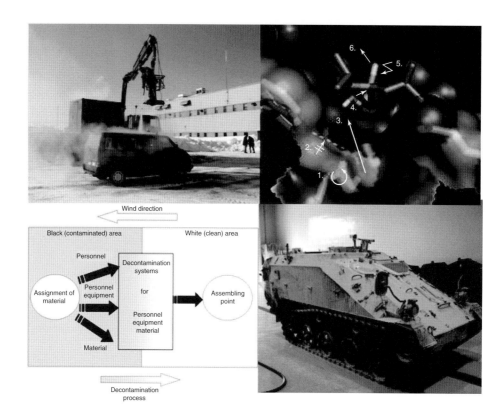

13
Decontamination of Chemical Warfare Agents – What is Thorough?

Hans Jürgen Altmann, Martin Jung, and Andre Richardt

Since the first massive use of toxic chemicals as chemical warfare agents during World War I effective decontamination has been a major part in any effective countermeasure strategy against the threat of chemical warfare agents. We have seen the application of different hazardous chemicals like organophosphates as chemical warfare agents, which are similar in their chemical properties to several products of industrial processes (e.g., pesticides). Therefore, besides the threat related to chemical attacks by terrorist or during military operations these common materials represent a constant risk in modern society. For this reason, efforts in the field of decontamination research were increased over recent decades. Our knowledge about the design of effective decontamination systems is based upon the scientific findings in the field of the kinetics and mechanism of chemical, physical, and enzymatic detoxification of toxic chemical compounds, which has became more and more sophisticated. Nowadays, we understand the general principles about the chemical and environmental fate of chemical warfare agents. Based on this knowledge we are able to design new decontaminants for specific requirements and an effective destruction of toxic chemicals. For the future we can expect more specific decontamination procedures designed for special tasks.

13.1
What Is Decontamination?

To understand the specifications in the design and development of decontamination systems we have to answer several questions:

1) What is the aim of the selected decontamination (e.g., complete cleaning or just prevention of any carry-over of the chemical warfare agent)?
2) Is the selected decontamination process linked to the detection process (if we know the contaminant, can we choose a designed decontaminant for the best decontamination result under the given conditions)?

CBRN Protection: Managing the Threat of Chemical, Biological, Radioactive and Nuclear Weapons,
First Edition. Edited by A. Richardt, B. Hülseweh, B. Niemeyer, and F. Sabath.
© 2013 Wiley-VCH Verlag GmbH & Co. KGaA. Published 2013 by Wiley-VCH Verlag GmbH & Co. KGaA.

3) What type of material has to be decontaminated (e.g., porous or non-porous surfaces, agent sorbing or non-sorbing material, electronics, or skin)?
4) Do we understand the fundamental physical and chemical processes during contamination [35, 36] and decontamination?

To answer these questions properly we have not only to understand the fate of selected chemical warfare (CW) agents in the environment. The main important chemical detoxification reactions of chemical warfare agents and the mass transport dynamics of the chemicals are also of basic interest for designing new enhanced decontaminants and, furthermore, the reader will obtain an introduction to the solvation process of the chemical warfare agents into the basic media. Based on these fundamentals we will give a summary of existing decontamination technologies and an outlook for future developments.

Before we start our tour through the world of decontamination of chemical warfare agents we have to define different terms for decontamination related topics:

- **Decontamination**: is the process of removal of hazardous material (the contaminant) from equipment, buildings, personnel, and the environment and is based on one or more of the following principles:
  - to physically screen-off the CW agent so that it causes no further damage to personnel;
  - to physically remove CW agents by evaporation, washing, or absorption;
  - to convert CW agents into less harmful or harmless products by chemical reaction (e.g., destruction).
- **Persistency**: describes the duration of an area's toxic contamination.
- **Volatilization**: is a function of the vapor pressure of the agent.
- **Thickener**: increases the viscosity of a given chemical warfare agent or a decontaminant without substantially modifying its other properties.

13.2
Dispersal and Fate of Chemical Warfare Agents

To understand the need for effective decontamination technologies we have to examine another important point concerning chemical warfare agents. The dispersal and fate of chemical warfare agents in an environment unterlies several processes and is a complicated process. Once a chemical weapon has been deployed (e.g., by detonation), the chemical agent payload is dispersed over a wide area and can undergo degradation processes such as hydrolysis, dehydration, photolysis, and oxidation [1]. A solid or liquid aerosol cloud is created, the so-called "primary" cloud. We have high ground contamination at the point of detonation and contaminated ground along the track of the aerosol-/vapor cloud, due to movement and settling of aerosols. The lifetime of the ground contamination is based on the characteristics of the dispersed chemical agent by local weather and terrain (Chapter 3). During this lifetime ground contamination can lead to casualties by direct contact of personnel with liquid agents and due to the vapor hazard (secondary cloud) over contaminated areas over a long period of time (Figure 13.1), whereas the primarily produced cloud of aerosols and vapor drifts with the wind and diluted by air. The danger from the primary cloud, ground contamination, and secondary cloud is related to local weather, humidity, soil, fauna and the chemical agent used [1]. For example, weather conditions are able to increase (increased vapor concentrations due to hot weather) or diminish (rain out) the risk for personnel or equipment and have to be taken into consideration during the process of selecting the right decontamination strategy beside the operational requirements (Figure 13.1, Table 13.1).

Another important point we have to consider is the persistency of chemical warfare agents. In general, a chemical agent is considered to be persistent when contamination lasts over a minimum of 24 h. Non-persistent or volatile agents are considered to be non-persistent when the contamination dissipates in minutes or hours. The fate of dispersed chemical warfare agents in the environment can be influenced by natural environmental conditions such as wind, rain, humidity, and temperature [1]. One of the major factors concerning the fate of dispersed chemical warfare agents is the change of their aggregate state by volatilization, which is strongly temperature dependent. In addition, photochemical reactions with sunlight lead to decreased concentrations of the chemical warfare agent in the environment [1]. To calculate the precise lifetime of a chemical warfare agent on

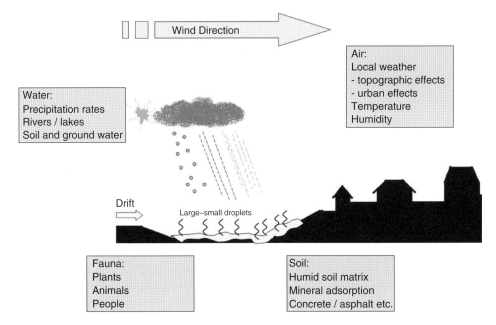

Figure 13.1 Fate of a chemical warfare agent in an outdoor environment. Factors influencing the fate are i) air, ii) water, iii) soil and iv) fauna.

Table 13.1 Factors affecting the fate of chemical warfare agents in an outdoor environment.

Air	Water	Soil	Fauna/Flora
Transport			
Local weather	Rain	Humid soil matrix	People
Topographic effects	Soil and ground water	Asphalts	Animals
Urban effects	Precipitation rates	Concrete	Plants
	Rivers/lakes	Plastic	
Humidity	Urban systems		
Fate			
Temperature	Solubility	Mineral adsorption	Life-times
Water	Adsorption	Matrix effects	Toxic effects
UV	Oxidation	Biodegradation	Surface chemistry of plants
Ozone	Catalysis	Surface catalysis	Reversibility
	Chlorine reactivity		

contaminated ground or equipment information about the physical and chemical properties of the agent (Chapter 3), the local climate conditions, the existing fauna and soil matrix effects are necessary. But we have to confess that the chemical transport and fate of chemical warfare agents is a complicate process and not fully understood yet. The transition of every chemical compound among different physical states (gas, liquid, or solid) depends on the surrounding environmental conditions (table 13.1) and there is a driving need for a better understanding of the transport and fate of chemical warfare agents in the environment. However, based on this information and knowledge of evaporation and transport effects in the atmosphere we should be able generate algorithms that are able to calculate lifetimes for the chemical warfare agents under defined weather conditions (Figure 13.2).

13.3
Decontamination Media for Chemical Warfare Agents

There is a multitude of possibilities through which to cope with contamination by toxic chemicals and the present section is far from a complete listing of all possible decontaminants and procedures. But what are the challenges for modern C-decontaminants? Based on current discussions about new and modern decontamination media we can identify several main properties (Figure 13.3).

Decontamination media can be divided into liquid, solid and vapor based systems. In this chapter we focus first on the basic principles of liquid systems and later on a more in-depth discussion of this topic. Solid and vapor based decontaminants

13.3 Decontamination Media for Chemical Warfare Agents | 357

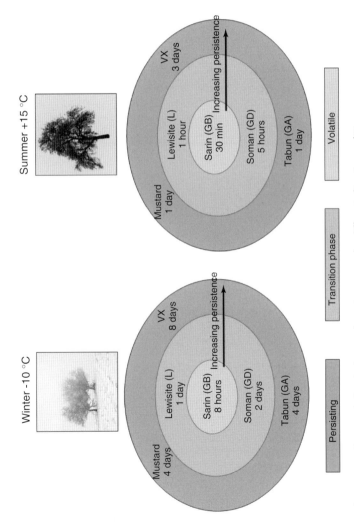

Figure 13.2 Estimated chemical agents' persistence under different climatic conditions.

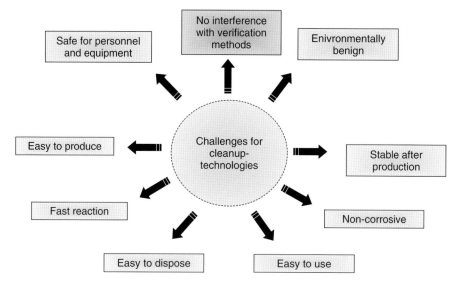

Figure 13.3 Challenges for modern C-decontamination media used in cleanup technologies.

will only be discussed briefly. Liquid systems can be divided into aqueous and non-aqueous media. The procedures and decontaminants available can be broken down into four different groups:

1) aqueous-based decontaminants,
2) non-aqueous decontaminants,
3) heterogeneous liquid media,
4) foams and gels.

We will discuss the advantages and disadvantages of all these groups with respect to decontamination efficiency, costs, material compatibility, and their influence on the environment.

13.3.1
Aqueous-Based Decontaminants

Aqueous based decontaminants have the great advantage of using water as the main ingredient from natural resources. Water would be the ideal medium to remove contaminants from a surface. It is inexpensive, not harmful to humans and the environment, and available in large amounts.

13.3.1.1 Water
If we want to use water as a decontaminant we have to consider its poor ability to dissolve hydrophobic substances such as most chemical warfare agents and its very poor ability to degrade these substances rapidly at neutral pH. Thickened chemical warfare agents are even more difficult to dissolve in water. The ability of

water to dissolve chemical warfare agents and decompose them by hydrolysis can be improved by adding "wetting substances" to the water. These "surfactants" improve the ability of water to remove chemical warfare agents from surfaces and also enable water to solubilize them more rapidly. This increases the rate of degradation in such a way that higher concentrations of dissolved material are available for the decontamination reaction. However, in most cases, even with the improved ability of water–surfactant mixtures to solubilize the hazardous material, the reaction velocity is not high enough to "clean up" a surface within acceptable time limits [2].

13.3.1.2 Water-Soluble Decontamination Chemicals

Owing to the difficulties with water as decontaminant, reactive components are needed to improve the effectiveness of water-based decontamination systems and to accelerate the decontamination process. We can identify some common water-soluble chemical products that are able to accelerate the reaction velocity of aqueous solutions and improve the ability of water to decontaminate surfaces (Table 13.2).

The main drawback of water-based decontaminants is their inability to penetrate hydrophobic paint and plastic layers. Hence, waterborne systems cannot extract the chemical warfare agents dissolved in the surface layer. Moreover, the thickeners used are often hydrophobic and, therefore, thickened agents are also hardly dissolved using water based decontaminants. In employing Table 13.2 we should be aware that in most applications residual agent will remain in the varnished surface; the remaining amount still represents a significant hazard for personnel. Dependent on surface material and temperature the agent will desorb and evaporate from the surface more or less rapidly and at higher temperatures the concentration of the toxic compound might again reach dangerous levels, at least close to the surface. The decontamination efficiency of waterborne systems can only be increased by using mechanical force, such as rubbing or scrubbing, and by combining the use of mechanical cleaning with surface active substances to improve solubilization. This means that the operator is highly involved and there is a high logistical burden to ensure the utilizability of the decont system. More work performed in chemical protection suits automatically means a need for more decontamination personnel resources. In addition, the high physical stress placed on the personnel will lead to strong signs of fatigue at the end of the decontamination procedure.

13.3.2
Non-aqueous Decontaminants

Non-aqueous decontamination systems combine the advantage of a smaller logistical burden for the user with a better solubilization for the most relevant CW agents. Non-aqueous decontaminants can be prefabricated and used immediately after a chemical or biological attack. Some of the relevant systems are commercially available (Table 13.3).

Table 13.2 Water-soluble chemical products suitable for speeding up the reaction velocity of aqueous solutions (from variable sources).

Decontaminant	Active ingredient	Application	Material compatibility
High test hypochlorite (HTH)	Ca hypochlorite	Terrain, equipment	Highly corrosive
Chlorinated lime	Ca hypochlorite	Infrastructure, equipment	Corrosive
Super tropical bleach (STB)	Ca hypochlorite	Infrastructure, equipment	Corrosive
COTS[a] – aqueous solution of sodium hypochlorite	Na hypochlorite	Infrastructure, equipment	Corrosive
Peroxodisulfate/ sodium chlorite	ClO_2	Water, B-decont	Corrosive, less corrosive than hypochlorite
Isocyanuric acid, sodium and potassium salts	Hypochlorite; in combination with organic solvents and as aqueous solution	Equipment, water, ointments	Slightly corrosive
Sodium hydroxide	Alkaline pH	Equipment	Non-corrosive, may be harmful to synthetic material
Lithium hydroxide	Alkaline pH	Equipment	Destroys alkyd paints, difficult to handle
Hydrogen peroxide	Oxygen in statu nascendi,	Equipment	Destroys dyes
COTS[a] peroxide contaminant household detergents	Oxygen in statu nascendi, alkaline pH	Equipment	Destroys dyes, harmful to skin
COTS[a] surfactants	Improvement of agent solubility	Equipment, clothing, human skin	May be harmful to skin
COTS[a] household detergents	Improved solubility of agent, slightly alkaline pH	Clothing, equipment, as hasty decontaminant human skin	May be harmful to skin
BX 24	Hypochlorite (Fichlor based)	Equipment	Corrosive
Decon Green	Peroxide based	Equipment	Low corrosivity

[a]COTS = commercially off the shelf.

Table 13.3 Non-aqueous decontamination systems.

Decontaminant	Active ingredient	Application	Material compatibility
Dichloramine T	Chlorine; only in combination with organic solvents like dichloromethane	Equipment, clothing, human skin in ointments	Corrosive
Monochloramine T	Hypochlorite; not applicable for GB,VX,GD; soluble in water and alcohol	Clothing, human skin in ointments	Less corrosive than STB and relatives
Dichloramine B	Chlorine; only in combination with organic solvents like dichloromethane	Equipment	Corrosive
Hexachloro-melamine	Chlorine; only in combination with organic solvents like dichloromethane	Equipment	Corrosive
DANC (chlorinated hydantoin in $C_2H_2Cl_4$)	Chlorine; not applicable for GB decontaminant	Sensitive equipment	Less corrosive than hypochlorite
DS2	2-Methoxyethanol diethylenetriamine and NaOH	Equipment	Can be harmful to paints, plastics, and elastomers
GDS 2000	Diethylenetriamine, several amino alcohols and Na alcoholates	Equipment	Can be harmful to alkyd paints, plastics, and elastomers
GD5	2-Amino-ethanol, KOH	Equipment	Can be harmful to alkyd paints, plastics, and elastomers
GD6	2-Amino-ethanol, potassium amino-ethoxide	Equipment	Can be harmful to alkyd paints, plastics, and elastomers
RSDL	Sodium (2,3-butanedione-monooximate) combined with poly(ethylene glycol) and tetraglyme	Human skin, wounds; sensitive equipment	Slower reaction with HD than with nerve agents
Phase-transfer catalysts	For example, N,N-hexadecane-methyl-diethanolammonium bromide	Sensitive equipment, human skin	Very slow reactions with HD

The average surface concentration necessary to decontaminate with non-aqueous systems is lower than with aqueous decontaminants. Compared with a 10% aqueous hypochlorite solution, which is applied in amounts of $6-10\,\text{l m}^{-2}$, organic decontaminants like DS 2 or GDS 2000 are applied in $0.5-1\,\text{l m}^{-2}$, which is only a tenth of the amount for an aqueous system. Non-aqueous decontaminants and formulations such as DS2, GD5, GD6, and GDS 2000 can cope with greasy surfaces and have an improved ability to dissolve and decontaminate thickened CW agents [3]. Scrubbing the surfaces helps to make the formulations more effective and is necessary to decontaminate strongly contaminated surfaces with thickened agent. The use of non-aqueous decontaminants is directly linked to problems resulting from the use of large amounts of organics, such as:

- only usable in combination with organic solvents [2],
- flammability,
- for some formulations upper and lower explosive limits are established,
- use of toxic or harmful organic solvents,
- environmental problems after use,
- toxic waste remaining after decont procedure,
- storage and handling problems,
- commercially off the shelf (COTS) products could be harmful to paints, plastics, and elastomers,
- fairly high costs of the COTS-available products in comparison to water-soluble decont chemicals.

An important reason for the use of "water free" chemicals and formulations is the decrease in the logistic burden arising from transport organization due to not needing large amounts of water for mixing the decontaminant solutions. Non-aqueous decontaminants are normally ready to use or premixed, so that only small handling operations are necessary immediately before use. The decontaminants or decontaminant systems shown in the Table 13.3 are mainly COTS products and are used by several armies worldwide.

13.3.3
Heterogeneous Liquid Media

We know that CW agents are organic compounds of low polarity and most reactants have polar character. Based on these facts mixtures of water and oily ingredients should be an ideal combination for the design of liquid decontamination media.

13.3.3.1 Macroemulsions (Emulsions)
Macroemulsions (or simply "emulsions") are mixtures of water, oil, and small amounts of an emulsifier (surfactants, amphiphilic block copolymers, or lipids). Formation of the emulsion usually occurs under vigorous stirring by means of a high-speed stirring device. Macroemulsions are not in thermodynamic equilibrium and, hence, are unstable with respect to a macroscopic phase separation due to Oswald ripening and coalescence of the droplets forming the emulsion.

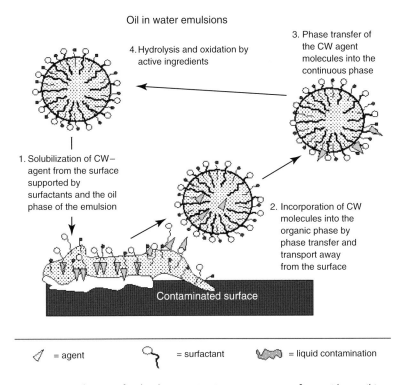

Figure 13.4 Schematic for the decontamination process on surfaces with an oil-in-water emulsion.

The droplet size is usually in the range of micrometers. Hence, based on the characteristic length scale in these systems, they should be called microemulsions. However, for historical reasons the term microemulsion is used for another type of system (Section 13.3.3.2). Emulsions combine the advantages of the aqueous and non-aqueous decontaminants. Two forms are possible:

- water in oil emulsions;
- oil in water (o/w) emulsions.

In a water-in-oil emulsion, water droplets are suspended in a continuous oil phase (Figure 13.5 below). In an o/w system the continuous phase is formed by water (Figure 13.4). The preferable form for decont applications is the water-in-oil form because the continuous phase is oil (organic solvent). The used solvent type has to match the solubility parameters or Hildebrand solubility parameters of the CW agents as closely as possible. The Hildebrand solubility parameter [4] is a dimensionless number that is available for many solvents and other organic compounds and can be calculated from the Hildebrand expression [4]:

$$\delta = \left[\frac{\Delta E}{V}\right]^{\frac{1}{2}} = \left[\frac{\rho(\Delta H v - RT)}{M}\right]^{\frac{1}{2}}$$

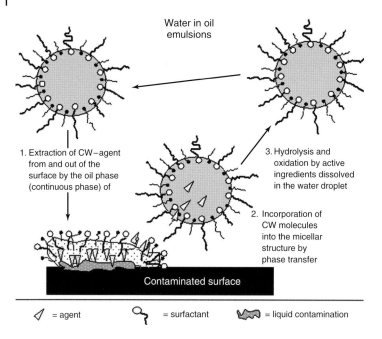

Figure 13.5 Water-in-oil emulsions used for decontamination of surfaces.

where

δ is the Hildebrand parameter,
ΔH_v the enthalpy of vaporization of the respective compound,
M is the molecular weight of the compound,
ρ is the density,
R is the gas constant (8314 J K^{-1} mol^{-1})
T is the absolute temperature.

Based on this expression the value of δ can be used to predict the solubility of most solutes (here the chemical warfare agent) in most solvents (Table 13.4). A further possibility to calculate the solubility of chemical warfare agent, are the Hansen solubility parameters, which are based on several empirical parameters. For further information on this parameters the interested reader is referred to the relevant literature [The determination of solubility parameters of solvents and polymers by means of correlations with other physical quantities.

Figure 13.5 shows a schematic decontamination process using water-in-oil macroemulsions [5].

During the decont process the oily phase (continuous phase) is in direct contact with the contaminated hydrophobic surface and has to dissolve CW agents and thickeners, to remove them from the surface and to penetrate the surface so as to extract CW agent from paint layers or plastic material. Decomposition of the agent takes place inside the water droplet or at the interface between the aqueous and oily phase [6, 7]. Therefore, the dissolved agents have to be transported to the

Table 13.4 Hildebrandt parameter of selected chemical warfare agents, solvents, and thickeners.

Chemical warfare agent	Solubility parameter $[(\text{cal cm}^{-3})^{\frac{1}{2}}]$	Solvent	Solubility parameter $[(\text{cal cm}^{-3})^{\frac{1}{2}}]$	Thickener	Solubility parameter $[(\text{cal cm}^{-3})^{\frac{1}{2}}]$
GA	9.0	Hexane	7.3	Poly(methyl methacrylate) (PMMA)	8.9–12.7
GB	9.0	Heptane	7.4	Poly(ethyl methacrylate)	8.5–11.1
GD	8.6	Jet fuel	≈7.4	Poly(n-butyl methacrylate)	7.4–11.1
GF	9.8	Gasoline	≈7.5	–	–
VX	8.8	Diesel	≈7.6	–	–
HD	10.6	Tetrachloroethylene	9.3	–	–
HN1	9.0	Dichloroethane	8.5	–	–
HN2	9.3	Ethylene glycol monomethyl ether	10.5	–	–
HN3	9.3	Ethanol	12.7	–	–

boundary layer between oil and water phases, and surfactants (chemicals with a phase-transfer capacity) could carry them into the water droplet.

In contrast to w/o emulsions the extraction effectiveness of o/w emulsions is limited because water is the contact media to the surface. Figure 13.4 shows a schematic decontamination process operating with o/w emulsions [5]. Figure 13.4 above shows a schematic decontamination process operating with o/w emulsions [5]. The extraction effectiveness of o/w emulsions is limited because water is the contact media to the surface. The surfactants used in o/w emulsions as emulsifiers must have phase-transfer properties to improve the solubilization of contaminants. If the surfactants used do not have phase-transfer abilities, the reaction rates are low despite the dominant rate of solubilization. This means that the overall effectiveness will be no better than when using a fully aqueous system such as bleach or high test hypochlorite (HTH) solutions.

A few decont emulsions have been introduced into the military inventory. One example is the German Emulsion, introduced into the German Army, and a relative of the German Emulsion, namely, the Xylene Emulsion of the Austrian Army (Table 13.5). Both decont formulations are water-in-oil emulsions. They use tetrachloroethylene or xylene as solvents. In both emulsions HTH or sodium diisocyanurate (Fichlor) is employed as active component. Both solvents have advantages and disadvantages. Tetrachloroethylene is the better choice as solvent for CW agents because it is not flammable. Unfortunately, it causes environmental problems because its half-life in soil is up to six years and some of its metabolites are toxic [8]. In contrast, xylene is a flammable liquid with a fairly low flash point of 21 °C a half-life time under solar radiation of two to four weeks.

Table 13.5 Examples of macroemulsions.

Emulsion type	Type and concentration of active substance	Type and concentration of surfactant	Type and concentration of solvent (%)	Amount of water (%)
C8 emulsion (German Emulsion)	HTH^a 7.5%	Anionic/nonionic 50/50, 1%	15	76.5
Xylene emulsion	HTH 1% or Fichlor 1–2%	Anionic/nonionic 50/50, 1%	10	77

^{a}HTH = high test hypochlorite.

The effectiveness of both emulsion types against CW agents is well tested; the formulations are used in the German and Austrian Armies. The formulations are suitable for decontamination equipment even under difficult conditions such as extremely dirty vehicles, thickened CW agent contamination, absorbing paints, and so on.

13.3.3.2 Microemulsions

It is well established that two immiscible liquids can form a macroemulsion by using surfactants or surfactant mixtures and strong mixing. However, as mentioned above, macroemulsions (or simply emulsions) are thermodynamically unstable and will phase separate due to Ostwald ripening and coalescence of droplets. The lifetime of emulsions can be improved by adding block copolymers and other rather expensive additives, but eventually they always segregate.

Microemulsions are based on a different concept compared to emulsions. Microemulsions are also obtained by mixing the immiscible liquids water and oil in the presence of surfactant and in some cases of cosurfactants (often alcohols). However, due to the chosen amounts of the components microemulsions are in thermodynamic equilibrium and therefore form spontaneously. This unique class of optically clear, or opaque, thermodynamically stable and normally low viscous liquids has been subject to intensive research due to its properties and technological importance [9, 10].

Microemulsions have the advantage over macroemulsions of having up to 1000 times larger interfacial surface areas, which leads to a much faster reaction between water-soluble active decontamination chemicals and oil-soluble CW agents (Table 13.6). This is simply related to the enormous internal surface and the improved contact between the solvents. Owing to this reason, microemulsions are used in the chemical industry to replace expensive phase-transfer catalysts. Moreover, they can also be used to reduce the necessary amount of harmful organic solvents. When the composition is chosen appropriately they still have a high capacity to solubilize both oil- and water-soluble compounds.

Moreover, it is possible to introduce less corrosive and less environmentally harmful chemicals into the decontaminant formulation. In addition, bicontinuous

Table 13.6 Comparison of macro- and microemulsions.

	Emulsion	Microemulsion
Appearance	Milky-white	Transparent
Droplet (μm radius)	0.15–100	0.0015–0.15
Production	Mechanically or chemically	Spontaneous
Thermodynamically stability	No	Yes
Internal surface	Medium	High
Reaction velocity	Medium	High

microemulsions with hydrophobic and hydrophilic continuous phases are able to wet both hydrophilic and hydrophobic surfaces. This also leads to enhanced contact to the contaminants on the surface. The properties of microemulsion systems and their increased effectiveness can lead to a minimization of reactive compounds in decontaminant formulations and a decrease in the amount of decontaminant necessary to clean up a given surface area. This lowers the costs and the logistical burden of decontamination procedures in comparison to systems based on aqueous solutions or even macroemulsions.

In decontamination research several developments of microemulsions for decont purposes have taken place [7, 11, 12]. One interesting development is a microemulsion formulation to which different reactive compounds, even enzymes, can be added [10]. This makes a decont system based on microemulsions very flexible and offers the possibility of developing tailored decont solutions for different applications. Laboratory tests and published reports show that the implementation of enzymes active against G-agents is possible and effective [13].

13.3.3.3 Foams and Gels

We can find several reasons for the development of foams and gels. They can be non-corrosive and environmentally friendly. Another reason can be the visibility rendered as to where the decontaminant has been applied. Foams and gels can decompose the contaminant by hydrolysis and oxidation in combination.

Foams The most common definition of foam is that it is a substance consisting of many gas bubbles that are encased in a liquid or solid. Consequently, foams are extremely complex systems consisting of polydisperse gas bubbles separated by a network of interconnected draining films called lamellas. Idealized foams are closely linked to mathematical problems like space-filling and minimization of surfaces. Several conditions are needed to produce foam. There must be mechanical work to produce the foam, surface active substances (surfactants) to reduce the surface tension, and the formation of foam must be faster than its breakdown. Real-life foams are typically disordered and have various bubble sizes. A big advantage over "conventional" decont solutions is that it can be easily tailored to the needs of the user and can thereby also be used for different systems like fire fighting foam systems. In addition to this advantage foam has some other interesting properties, such as:

1) sticking on nearly any kind of surface;
2) slowly decomposing under bleeding out fresh decont solutions;
3) less use of consumables due to the low density;
4) adjustable residence time on a surface.

These properties are preferable for decont systems, but difficult to realize with decont solutions on the market. A foam system can produce "wet" or "dry" foams depending on the type and concentration of surfactant or surfactant mixture. The preferable form for decontamination purposes is foam that releases permanently liquid. Owing to bleeding of the foam the surface is steadily exposed to fresh solution and decontamination can take place. Wet foam sticks on every kind of surface independent of its orientation in space, which means it sticks on horizontal or vertical and even on overhanging surfaces. The ideal foam system should combine the properties of foam with the positive properties of emulsions or microemulsions, such as:

1) higher viscosity than aqueous solutions;
2) integrated solvent to improve extraction capacity from non-resistant surfaces;

Combination of these properties leads to a three-phase foam, where an emulsion or microemulsion builds the foam cells. The bleeding effect can take place immediately when this decontaminant is sprayed on a surface. The first commercially available system is the so-called C8 emulsion [14], which brings the decontaminant consumption from, formerly, $3 \, l \, m^{-2}$ pure emulsion down to $0.5-1 \, l \, m^{-2}$ foamed emulsion. Another well-tested system is the Canadian CASCAD foam system. Originally designed for decontamination of military equipment it is one of the eight decontaminants identified in a study of the Joint Fixed Site Decontamination program of the US Government [15] as preferable decontamination formulation for the decontamination of infrastructure. The list of formulations presented in Table 13.7 is far from being a complete compendium of all foam based decontamination systems but will lead the reader to further literature.

Gels A gel is a three-dimensional network made of crosslinked polymer chains or of aggregated colloids. If the network is only filled with gas the gel is called an aerogel. Typical aerogels are often made by sol–gel chemistry using silica precursors. These systems have exceptional properties with respect to heat conductivity, but for the purpose of decontamination aerogels are not useful. If the 3D network is swollen by a solvent, the system is called a lyogel. A common case is so-called hydrogels with water as the solvent component. Lyogels can be seen as intermediates between solids and liquids. They could be called viscoelastic solids. Both by weight and volume, gels are mostly liquid in composition and thus exhibit densities similar to liquids; however, they have the structural coherence of a solid. An example of a common gel is edible gelatin (here the crosslinks are entanglements and the gel is a physical gel). The application of a finite shear to a

Table 13.7 Foam-based decontamination systems.

Decontaminant	Active ingredient	Application	Material compatibility
CASCAD	Hypochlorite (Fichlor) and surfactants	Equipment decontamination, blast reduction	Less corrosive than HTH-solutions
Sandia Foam	Peroxide based	Equipment, buildings	Low corrosivity
Easy Deon	Peroxide based, formulation equals Sandia Foam	Equipment, buildings	Low corrosivity
MDF 200	Peroxide based, formulation equals Sandia Foam	Equipment, buildings	Low corrosivity
Decon Schaum	Tensides plus active ingredients	Equipment, buildings	Low corrosivity
Foamed C8 emulsion	Hypochlorite macroemulsions foamed	Equipment	Corrosive

Table 13.8 A commercially available gel.

Decontaminant	Active ingredient	Application	Material compatibility
L-Gel	Oxone ($2KHSO_5 \cdot KHSO_4 \cdot K_2SO_4$)	Equipment, paints surfaces	Low corrosivity

gel system after a long rest may result in a decrease of the viscosity. If the decrease persists when the application of shear forces is discontinued, this behavior is called work softening (or shear breakdown), whereas if the original viscosity is recovered the behavior is called thixotropy. The fact that gels have an internal liquid phase makes them interesting for the development of decontaminants. There are a few developments found in the open literature (e.g., L-Gel, Table 13.8) [16]. This might be an area of research for the future as much as the microemulsion area is in the present.

13.4
Selected Chemical Warfare Agents and Decont Reaction Schemes

Most of the common chemical warfare agents will break down via oxidation or hydrolysis. However, the sulfur atoms of VX and sulfur mustard (HD) are readily subjected to oxidation. In addition, the central phosphorus atom of VX and the

G-agents may be subjected to hydrolysis. VX hydrolysis can result in a toxic breakdown product [1]. Knowledge about the decont reaction schemes of different chemical warfare agents has lead in consequence to the design of decontaminants that are able to the design of decontaminants that are able to target systematically the weaknesses of the chemical warfare agents towards hydrolysis or oxidation.

13.4.1
Sulfur Mustard (HD)

Mustard, also known as *HD*, *Yperite*, or *Lost*, is a colorless and odorless liquid with the chemical name 1,1-dichlorodiethyl sulfide. It is a strong vesicant [1, 17] and is persistent in the environment. Especially under cold weather conditions a sulfur mustard contamination can cause injuries up to weeks or even years after release, depending on the environmental conditions of the contaminated area.

13.4.1.1 Hydrolysis
The first step in the hydrolysis of HD (Ia) is the formation of a positively charged sulfonium ion (Ib) through a molecular internal ring-closure reaction with a release of chloride ion as negatively charged counterpart. A water molecule is now able to substitute a ring carbon atom by opening the ring structure and so form hemi-mustard (IIa) and hydrochloric acid (Scheme 13.1). Owing to the chemical equilibrium between Ia and Ib, the amount of hemi-mustard increases during the hydrolyzation process. However, hemi-mustard is also a vesicant, which reacts with water similarly to the original mustard to form thiodiglycol (III) (TG) and hydrochloric acid. The cyclic intermediate (IIb) product reacts in an analogous reaction by internal displacement and forms 1,4-thioxane and hydrochloric acid. The ratio between TG and thioxane produced by the hydrolysis reaction is given as 4 : 1 in the literature [3]. The rate of the full hydrolysis reaction of mustard is fairly high. The limiting factor clearly is the amount of mustard dissolved in water. Mustard is very persistent in an aqueous environment and the dissolution velocity of mustard is low. This has consequences for decontamination processes using plain water. Thorough decont of mustard contaminated surfaces with plain water and even with water plus detergents is not very effective [2, 3, 18]. Further information on chemical reactions of mustard is given in Scheme 13.1 and can be drawn from the literature [2, 18].

13.4.2
Sarin (GB)

Sarin is a colorless, odorless, and very toxic liquid. Its chemical name is *O*-isopropyl methylphosphonofluoridate and like other nerve agents it acts as a strong cholinesterase inhibitor. Decontamination is relatively simple because hydrolysis can be accelerated by using basic aqueous solutions such as Na_2CO_3, NaOH, or KOH solutions [2, 14, 19].

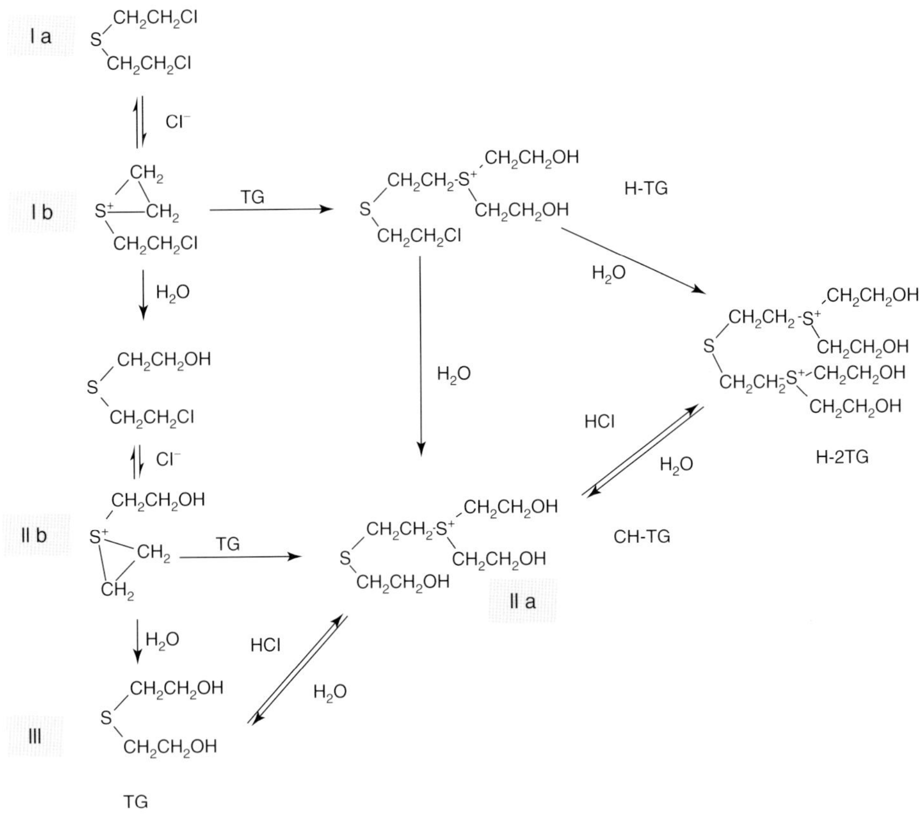

Scheme 13.1 Hydrolysis of mustard (HD).

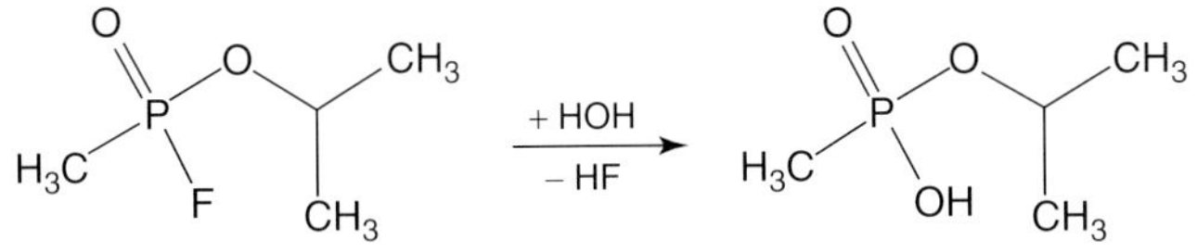

Scheme 13.2 Hydrolysis of sarin.

The reason for this is that the hydrolysis of sarin is strongly dependent on the pH of the solution. The half-life time at pH 1 is 15 min., while at pH 5 it is 165 h, and at pH 13 it is 0.3 s. Sarin is miscible with water in any ratio and at neutral pH the rate of hydrolysis is $50\,h^{-1}$. The decomposition reaction with water at neutral pH follows Scheme 13.2 [2].

Hydrolysis can be accelerated by increasing the pH, adding anionic-, nucleophilic catalysts, or metal chelate complexes [2, 19].

13.5
Soman (GD)

Soman is one of the toxic compounds most used as a CW agent. It is a colorless liquid with a fruity odor; the industrial product is yellow–brown with a camphor-like odor. Soman is degraded by hydrolysis in acidic, neutral, and basic media (Scheme 13.3). The end products of hydrolysis are hydrogen fluoride and pinacolyl methylphosphonate [2, 18].

Owing to its chemical similarity to sarin (GB), the hydrolysis of soman can be catalyzed by the hypochlorite anion and even heavy metal complexes such as copper(II) accelerate the soman hydrolysis to, for decont purposes, useful rates [20]. Na_2CO_3, KOH, and NaOH solutions are also used in decontamination procedures for the fast decomposition of soman, with half-life times of approximately less than 1 min [2, 18, 19, 21].

13.6
VX

Although VX is many times more persistent than the G-agents, it is very similar to GB in mechanism of action and effects. The substance has a very low volatility. The attack of a hydroxide on the P atom is the first step in the basic hydrolysis of VX (I) (Scheme 13.4). It forms a phosphorus intermediate (II). The phosphorus intermediate can decompose in two ways:

1) The anion of diisopropylaminoethane thiol is expelled and as a result the ethyl ester of methylphosphonic (III) acid is formed.
2) Ethoxide is expelled to form a compound called *EA 2192 (IV)* [22–24].

The ratio between EA 2192 and the reaction product of pathway 1 is 13% EA 2192 to 87% methylphosphonic ester. EA 2192 is stable against further hydrolysis (the hydrolysis rate of EA 2192 is about 1000 times slower than that of VX) and is just as toxic as VX. Therefore, the formation of this compound has to be

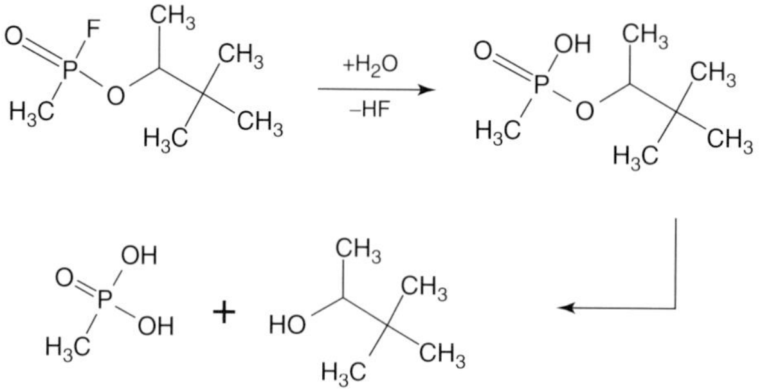

Scheme 13.3 Hydrolysis of soman.

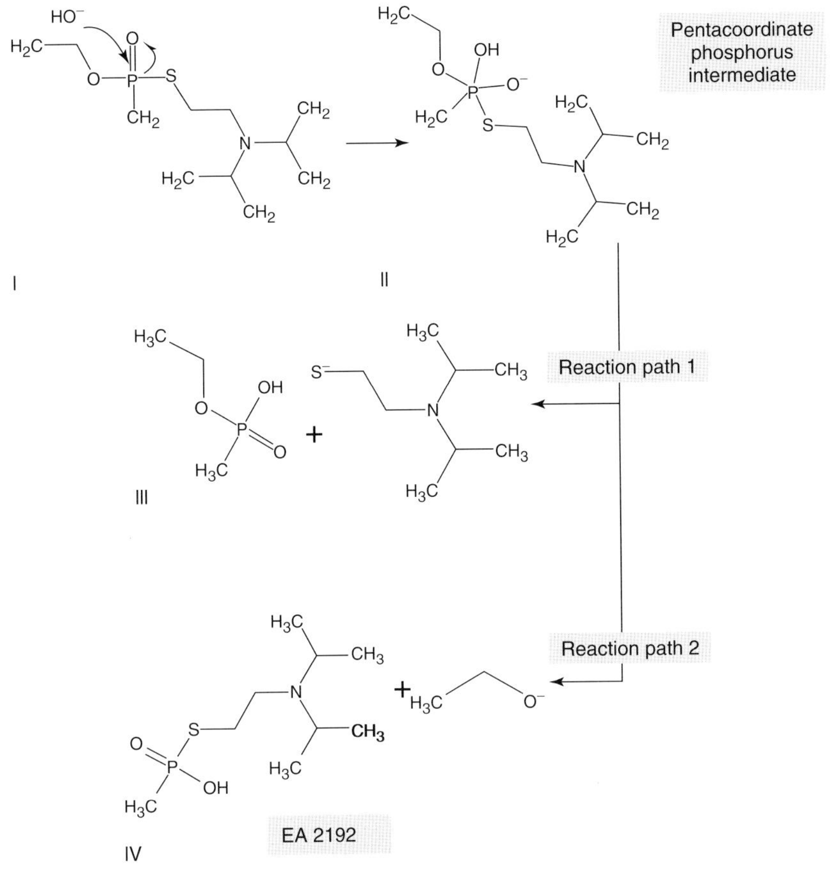

Scheme 13.4 Two possible pathways for the hydrolysis of VX.

avoided by running the hydrolysis process during operations as shown in pathway 1 (Scheme 13.4).

A third way to destroy VX during hydrolysis is by displacement of the thiophosphonate anion from the carbon atom (Scheme 13.5) [22].

The hydrolysis of VX is a slow reaction and, as shown in Scheme 13.4, EA 2192 is formed (at a ratio of 13%) [19]. Therefore, for the decontamination of VX, the alkaline-catalyzed hydrolysis in combination with oxidation reactions with commonly available bleach or HTH-solutions is a better choice [23–26].

13.7
Catalysis in Decontamination

The use of catalysts in liquid or solid form could have several advantages compared to stoichiometric based decontamination of chemical warfare agents [13]. Catalysts

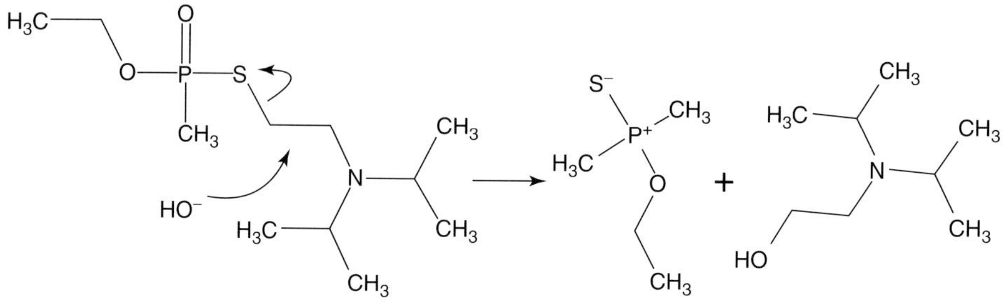

13 Decontamination of Chemical Warfare Agents – What is Thorough?

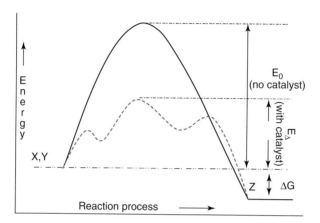

Scheme 13.5 Hydrolysis of VX – hydrolytic attack of the S–C bond.

Figure 13.6 Generic potential energy diagram showing the effect of a catalyst in a hypothetical exothermic chemical reaction $X + Y \rightarrow Z$. The presence of the catalyst opens up a different reaction pathway (– – –) with lower activation energy. The overall thermodynamics stay the same.

are not consumed in the decontamination process and can therefore be used in relatively small amounts compared to the reactants. In addition, catalysts can increase the rate of a reaction due to a reduction of the activation energy (Figure 13.6).

The relatively small amounts of catalysts can led to a reduction of the logistic burden. We can divide the catalytic processes usable for decontamination process into three different basics:

- metal-ion catalyzed hydrolysis and metal-catalyzed alcoholysis reactions,
- biotechnology based decontamination (hydrolysis or oxidation),
- catalytic oxidation.

For all three basic principles a general introduction will be given. The use of metal ions as catalysts in industrial processes is widely known. For this reason, the ability of several metal ions to catalyze the hydrolysis or alcoholysis of chemical warfare agents has been investigated [27]. For example, copper(II) is a potent catalyst for the hydrolysis of GB [19]. In addition, metal-ion catalyzed alcoholysis of neutral organophosphate chemical warfare agents and pesticides proceeds very rapidly [28]. Products are non-toxic and can be simply disposed by conventional means. Two main metal ions are the focus of research:

1) lanthanide-containing catalysts for decomposition of P=O based materials – very effective for the destruction of nerve agents [29];
2) palladacycle catalysts for P=S based pesticides [30].

There are certain advantages to the alcoholysis processes (rate, substrate solubility, no EA2192 when used for VX decontamination).

Biotechnology based decontamination methods have the charm of being environmentally friendly and not being harmful for personnel and equipment [13]. Over the years, many enzymes have been studied that show some ability to hydrolyze or to oxidize chemical warfare agents. At present, several enzymes have entered a stage of development beyond the laboratory. They show the required activity, stability, and can be produced in large quantities. The enzymes diisopropyl fluorophosphatase (DFPase) from the squid *Loligo vulgaris*, organophosphorus hydrolase (OPH) from *Pseudomonas diminuta*, and organophosphorus acid anhydrolase (OPAA) from *Sphingomonas* are examples of hydrolytic enzyme systems. As an example of oxidation enzyme systems oxidoreductases with oxidative and broad substrate specificity might be useful for destroying a wide range of chemical warfare agents simultaneously. The improvement of several enzyme systems is under way and in the future we can expect optimized systems. Liquid systems with granulated enzyme powder and immobilized enzymes could be integrated into equipment and fixed on fabrics [13].

Photocatalysis has been widely applied to solar-energy conversion and environmental purification [37]. Different photocatalysts, based typically on titanium dioxide (TiO_2), produce active oxygen species under irradiation of different wavelengths and decompose not only conventional pollutants but also different types of hazardous substances such as chemical warfare agents at mild conditions (Figure 13.7) [31]. The currently available data indicate that photocatalysis, which may not always have the striking power of decontaminants like calcium hypochlorite, certainly helps detoxification of hazardous compounds such as chemical warfare agents [32].

13.8
Decont Procedures

The next subsection gives an overview of decontamination procedures and equipment as a guide. The procedures and substances presented here are not an exhaustive evaluation of all available decont technologies and decont equipment, but it will give the reader an impression of the different procedures and why an overall concept not only for decontamination but also for the whole sector of CBRN-defense is necessary.

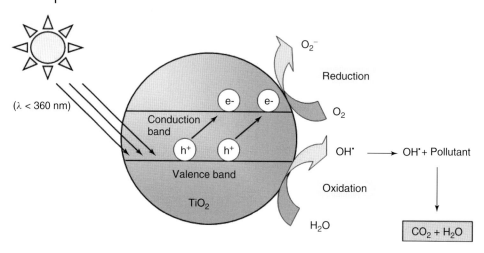

Figure 13.7 Photooxidation of chemical warfare agents by photocatalysts.

13.8.1
Generalities

Decontamination of personnel and equipment means in general the division of the decont side into two parts, a black (contaminated) and a white (clean) part (Figure 13.8). In the black part contaminated personnel and material are assigned to appropriate decont procedures and in the white part personnel gear, vehicles, and equipment without the need to wear NBC-protection are reassembled. One of the most important parts is the design of a decont site under consideration of the wind direction (Figure 13.8).

13.8.2
Equipment Decontamination

The decontamination of equipment can be divided into two major processes:

1) wet procedures, using liquid decontaminants on an aqueous or non-aqueous base;
2) dry procedures using adsorbents.

Most of the decont procedures used by armed forces worldwide are based on "wet procedures," with aqueous or non-aqueous decontaminants for the thorough decontamination of equipment.

13.8.2.1 Wet Procedures
As mentioned above "wet procedures" use water or solvent based decontaminants (emulsions/microemulsions are a combination of solvent and aqueous solutions). In general, they can be used with the same type of spray equipment as used for the water-based systems.

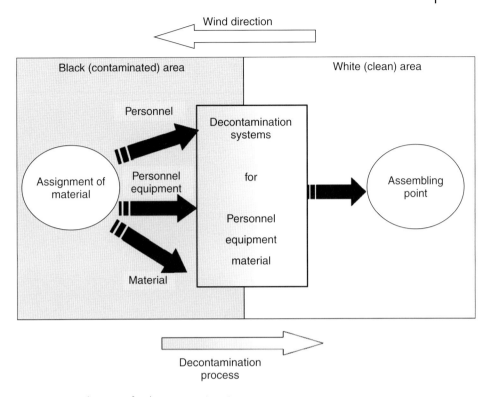

Figure 13.8 Schematic of a decontamination site.

Aqueous Decontaminants This group of decontaminants is still widely used by different armies worldwide and has been well-established since World War I. Most systems for large equipment decontamination use a mixing tank, where a solid decontamination product is dissolved in water. The ready to use solution is sprayed on the object to be decontaminated and after a dwell time of several minutes the "film" is rinsed off by using water jets with elevated pressure. For the decontamination of smaller numbers of vehicles, vehicle parts, or equipment most armies use small size decont systems operating with a pressurized storage container for the decontaminant. The decont solutions can be sprayed on the surface; sometimes a disposable brush is integrated to increase efficiency.

Non-aqueous Decontamination The main parts are the storage container or spray canister with integrated pump and brush. The decontaminant is pumped out of the storage vessel and simultaneously spread on the surface to be decontaminated. Brushing the surface helps to increase the speed of reaction due to the faster mixing of agent and decontaminant. Therefore, the decontamination of large surface areas is labor intensive and the user has to climb on higher parts of a vehicle, which makes decont operations due to DS2 on the surface quite dangerous. More modern systems operate with crane elevated platforms and spray guns, which makes decont of large vehicles easier.

13.8.2.2 Dry Procedures

Dry decontamination procedures use the exhaust energy of combustion engines, such as gas turbines or pulse jet engines. The thermal energy in form of heat and high velocity gas flow evaporates and shears off CW agents from the surface. The principle was and is still in use in Russia, where a truck mounted jet engine is used to decontaminate vehicles of any kind. The advantage of this system over wet processes is the dry surface after decont and the saving of large amounts of decontamination chemicals. Most of these systems have the additional possibility to inject aqueous solutions of decontaminants or water into the plume and use the increased reaction velocity of hot decont solutions.

13.8.2.3 Clothing and Protective Clothing

Contaminated clothing or protective clothing must be discarded and replaced by clean material. This leads to higher costs and an increased logistic burden. Therefore, armies worldwide are trying to find procedures and equipment that enable them to clean up contaminated clothing in a minimum of time with a maximum of effectiveness. Several decont procedures and processes were introduced in different armies and then replaced by "new" procedures, starting from simple washing processes over dry cleaning to treatment with adsorptive material. Especially when using protective suits, such as the well-known over- and undergarments, where the protective capacity is assured by integrating adsorptive charcoal layers, decontamination is difficult. Washing processes harm the adsorptive capacity of the charcoal layers or remove at least a good part of the integrated charcoal. Additional problems are caused by the high heat for evaporation of water, which is responsible for the slow and time consuming drying of wet garment material. Owing to the limited drying temperatures water will be adsorbed on the charcoal, blocking its adsorptive capacity. This decreases the protective capacities of the permeable protective suit down to a not-tolerable level.

Solvent treatment or dry cleaning of protective suits with integrated charcoal layers is carried out with solvents like FC113 (Fluoro Carbon 113). Tetrachloroethene or relatives cause problems by building up monolayers of solvent on the charcoal surface, which reduces the adsorptive capacity for CW agents. These systems are much faster than washing machines due to the shorter drying times, but solvent residues within the garments can cause health and environmental problems.

Another way to decontaminate garment material or protective clothing is the use of "dry processes." These dry processes operate with hot air or mixtures of overheated water vapor and hot air. The advantage over the washing process is that no drying times have to be taken in consideration because the garment material comes out of the process dry.

13.8.2.4 Decontamination of Personnel

Persons suspected of being contaminated need immediate decontamination to minimize casualties. To achieve this goal, personnel can be decontaminated by two major processes:

1) rapid (operational) decontamination to remove CW agents from skin, personal equipment, and protective gear as soon as possible;
2) thorough decontamination of personnel and personal equipment.

We now describe some of the most important and available procedures and systems for the decontamination of personnel and personal equipment.

13.8.2.5 Rapid Decontamination of Personnel and Personal Gear

In a chemical, biological, radiological, and nuclear (CBRN) attack, soldiers and civilians are likely to become contaminated. Every soldier should be well educated in terms and methods of hasty decontamination. They have to know how to react in dangerous situations and should be able to choose the right solutions. Civilians do not have this background and therefore the first responders have to react in the right way. The first responders have to follow the overarching concept for the decontamination of civilians. However, based on military experience the civilian side could learn and could adapt the military guidelines for their own purposes. The time it takes for a definitive (complete) decontamination is not given and the soldier, based on his training, should react as soon as possible to minimize agent penetration through protective or non-protective clothing causing massive casualties. One must remember that the first 10 min after the incident are critical to a favorable outcome. Under these circumstances hasty decontamination means to remove *every*, and the emphasis lies on every, reachable smear or droplet that might be CW agent. Many armies have therefore developed detailed standard operation procedures and decontamination chemicals for hasty decont. Some armies use Fuller's earth, a solid absorbent based on clay material. Fuller's earth is composed mainly of alumina, silica, iron oxides, lime, magnesia, and water, in extremely variable proportions, and is generally classified as sedimentary clay. It is a common name for several fine-grained, earthy materials that plastify after wetting with water. Chemically, clays are hydrous aluminum silicates, ordinarily containing impurities, for example, potassium, sodium, calcium, magnesium, or iron in small amounts [33].

These absorbents are able to bind liquid CW agents and make them removable from cloth surfaces or personal equipment such as weapons, gas masks, and so on. When solid absorbent materials like Fuller's earth, soil, or diatomaceous earth are used, the contaminant is usually not degraded. For example, petroleum products are readily absorbed but are not changed in their character. Thus, the sorbent material becomes as toxic and has to be collected and then disposed of. Caution needs to be taken during this collection process, because fine dust or particles can be inhaled or stick to exposed skin.

Examples of the use of adsorbents combined with reactive capacity are the M 291 Skin decont kit of the US Army and RSDL. Newer developments are the Canadian RSDL, an oxime-based liquid decontaminant with FDA approval for use on human skin.

RSDL is a topical decontamination solution that has been tested and shows the ability to reduce toxic effects from exposure to chemical warfare agents like VX and HD and T-2 toxin. RSDL contains Dekon 139 and a small amount of 2,3-butadiene monooxime (DAM). These compounds are dissolved in a solvent composed of poly(ethylene glycol monomethyl ether) (MPEG) and water. This solvent system is particularly important as it promotes the decontamination reaction by actively desorbing, retaining, and sequestering the chemical agent, while the active ingredient chemically reacts with, and rapidly neutralizes, the vesicant chemical or the organophosphorous nerve agent. This reaction starts immediately and neutralization is usually complete within seconds or a few minutes.

Future developments for hasty decontamination could move in the direction of microemulsions with catalytically active ingredients that are effective against chemical warfare agents. The advantage of these systems is the very low active ingredient concentration necessary to destroy chemical warfare agents safely in combination with skin friendly microemulsions. These systems could be usable on human skin without thinking about medical aspects such as burns due to the use of aggressive chemical compounds in the decont solution and they will be friendlier to the environment than established decontaminants.

13.8.2.6 Thorough Decontamination of Personnel

Persons suspected of being contaminated have to go through a thorough decontamination process even after hasty decont of protective gear and personal equipment. They are normally separated by gender and led into a decontamination system designed for the decontamination of personnel. The decont unit is strictly separated into a "black" or contaminated and a "white" clean area (Section 13.8.1). After passing a strip down room, where the contaminated clothing is removed, they enter a wash down room, where they are showered. After the washing procedure they enter a re-dressing room, where new or freshly decontaminated clothing and personal gear is issued, and after re-dressing they leave the decontamination unit on the "white" side. The decont unit has to be placed in the environment in such a way that no contact- or gas-hazard can occur on the white side.

13.9
Conclusions and Outlook

In this chapter we have given the fundamentals of decontamination reaction pathways. Furthermore, we have given an introduction to various decontamination media. Nowadays, we know the basic principles of how decontamination processes work and we can appraise how difficult it is to design decontamination media with tailored chemistry for specific missions. Looking ahead, we can assume that the highest research priority will be in the field of reactive decontamination and of the clearance of decontaminated equipment. The on-going research tries to identify both liquid and solid decontaminants that are easy to handle, have no adverse or corrosive effects on equipment or the environment, and are able to lower the logistic burden. Therefore, the search for catalysts will continue, especially for those catalysts that are pH independent, have a broad range of possible use, and can catalyze the hydrolysis of the ester groups in G agents as well. Another area of interest is the interaction of agents with solid decontaminants. Based on current research, many of the results may be applied to the safe destruction of chemical weapons. Especially for the development of oxidative paint systems, based on titanium dioxide, we can expect major breakthroughs. Also, a fundamental understanding of the different processes as mass transport, evaporation and absorption is essential for the evaluation of decontamination of processes. However, an understanding of the technical aspects of decontamination is only the basic for a concept to proof the success of a decontamination procedure. At the end of each decontamination procedure should stand the clearance of the decontaminated equipment and that it can be used without any healthy risk.

References

1. Talmage, S.S., Munro, N.B., Watson, A.P., King, J.F., and Hauschild, V. (2007) in *Chemical Warfare Agents in the Environment: Toxicology and Treatment*, 2nd edn, ch. 4 (eds R.C. Marrs, R.L. Maynard, and F.R. Sidell), John Wiley & Sons, Ltd, Chichester, pp. 89–125. ISBN: 978-0-4700-1359-5.
2. Franke, S. *et al.* (1977) *Lehrbuch der Militärchemie*, vol. 2, Militärverlag der DDR, Berlin, 206 ff.
3. Bartlett, P.D. and Swain, C.G. (1949) *J. Am. Chem. Soc.*, **71**, 1406–1415.
4. Weast, R.C. (ed.) (1978) *CRC Handbook of Chemistry and Physics*, 59th edn, CRC Press, p. 726.
5. Richardt, A. and Mitchell, S. (2006) *J. Def. Sci.*, **10**, 261–265.
6. Clark, J.H. (1995) in *Chemistry of Waste Minimization*, (ed. J.H. Clark), Chapman & Hall, London, ch. 5, pp. 116–140. ISBN: 978-0-7514-0220-9.
7. Fallis, I.A. *et al.* (2009) *J. Am. Chem. Soc.*, **131** (28), 9746–9755.
8. McConnel, G., Ferguson, D.M., and Pearson, C.R. (1975) *Endeavor*, **121**, 13–18.
9. Kahlweit, M. and Strey, R. (1985) *Angew. Chem. Int. Engl.*, **24**, 654–668.
10. Hellweg, T., Wellert, S., Altmann, H.-J., and Richardt, A. (2008) Microemulsions as decontamination media for chemical weapons and toxic industrial chemicals, in *Microemulsions: Properties and Applications*, ch. 14 (ed. M. Fanoun), CRC Press. ISBN: 978-1-420-08959-2.
11. Gäb, J., Melzer, M., Kehe, K., Wellert, S., Hellweg, T., and Blum, M.M. (2009) *Anal. Bioanal. Chem.*, **396** (3), 1213–1221.
12. Menger, F.M. and Rourk, M.J. (1999) *Langmuir*, **15**, 309–313.
13. Richardt, A. and Blum, M.M. (2008) *Decontamination of Warfare Agents – Enzymatic Methods for the Removal of B/C-Weapons*, Wiley-VCH Verlag GmbH, Weinheim. ISBN: 978-3-527-31756-1.
14. Wagner, C.W. and Yang, Y.C. (2002) *Ind. Eng. Chem. Res.*, **41** (8), 1925–1928.
15. Science Applications International Corp (2005) Compilation of Available Data on Building Decontamination Alternatives, EPA600-R-05-036, U.S.

Environmental Protection Agency, Washington, DC. http://www.epa.gov/NHSRC/pubs/600r05036.pdf (accessed 30 December 2010).
16. Sandia (2002) Sandia Decon Formulation, Publication Nr.: SAND2000-0625, Sandia National Laboratory.
17. Seidel, A. (ed. in chief) (2006) *Kirk Othmer Encyclopedia of Science and Technology*, 4th edn, vol. 5, John Wiley & Sons, Inc., Hoboken, NJ, pp. 795–802.
18. Albrizo, J.M. and Ward, J.R. (1991) *J. Mol. Catal.*, **66** (2), 191–194.
19. Yang, Y.C., Baker, J.A., and Ward, J.R. (1992) *Chem. Rev.*, **92**, 1729–1743.
20. Katritzky, A.R. et al. (1989) *J. Fluorine. Chem.*, **44** (1), 121–131.
21. Ward, J.R., Yang, Y.C., Wilson, R.B., Burrows, W.D., and Ackerman, L.L. (1988) *Bioorg. Chem.*, **16** (1), 12–16.
22. Yang, Y.C., Szafraniec, L.L., Beaudry, W.T., and Bunton, C.A. (1993) *J. Org. Chem.*, **58**, 6964–6965.
23. Yang, Y.C. (1999) *Acc. Chem. Res.*, **32**, 109–115.
24. Yang, Y.C. et al. (1990) *J. Am. Chem. Soc.*, **112** (18), 6621–6627.
25. Epstein, J. et al. (1974) *Phosphorus*, **4**, 157–163.
26. Szanfraniec, L.J. et al. (1990) On the stoichiometry of phosphonothiolate ester hydrolysis. CRDEC-TR.212, July, AD-A225952.
27. Melnychuk, S.A., Neveroy, A.A., and Brown, R.S. (2006) *Angew. Chem. Int. Ed.*, **45** (11), 1767–1770.
28. Neverov, A.A. and Brwon, R.S. (2004) *Org. Biomol. Chem.*, **2**, 2245–2248.
29. Lewis, R.E., Neverov, A.A., and Brown, R.S. (2005) *Org. Biomol. Chem.*, **3**, 4082–4088.
30. Lu, Z., Neverov, A.A., and Brown, R.S. (2005) *Org. Biomol. Chem.*, **3**, 3379.
31. Ménesi, J., Kõrösi, L., Bazsó, É., Zöllmer, V., Richardt, A., and Dékány, I. (2008) *Chemosphere*, **70** (3), 538–542.
32. Kõrösi, L., Papp, S., Menesi, J., Illes, E., Zöllmer, V., Richardt, A., and Dékány, I. (2008) *Colloids Surf. A: Physicochem. Eng. Aspects*, **319**, 136–142.
33. Menesi, J., Kekesi, R., Kõrösi, L., Zöllmer, V., Richardt, A., and Dékány, I. (2008) *Int. J. Photoenergy*, 9 pp, doi: 10.1155/2008/846304, article ID 846304. http://www.hindawi.com/journals/ijp/2008/846304/
34. Koenhen, D.M., Smolders, C.A., (1975) *J. Appl. Polym. Sci.*, Vol. **19** pp. 1163–1179.
35. Willis, M.P., Mantooth, B.A., and Lalain, T.A, (2012) *J. Phys. Chem*, **116** (1), 546–554.
36. Willis, M.P., Mantooth, B.A., and Lalain, T.A, (2012) *J. Phys. Chem*, **116** (1), 538–545.
37. Ram, M.K., Andreescu, E.S., and Hanming, D., (2011) *Nanotechnology for Environmental Decontamination*, McGraw-Hill Professional, ISBN 978-0071702799

14
Principles and Practice of Disinfection of Biological Warfare Agents – How Clean is Clean Enough?

Andre Richardt and Birgit Hülseweh

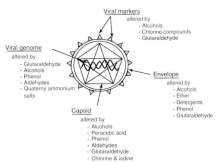

Hygienic concepts have been applied throughout human history and ancient reports already recommend (i) the passing of metal objects through fire to cleanse them and (ii) the addition of silver or copper to make water storable. Over the centuries, more effective biocides were developed and concepts for disinfection and sterilization improved. In recent decades disinfection and sterilization followed the increasing knowledge of biological agent inactivation. Different technologies for disinfection are today at our disposal. This chapter shows that disinfection is a complex field

CBRN Protection: Managing the Threat of Chemical, Biological, Radioactive and Nuclear Weapons,
First Edition. Edited by A. Richardt, B. Hülseweh, B. Niemeyer, and F. Sabath.
© 2013 Wiley-VCH Verlag GmbH & Co. KGaA. Published 2013 by Wiley-VCH Verlag GmbH & Co. KGaA.

and that the various disinfectants have different strengths and weaknesses. Which kind of disinfectant is used depends on the circumstances and many other criteria. Moreover, understanding the mechanism(s) of action of a biocide has become an important issue with the emergence of resistance to biocides. For the future we await the combination of different chemicals to increase their antimicrobial activity since synergistic effects of biocides have long been known.

14.1
General Principles of Disinfection and Decontamination

Key questions for decontamination and disinfection of biological agents are: "What is sufficient?" and "Is clean clean enough?" Is it possible to determine a necessary inactivation level that meets regulatory needs in the case of a biological incident or is an estimation of the reduction level sufficient?

As the case may be, it is important to realize that most of our knowledge and experience about the susceptibility of potential biowarfare agents to germicides relies on studies with related pathogens or surrogates. For example, the susceptibility of Variola virus to certain germicides is expected to be similar to Vaccinia virus, and *Bacillus anthracis* behaves in a similar manner to disinfectants as *Bacillus atrophaeus* does. Thus, we can extrapolate from the larger databases available on the susceptibility of genetically similar organisms.

In this chapter we give an idea of the complexity of disinfection and focus on it from the practical point of view. We present an overview of the different chemicals that inactivate biological agents and provide information on their mode of action. We critically evaluate different disinfection technologies and point out factors influencing the efficacy of biocides. Moreover, we discuss the emergence of increasing microbial resistance.

14.1.1
Definition of Terms

The term **decontamination** generally refers to the removal and reduction of microbial contamination. It does not necessarily mean the destruction of biological agents. In contrast, disinfection means the destruction of hazardous and pathogenic biological agents by physical or chemical methods, although the process does not necessarily kill all microorganisms. However, it leaves an object or sample safe to handle. Disinfection is less effective than sterilization and first **sterilization** means the destruction of all biological agents. By definition, sterile is the state of being free from all living or viable microorganisms. In contrast **cleaning** means only the removal of all foreign material from objects either by mechanical action or washing and must precede disinfection and sterilization procedures.

In general, terms with the suffix -cide or -cidal indicate *a* killing action and are commonly used. For example, a **germicide** is an agent that kills microorganisms, particularly pathogenic organisms ("**germs**"), while a **biocide** is an agent that inactivates biological agents.

14.1.2
Physical Methods of Disinfection

The physical processes of disinfection always have smooth transitions to sterilization and they are mentioned here to complete the chapter. Important physical disinfection methods include gamma and ultraviolet (UV) radiation as well as boiling. Moreover, ultrasonic waves, heat, and pressure are applied as physical principles for disinfection. Although we consider physical and chemical disinfection methods separately, it must be pointed out that both modes of action can be combined what is common in practice.

14.1.3
Chemical Methods of Disinfection

Generally, biocides are chemical agents capable of disinfection under defined conditions. They can be divided into chemicals used for liquid or vapor disinfection; however, some are used for both applications. Disinfection with liquids is preferred for solid surfaces and equipment where the size allows a dipping process. Gaseous disinfectants are frequently applied to decontaminate large rooms and areas and susceptible instruments.

Biocides are available under various trade names and typical chemical compounds applied for liquid disinfection are phenols, alcohols, and various aldehydes while vapor phase biocides frequently include substances like ethylene oxide, ozone, and formaldehyde-releasing agents. In addition, some disinfectants are formulated in combinations to potentiate their effects by synergisms.

Modern disinfectants should fulfill plenty of criteria. For example, they should be efficient in decontamination but simultaneously environmentally friendly, easy to prepare, stable, and cheap to produce. By- and end-products of biocides should be biologically degradable and harmless. In addition, they should be effective against all types of infectious agents and serve for various disinfection purposes of interiors of buildings, vehicles, sensitive equipment, and clothing. The ideal disinfectant should be fast acting even in the presence of organic substances and matrices like soil, serum, or body fluid.

14.2
Mechanisms of Action of Biocides against Microorganisms

The efficacy of biocides toward microorganisms is influenced by diverse factors, some of which are inherent to the agent itself while others depend on the type of microorganism and the applied environmental conditions.

Most of the subsequently mentioned biocidic reagents act upon multiple sites within the cell, bacteria, spore, or virus. Their interactions responsible for death are in many cases still under investigation. Moreover, the site of lethal action depends upon the employed concentration.

> According to their mode of action we can divide biocides into:
> - oxidizing agents,
> - alkylating agents,
> - nucleic acid binding agents,
> - protein denaturants
> - agents interacting with lipids,
> - metal-ion binding agents.

14.2.1
Chemicals for Disinfection

Oxidizing agents consist of substances like hydrogen peroxide, peracetic acid, and hypochlorite. All these substances release free oxygen radicals and damage microorganisms primarily by oxidizing sulfhydral and sulfur bonds in essential proteins, enzymes, and metabolites. Thereby, they denature proteins and disrupt cell-wall permeability. They inactivate Gram-positive and Gram-negative bacteria, viruses, fungi, and yeast as well as spores.

In contrast, **alkylating agents** like ethylene oxide and aldehydes modify microbial metabolites irreversibly and denature proteins, enzymes, and nucleic acids by adding alkyl groups to either essential amino acid residues or purines and pyrimidine bases. The modifications result in loss of function of proteins and mutations that interfere with replication and gene expression.

Ethylene oxide (C_2H_4O), for example, has bactericidal, fungicidal, virucidal, and sporicidal properties and converts hydroxyl, sulfhydryl, carboxyl, and amino groups into hydroxyethyl adducts (Scheme 14.1).

In addition, alkylating substances such as formaldehyde and glutaraldehyde display crosslinking activities and cause irreversible modification of protein structures.

R-OH + C_2H_4O → R-O-CH_2-CH_2-OH

R-SH + C_2H_4O → R-S-CH_2-CH_2-OH

R-COOH + C_2H_4O → R-COO-CH_2-CH_2-OH

R-NH_2 + C_2H_4O → R-NH-CH_2-CH_2-OH

Scheme 14.1 Modifications of different chemical side chains by ethylene oxide (C_2H_4O).•

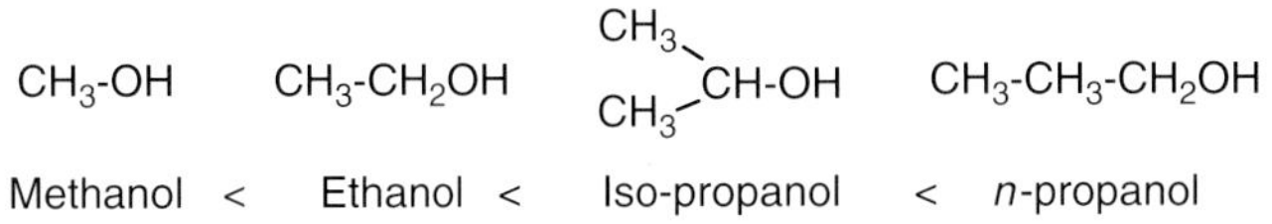

CH$_3$-OH   CH$_3$-CH$_2$OH   (CH$_3$)$_2$CH-OH   CH$_3$-CH$_2$-CH$_2$OH

Methanol < Ethanol < Iso-propanol < n-propanol

Figure 14.1 Increasing effectiveness of alcohols for surface disinfection.

Formalin is a 37% (w/v) solution of formaldehyde in water and dilution of formalin to 5% (v/v) results in an effective disinfectant.

Nucleic acid binding agents like acridine dyes bind to DNA by intercalation between base pairs in the double helix. Thereby, they inhibit DNA replication, gene expression, and protein synthesis.

Agents that interact with lipids like cationic detergents have strong surface activity. At low concentrations they damage the cytoplasmic membrane of Gram-positive bacteria and lipid-containing viruses. Owing to re- and disarrangements of lipids and proteins in the membrane structure they cause leakage of cytoplasmic constituents. At high concentrations cationic detergents cause coagulation of the cytoplasm and denaturation of proteins. They are less active against Gram-negative bacteria and are not active against non-lipid-containing viruses.

Phenols and alcohols like ethanol, benzyl alcohol, and chlorobutanol act as **protein denaturants** and displace water molecules from biological agents. Ethyl or isopropyl alcohol are effective in concentrations of 70–90% (v/v) but they evaporate rapidly and therefore have limited exposure time. They are less active against non-lipid viruses and ineffective against bacterial spores. The denaturing effect of alcohols increases with their chain length. Primary alcohols are more effective than iso- or secondary and tertiary alcohols. Figure 14.1 shows a ranking in order of increasing effectiveness.

Phenol-based disinfectants come in various concentrations ranging from 5 to 10%. They are mainly used for disinfection of contaminated surfaces like walls, floors, and bench tops. They effectively kill lipid-containing viruses, bacteria and fungi but are not active against spores or non-lipid viruses.

Halogens like chlorine and iodine are applied for disinfection, while chlorine-containing solutions have the broadest spectrum activity. Sodium hypochlorite is the most common base for chlorine disinfectants. Common household bleach (5% available chlorine) can be diluted 1/10–1/100 with water to yield a satisfactory disinfectant solution. Diluted solutions may be stored for extended periods if kept in a closed container and protected from light. Chlorine-containing disinfectants are inactivated by excess organic materials. They are also strong oxidizers and very corrosive. Always use appropriate personal protective equipment when using these compounds. At high concentrations and extended contact time, hypochlorite solutions are considered cold sterilants since they also inactivate bacterial spores.

Iodine has similar properties to chlorine, and iodophors (organically bound iodine) are recommended clinical disinfectants. They are most often used as antiseptics and in surgical soaps and are relatively non-toxic to humans.

14.2.2
Fumigation – Well-Known for Decontamination of Objects

Smoke or vapor-generating systems generally disperse an airborne biocide to decontaminate an area or object. This normally poses an unprecedented challenge since hundreds to thousands of cubic meters have to be decontaminated. Currently, only limited experience exists regarding the biological decontamination of large buildings and spaces. Published experience and knowledge is mainly restricted to US examples like the decontamination of the:

- Postal Service Curseen-Morris Processing and Distribution Center (Brentwood Post Office),
- Hart Senate Office Building (HSOB),
- American Media Inc. office building in Boca Raton.

The most commonly used gases for decontamination of large areas, spaces, or objects are formaldehyde and ethylene oxide, but the airborne forms of chlorines and hydrogen peroxide also possess germicidal properties. Generally, all gases are applied in closed systems under controlled conditions with strict regulations, and monitoring the specific process parameters is essential. Moreover, the effectiveness of the fumigation prior to a clearance and reopening has to be demonstrated by using either spore strips containing surrogates or by environmental sampling. In the following we introduce four of the frequently taken fumigation methods and discuss their advantages and disadvantages.

14.2.2.1 Fumigation with Ethylene Oxide
Ethylene oxide is an odorless gas at room temperature and has been widely used for hospital and biomedical sterilization applications. The gas can be delivered from bulk sterilizers or in prepared packages that contain measured volumes. The gas penetrates surfaces and is often applied for critical items if re-use is requested. However, its application is critical since it is a highly reactive gas with flammable and explosive vapors [1]. It is rated as a Group B1 agent, which means it is a probable human carcinogen and it is hazardous to health in case of unprotected exposure. Acute exposure to ethylene oxide can cause nausea, vomiting, and death and chronic exposure is responsible for irritation of skin, eyes, mucous membranes, cataracts, and problems in brain function [2].

14.2.2.2 Fumigation with Chlorine Dioxide Gas
Chlorine dioxide (ClO_2) is an oxidizing agent and is produced from either chlorite (ClO_2^-) or chlorate (ClO_3^-). Its use as a decontaminant is a mature technology.

The gas is normally generated directly at the decontamination site (on-site production) and is injected into the sealed building. Incubation with chlorine dioxide takes 12–24 h and is followed by neutralization with sodium sulfite or bisulfite.

The main concerns for the application of chlorine dioxide are the amount of needed gas and the toxicity of its precursors and by-products [2].

14.2.2.3 Fumigation with Formaldehyde Gas

Since pure formaldehyde is unstable at ambient conditions, paraformaldehyde, the polymer $(CH_2O)_n$, and the stable crystalline form of **formaldehyde** is employed as the ready-to-use formula. Usually, the gas is generated by heating on site. The gas is traditionally used as a fumigant to decontaminate laboratories of the biosafety level 3 or 4. Like chlorine dioxide fumigation vaporization of paraformaldehyde is a mature and commercialized technology for disinfection [3].

For sterilization, gas exposure for 16 h at a concentration of 1.0 mg l^{-1} at relative humidity (RH) and ambient temperature of 24 °C ± 5% is recommended since these parameters have an enormous influence of on spore disinfection [4, 5].

However, the use of formaldehyde gas is subject to strict regulations. Moreover, it has to be neutralized with ammonium bicarbonate after fumigation and outgassing over weeks and months besides neutralization from porous surfaces is still a major problem. In addition, formaldehyde is a flammable gas, has a pungent odor, and it has been identified as a potential carcinogen.

14.2.2.4 Vaporized Hydrogen-Peroxide (VHP)

Vaporized hydrogen-peroxide (VHP) has a broad antimicrobial efficiency, as shown in Figure 14.2, although the susceptibility of these organisms towards vaporous disinfectants varies enormously [6, 7].

Bacillus spores, small viruses without envelope, and mycobacteria are more resistant to VHP disinfection than viruses with a lipid-envelope and therefore require a longer decontamination time. VHP is a material-compatible disinfectant and several hydrogen peroxide vapor generation systems are available commercially although only some of them have been adapted for large-scale applications. Usually, the hydrogen peroxide vapor is generated from a concentrated 30–35% H_2O_2 aqueous solution of hydrogen peroxide by controlled heating (Figure 14.3).

Since hydrogen peroxide gradually decays fresh peroxide is continuously supplied into the space. Typical H_2O_2 vapor **working concentrations** are about 0.3 mg l^{-1} and complete inactivation requires a minimum of 2–6 h of contact time to destroy

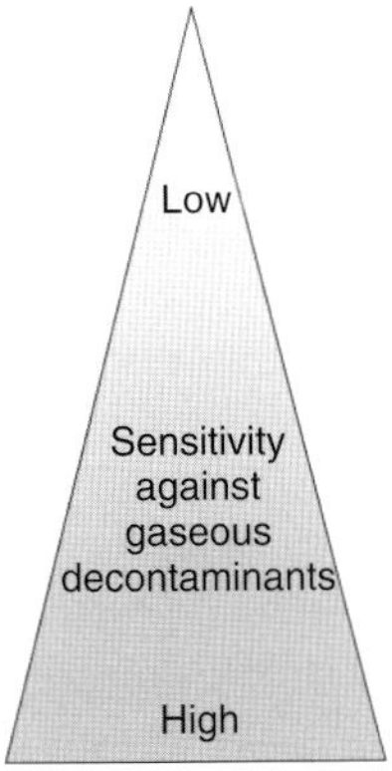

Sensitivity against gaseous decontaminants		
Low	Bacterial spores	e.g.: *B. anthracis, B. subtilis*
	Small viruses without envelope	e.g.: Parvoviridae, Picomaviridae
	Mycobacteria	e.g.: *Mycobacterium tuberculosis*
	Fungi	e.g.: *Aspergillus niger*
	Gram-negative bacteria	e.g.: *Burkholdia cepacia, Proteus vulgaris*
	Large viruses without envelope	e.g.: Adenovirus
	Gram-positive bacteria	e.g.: *Legionella pneumophilia, Lactobacillus casei*
High	Viruses with lipid-envelope	e.g.: Herpesviridae, Orthomyxoviridae, Poxviridae

Figure 14.2 Increasing sensitivity against gaseous decontaminants [8–10].

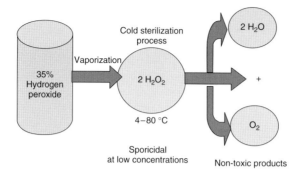

Figure 14.3 Generation of VHP. Adapted from Reference [9].

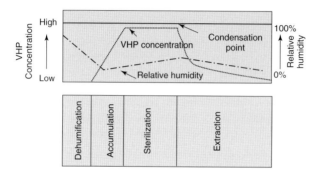

Figure 14.4 STERIS sterilization process for decontamination of spaces with VHP [9].

anthrax spores [11]. After turning off the generator the remaining of H_2O_2 vapor is usually converted enzymatically by catalase into water and oxygen. Thereby, no toxic by-products remain in the space [12].

Relative humidity is an important parameter in determining the performance of hydrogen peroxide vapor. For example, the STERIS process maintains humidity below 40% at the beginning of fumigation to keep the chemical in the vapor phase for improved penetration to substrate surfaces (Figure 14.4).

The main disadvantage of VHP technology concerns its chemical reactivity. Upon contact with certain materials and surfaces like galvanized steel or porous paper the vapor can break down since these convert the oxidizing gas into water and oxygen. Therefore the efficiency of the VHP technology was verified under worst-case conditions [12]. Table 14.1 summarizes the discussed advantages and disadvantages of the most commonly used chemicals for fumigation.

14.3
Levels of Disinfection

Germicides differ significantly in their antimicrobial spectrum and their depiction as virucidic, fungicidic, bactericidic, and sporicidic defines the biological

Table 14.1 Advantages and disadvantages of standard disinfection technologies.

Fumigation with	Advantages	Disadvantages
Ethylene oxide	Is a mature technology Has a broad antimicrobial efficiency Odorless Can be applied for critical items if re-use is requested	Is a probable human carcinogen Is hazardous to health in case of unprotected exposure
Chloride dioxide	Is a mature technology Has a broad antimicrobial efficiency with proven sporicidal activity Effectiveness on porous and non-porous surfaces Rapid natural breakdown Solubility and stability in water Odor could be detected at concentrations ≥ 0.1 pM	Gas is corrosive and could damage equipment Instability of the gas On-site generation of gas Produces a large volume of liquid waste Relative humidity is critical
Formaldehyde	Is a mature technology Has a broad antimicrobial efficiency	Outgassing besides neutralization from porous surfaces Is a potential carcinogen Has a pungent odor Is flammable
Vaporous hydrogen peroxide	Is a mature technology – ready to launch Has broad antimicrobial efficiency End products after catalytic breakdown are harmless	Chemical reactivity of the vapor On-site generation of gas Relative humidity is critical Gas is corrosive and could damage equipment

effectiveness (Table 14.2). For example, a bactericide is an agent that kills bacteria, while a virucide is an agent that kills viruses. Unlike sterilization, disinfection is not per se sporicidal but a few disinfectants will kill spores with prolonged exposure times.

A bacterial spore requires a considerably longer application time for deactivation than, for example, a Gram-negative bacteria with a relatively thin cell wall. The Gram-positive tuberculosis bacterium is also relatively resistant and is able to survive small doses of disinfectant. Polioviruses, adenoviruses, and noroviruses are examples of highly stable viruses; non-enveloped viruses such as the influenza virus are less stable. *Aspergillus niger* is one of the most resistant fungi.

According to the Centers for Disease Control and Prevention (CDC) and the World Health Organization (WHO) four levels of treatment are defined.

Disinfectants of level 1 or **Low Level Disinfection** inactivate most vegetative bacteria, fungi, and some viruses. This level of treatment does not inactivate mycobacteria (bacteria causing tuberculosis) and bacterial spores. In addition, this level of treatment is inadequate for biomedical waste treatment and is not recommended.

Disinfectants of level 2 also called **Intermediate Level Disinfection** inactivates vegetative bacteria, all mycobacteria, viruses, and fungi. It does not include the inactivation of bacterial spores as required in biosafety laboratories of levels 3 and 4. The CDC defines this as the destruction of all microorganisms except high numbers of bacterial spores. These two definitions are essentially equivalent. Tests for intermediate level disinfection must show that a $6\log_{10}$ reduction of the microorganisms most resistant to the treatment is attained.

Disinfectants of Level 3 or **High Level Disinfection** kills all microbial life forms present in a medical waste load as evidenced by the inactivation of surrogate pathogens (bacterial spores) having death curves similar to the most resistant human pathogens. A minimum of $4\log_{10}$ reduction of spores of either *Bacillus stearothermophilus* or *Bacillus subtilis* by thermal inactivation technologies or by chemical treatment is accepted as indicating high level and intermediate level disinfection. A $4\log_{10}$ reduction is equivalent to a 99.99% reduction in spores.

Disinfection of Level 4 is similar to **Sterilization** and kills all microbial life as indicated by complete inactivation of specific concentrations of those organisms recognized as most resistant to the treatment process. Sterilization is evidenced by a minimum $6\log_{10}$ reduction in spores of *B. stearothermophilus*, an extremely heat resistant *Bacillus* species.

Table 14.2 Microbicidal activity of selected disinfectants.

Chemical agent	Microbicidal activity[a]							Comment
	Bacteria				Viruses	Fungi		
	Spores	Gram positive	Gram negative	Myco-		Yeast	Mold	
Peracetic acid	+++	+++	+++	+++	+++	+++	+++	pH-optimum 2–3, fast reactivity
Na-hypochlorite	+++	+++	+++	+	+++	+	+	pH-optimum 4–6
Formaldehyde	+++	+++	+++	+++	+++	+	+	pH-optimum 4–9
Glutaraldehyde	+++	+++	+++	+++	+++	+	+	pH-optimum 2–3
Phenols	–	+++	+++	+++	+++	+	+	pH-optimum 2–4
Alcohols	–	+++	+++	+++	+++	+	+	–

[a] +++: high efficiency, +: moderate efficiency, –: no efficiency.
Table modified according to Reference [13].

14.4
Biological Target Sites of Selected Biocides

As explained in Section 14.2 biocides vary in their chemical structures and in their mode of action. However, the final damage, when lethal concentrations are used, may show similarities. To be effective, biocides must reach and interact with their microbial target site(s). In general, these target sites include outer cellular components, the cytoplasmic membrane, cytoplasmic constituents, nucleic acids, and metabolic processes.

14.4.1
Viral Target Sites

As explained in Part 2, Chapter 4 viruses are much smaller and have simpler structures than other microorganisms. Furthermore, viruses do not have any metabolic activity. For these reasons it is generally accepted that viruses have fewer target sites to biocides than bacteria and fungi.

Principally, enveloped viruses are considered to be more sensitive to disinfectants than non-enveloped viruses, as indicated in Table 14.3 [14]. This means that poliovirus, as a representative for Picornaviridae, and noroviruses, as representative for Caliciviridae, are examples of highly stable, non-enveloped viruses. In contrast, influenza virus (Orthomyxoviridae), as a lipid-enveloped virus, is less stable.

As shown in Figure 14.5 potential viral target sites for biocides are (i) the viral envelope, (ii) viral envelope proteins (markers), (iii) the capsid, protecting the viral nucleic acid, and (iv) the viral genome itself.

The viral envelope derives from the membrane of the host cells and is highly lipophilic and negatively charged. Therefore, enveloped viruses are highly susceptible to membrane active biocides such as ether or chloroform and to non-ionic detergents such as Nonidet P40, Triton, and Tween that break down lipid–lipid and lipid–protein interactions. However, these biocides have only reduced or no inactivation capacity on non-enveloped viruses.

The viral surface proteins as well as the viral capsid are mainly proteinaceous and therefore susceptible to all biocides that either react with NH_2 or SH groups of amino acids or coagulate and denature proteins. While the destruction of viral

Table 14.3 Susceptibility of different viral families to biocides.

Target structure of viruses	Susceptibility to biocides
Small viruses without envelope, for example: Parvoviridae, Picornaviridae, and some rotaviruses	High-level disinfectants needed
Large viruses without envelope, for example: adenovirus	Medium-level disinfectants needed
Viruses with a lipid envelope, for example: Herpesviridae, Orthomyxoviridae, and Poxviridae	Low-level disinfectants sufficient

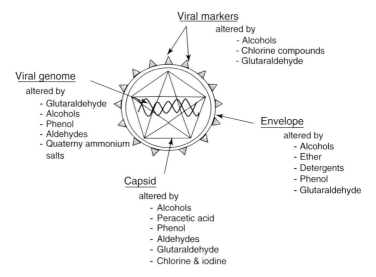

Figure 14.5 Potential viral biocidal target sites.

markers inhibits the infection process itself (no adhesion of the virus to the host cell receptor), destruction of the capsid sets the viral nucleic acid free.

Even if the capsid and the viral surface proteins are totally damaged, *some* viruses remain infectious. This specific disinfection problem arises from positive-sense RNA viruses such as alpha- and flaviviruses. Compared to other viral families these viruses can directly cause infection because their RNA contains all of the necessary genetic information for establishing its replication in the host cell. Thereby, they induce a natural infection, although the naked viral RNA is over 100 000-fold less infectious than the complete virus particle [15].

14.4.2
Bacterial Target Sites

Compared to viruses vegetative bacteria offer multiple target sites for biocides (Figure 14.6). Their basic structure and the composition of the cytoplasm and the cytoplasmic membrane (inner membrane) are conserved while the outer membrane differs widely. For example, compared to Gram-positive bacteria, Gram-negative bacteria possess an outer membrane that acts as a barrier to the uptake of disinfectants while mycobacteria have a waxy cell wall that hampers disinfectant entry.

The cytoplasmic membrane is, like the viral envelope, composed of a phospholipid bilayer with embedded proteins and regulates the transfer of solutes and metabolites in and out of the cytoplasm. Therefore, biocides like organic solvents, phenols organic acids, and detergents damage the bilayer irreversibly. They damage the lipid bilayer by either dissipation of the bacterial proton motive force (PMF) or denature and coagulate essential transport systems for amino acids and ions

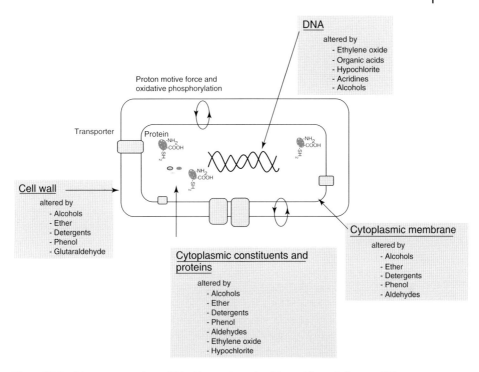

Figure 14.6 Primary target sites of biocides on bacteria. Adapted from Reference [16].

as well as virulence markers. Disintegration of the PMF leaves the bacterial cell without energy because active transport, oxidative phosphorylation, and adenosine triphosphate (ATP) synthesis are inhibited.

Components of the bacterial cytoplasm are not primary target sites of biocides, but once a biocide has penetrated the cell it could damage nucleic acids as well as ribosomes and various cytoplasmic enzymes. As a consequence, essential and vital processes like replication, translation, and metabolic activity of the bacterial cell die off.

For example, ethanol shows secondary effects in bacteria and inhibits DNA-, RNA-protein, and peptidoglycan synthesis. Other effects are the inhibition of enzymes involved in glycolysis and fatty acid and phospholipid synthesis.

14.5
The Spores Problem

Bacterial spores are among the most resistant of all living cells to biocides, although their response clearly depends on the stage of sporulation [17]. Dried spores on surfaces are often more resistant to inactivation than the same spores in aqueous suspension, clearly indicating that their susceptibility toward disinfection depends on their water-content [18].

As explained in Part 2, Chapter 4, spores are formed inside the vegetative cell in response to nutritional deprivation and enable the organism to survive without metabolism.

In electron microscopic pictures bacterial spores show a multilayered structure. The inner-most compartment, the spore core, contains all the necessary cellular components (DNA, RNA, and metabolic enzymes) to re-establish a vegetative bacterial cell after spore germination and outgrowth. The spore core is surrounded by an inner spore membrane and a spore cortex. While the inner membrane is a lipid bilayer without detectable fluidity, the cortex is composed of a thick layer of peptidoglycan. The cortex is followed by a complex proteinaceous coat and finally the spore is enclosed by an exosporium (Figure 14.7).

Since the spore coat and cortex act as a barrier to disinfectant entry, decontamination procedures for bacterial spores are often more rigorous than procedures used for vegetative bacteria. Moreover, incubation temperature, concentration, pH, and RH affect the sporicidal activity of various chemical agents.

Whereas agents like aldehydes, halogens, hydrogen peroxide, and peroxy acids are actively sporicidal, chemicals like organic acids and esters, quaternary ammonium compounds, and phenol-based agents and alcohols only inhibit germination or outgrowth of spores. They are unable to provide meaningful spore reductions. In Table 14.4 we compare different selected inactivation methods on different *Bacillus* spores. We evaluate their efficiencies and take the different reaction conditions into account.

Physical parameters that have been successfully evaluated for disinfection of spores are:

- high pressure,
- ultrasonic waves,
- UV_{254}
- gamma radiation,
- heat.

Compared to their vegetative cells, spores from *Bacillus anthracis* and their surrogates are extremely heat resistant. Therefore, their complete destruction by autoclavation frequently requires temperatures of 130–140 °C for 30–120 min. Moist heat is often more sufficient than dry heat; the efficiency of heat inactivation depends significantly on the water content and mineralization status of the spore [20].

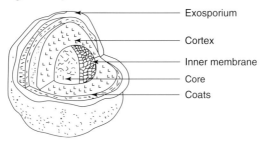

Figure 14.7 Schematic structure of spores.

Table 14.4 Efficiency of selected chemicals and gases for the inactivation of different Bacillus spores.

Method	Concentration	Inoculum size	Time	Efficiency
Chemical sterilization				
Free available chlorine	2.4–2.3 mg l^{-1} available; Cl$_2$, pH 7.2, 22 °C	1.1×10^5 spore suspension of B. anthracis	1 h	>99.99% killed
Sodium hypochlorite (NaOCl)	0.05%, pH 7.0, 20 °C 0.05%, pH 11.0, 20 °C	Spore suspension of B. subtilis, B. globigii, representing 1.6–2.2 $\times 10^9$ CFU ml^{-1}	30 min	99.99% killed 50% spores survived
Hydrogen peroxide (H$_2$O$_2$)	25.8%, 24 °C 25.8%, 76 °C 2.7%, pH 5.0 2.7%, pH 4.3	B. subtilis, B. globigii spore suspension (no concentration) 10^6 CFU ml^{-1} B. subtilis spore suspension 10 ml B. subtilis spore suspension coated onto stainless steel carriers	15 min <1 min 3 h 6 h	0.001% survived <0.0001% survived 100% killed 100% killed
Peracetic acid (CH$_3$COOOH)	0.13 mol l^{-1}, pH 5.0, 6.5, 8.0 0.39 mol l^{-1}, pH 4.0, 7.0, 9.0	10^6 CFU ml^{-1} B. subtilis 10 ml B. subtilis spore suspension coated on stainless steel carriers	<30 min 24 h	100% killed 100% killed
Formaldehyde (CH$_2$O)	4% in water 400 mg m^{-3}, 30% RH 280 mg m^{-3}, 50% RH 250 mg m^{-3}, 80% RH 400 mg m^{-3}, 98% RH	10^8 ml^{-1} B. anthracis 10^2–3×10^8 B. globigii NCTC 10073 dried on disks	2 h 22 min 31 min 16 min 9 min	10^4 inactivation factor 1 log$_{10}$ reduction, at 23.5–25 °C
Glutaraldehyde (C$_5$H$_8$O$_2$)	2% in water, pH 8.0 5%, 21.1 °C	10^8 ml^{-1} spores B. anthracis	15 min 3.6 h	10^4 inactivation factor 99% killed

(continued overleaf)

Table 14.4 (continued)

Method	Concentration	Inoculum size	Time	Efficiency
Gaseous sterilization				
Ethylene oxide (C_2H_4O)	Exposed to constant boiling HCl at 20 °C for 30 min before exposure to ethylene oxide at room temperature	B. globigii and B. anthracis dried onto suture loop carriers (no concentration)	1 h	100% killed
	500 mg l^{-1}, 30–50% RH, 54.4 °C	~10^6 spores B. globigii on non-hygroscopic surfaces	30 min	4 $\log_{10}$ reduction
		~10^6 spores B. globigii on hygroscopic surfaces		6 $\log_{10}$ reduction
Peracetic acid vapor (CH_3COOOH)	1 mg l^{-1}, 80% RH	6×10^5–8×10^5 B. subtilis, B. niger dried on filter-paper disks and glass squares	10 min	<1 spore remained on paper and glass
	1 mg l^{-1}, 60% RH			Two spores remained on paper; 38 spores remained on glass
	1 mg l^{-1}, 40% RH			Twenty-four spores remained on paper; 1530 spores remained on glass
Ozone (O_3)	1.0 mg l^{-1} generated in water pH 3	1.8×10^5 spores per ml B. cereus	5 min	<10^1 CFU ml^{-1} survived
	3.0 mg l^{-1}, preconditioned at 54% RH	10^8–2×10^8 B. subtilis dried on filter paper	1.5 h, 95% RH	<0.001% survived
		10^8–2×10^8 B. cereus dried on filter paper	1.5 h, 95% RH	<0.001% survived

RH, relative humidity; CFU, colony forming unit; mol l^{-1} = gram molecular weight per liter. Table adapted and modified from Reference [19].

Combinations of physical and chemical disinfection methods have been also successfully supplied to spore inactivation and are frequently more effective than individual methods. For example, Xu *et al.* inactivated *Bacillus anthracis* spores effectively in milk by a combination of heat and hydrogen peroxide [21], while Urakami and coworkers successfully combined ozone treatment with UV irradiation [22]. In addition, joint effects of heat and hydrostatic pressure for complete destruction of *Bacillus* species spores were observed.

14.6
Inactivation as Kinetic Process

Microbial inactivation is a kinetic process and is comparable to a chemical reaction. Frequently, this inactivation follows first-order kinetics and, therefore, survival curves are in their simplest form semi-log plots, where the ratio of the concentration of surviving organisms to the initial number of organisms is plotted versus time of incubation as illustrated in Figure 14.8.

The most commonly used model for inactivation of microorganisms by disinfectants has been derived from the work of Chick and Watson who in 1908 discovered that disinfectant concentration and contact time were the primary variables affecting microbial inactivation kinetics and efficiency [24]. Combining their expressions

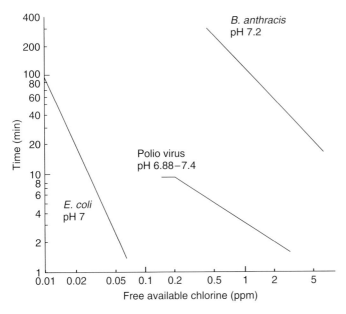

Figure 14.8 Disinfection (2 log$_{10}$) of viruses and bacteria by free available chlorine. Adapted from Reference [23].

Table 14.5 Ct-values for 4 $\log_{10}$ inactivation ($Ct_{99.99}$) of viruses by free chlorine, modified according to the EPA Guidance Manual *LT1ESWTR Disinfection Profiling and Benchmarking* [25].

Temperature (°C)	Chlorine concentration (mg l^{-1}) pH 6–9	pH 10
10	6	45
15	4	30
20	3	22
25	2	15

yields the equation that is known as the "Chick–Watson law:"

$$\ln (N/N_0) = -kCt$$

where

> N is the bacterial concentration at time t,
> N_0 is the bacterial concentration at time zero,
> C is the disinfectant concentration,
> k is an empirical constant, also known as the *Chick–Watson coefficient* of specific lethality,
> t is the contact time.

The Chick–Watson law is the basis for all other inactivation models, which can be considered as derivations of this formula. Normally, the mathematical product of concentration and contact time defines a disinfectant unequivocally but, although a lot of research has been performed on Ct values for various types of microorganisms and for different disinfectants, data on Ct values in the literature may differ strongly due to differences in experimental design (Table 14.5).

Generally, the level of inactivation of a biocide is expressed as $\log_{10}$ reduction and according to the biocide's classification level it has to reduce spores, bacteria, viruses, and fungi to a defined number in order to be called effective.

In addition, factors that influence the activity of biocides have been extensively reviewed [26] and depend on either environmental, chemical, or physical factors or they depend on the properties of the organism itself.

> Variables that we should take into account when disinfecting are:
> - the biocide's concentration,
> - pH,
> - humidity,
> - incubation time,
> - temperature.

The activity of most disinfectants increases as the temperature increases. Furthermore, too great an increase in temperature causes the disinfectant to degrade and weakens its germicidal activity and thus might produce a potential health hazard.

An increase in pH improves the antimicrobial activity of disinfectants like glutaraldehyde and quaternary ammonium compounds but decreases the antimicrobial activity of phenols, hypochlorites, and iodine. The pH influences the antimicrobial activity by altering either the disinfectant molecule itself or the cell surface of the organism to be inactivated.

Moreover, RH is an important factor influencing the activity of gaseous disinfectants/sterilants, such as ethylene oxide, chlorine dioxide, and formaldehyde. Excessive moisture can dilute the disinfectant or even neutralize it, and water hardness can reduce the rate of kill of certain disinfectants because divalent cations like magnesium and calcium interact with the disinfectant to form insoluble precipitates.

Morphology, structure, composition, and concentration of organisms also influence the efficacy of biocides. Higher concentrations of microorganisms either require a longer application time or/and a higher concentration of disinfectant. Moreover, many organisms can develop resistance to biocides, which results in a reduced susceptibility.

Also critical for the application is the nature of the surface being disinfected. The more porous and rough the surface is the longer a disinfectant will need to be effective because much of the reactive agent can be absorbed, bound to, or diffuse into the supporting material.

The presence of interfering substances of organic matter can limit or suppress the decontamination efficacy and can occur in at least two ways. Most commonly, interference occurs by a chemical reaction between the biocide and the organic matter, resulting in a complex that is less biocidal. This phenomenon applies in particular to oxidative biocides but also the activity of phenols is reduced by organic material like soil, serum, fecal material, and food residues. It has been shown that organic matter decreases the effect of hypochlorites against bacteria, viruses, and fungi [27]. Moreover, body fluids like blood serum can reduce biocide efficacy by neutralizing active agents. Alternatively, organic material can protect microorganisms from attack by acting as a physical barrier or by consuming the disinfectant.

Finally, pathogens could be protected from disinfectants by biofilms, which are microbial communities that produce thick masses of cells and extracellular materials. Bacteria within biofilms are up to 1000 times more resistant to antimicrobials than are the same bacteria in suspension. Moreover, capsules or slimes might play a role in bacterial insusceptibility to biocides.

14.7
Evaluation of Antimicrobial Efficiency

Monitoring the efficiency and success of disinfection presents a challenge but is an essential part of each initial investigation and is normally assessed on the basis of standardized protocols. Although an international test scheme for biocides does not exist, several national agencies like the American Association of Official Analytical Chemists (AOAC), the British Standards Institution (BSI), and the German Society for Hygiene and Microbiology (DGHM) regularly publish valid standards and define suitable test viruses, bacteria, and fungi.

Currently the greatest progress has been made in Europe. There, the Committee for European Standardization's (CEN's) technical work focuses constantly on the concept of disinfection quality, in order to harmonize the methods. A few of the standards are listed in Table 14.6.

Normally, a pass-fail criterion is the result of a disinfectant test. This means that the test either confirms or denies the ability of a biocide to inactivate a biological agent. Below we now explain and discuss some fundamental aspects of disinfectant validation. Further aspects can be found in reviews [28–30].

Table 14.6 Accepted European standards for disinfectant testing.

Norm	Test type	Valid since	Test organisms
EN 1276	**Quantitative suspension test** for the evaluation of **bactericidal activity** of chemical disinfectants and antiseptics used in food, industrial, domestic, and institutional areas	1997	*Staphylococcus aureus, Pseudomonas aeruginosa*
EN 13697	**Quantitative non-porous surface test** for the evaluation of bactericidal and/or fungicidal activity of chemical disinfectants used in food, industrial, domestic, and institutional areas – test method and requirements without mechanical action	2002	*Staphylococcus aureus, Pseudomonas aeruginosa, Escherichia coli, Enterococcus hirae, Candida albicans, Aspergillus niger*
EN 1650	**Quantitative suspension test for the evaluation of fungicidal or yeasticidal activity** of chemical disinfectants and antiseptics used in food, industrial, domestic, and institutional areas	1998	*Candida albicans, Aspergillus niger*
EN 13704	**Quantitative suspension test** for the evaluation of **sporicidal activity** of chemical disinfectants used in food, industrial, domestic, and institutional areas	2002	*Bacillus subtilis, Bacillus cereus, Clostridium sporogenes*
EN 1040	**Quantitative suspension test** for the evaluation of basic bactericidal activity of chemical disinfectants and antiseptics	1997	*Staphylococcus aureus Pseudomonas aeruginosa*
EN 14347	**Basic sporicidal activity**, test methods, and requirements	2005	*Bacillus subtilis, Bacillus cereus*

14.8
Carrier Tests versus Suspension Tests

Two main types of tests – carrier and suspensions tests – are differentiated to evaluate biocides used on contaminated surfaces and equipment.

Carrier tests are the oldest tests for disinfection and in these tests the carrier is artificially contaminated by a liquid culture of the test organism. It is then dried and afterwards immersed in the disinfectant for a certain period of time. After the exposure the carrier is checked for survivors and cultivated in a nutrient broth. No growth indicates activity of the tested disinfectant whereas growth indicates a failing.

The most widely used and reported carrier tests are those of the DGHM and the use-dilution test of the AOAC (Figure 14.9).

Carrier tests have their strengths and weaknesses and Table 14.7 lists the required reduction rates for different bacteria and the yeast *Candida albicans* for quantitative investigations according to the DGHM.

Carrier tests are particularly appropriate for dilutable chemical disinfectants but their application is limited because they are notoriously variable on the basis on statistics. The number of bacteria dried on a carrier is hard to standardize and the survival of bacteria on the carrier is inconstant during drying. To avoid false-positive results, the scientist performing the use-dilution test must have experience, and skill in handling the carriers under time pressure is essential.

Other tests to verify the effectiveness of disinfectants are suspension tests (Figure 14.10), where the bacterial culture is suspended in the disinfectant solution. After a defined exposure of the inoculum and disinfectant the culture is checked for survivors by subcultivation.

In general, suspension tests are preferred to carrier tests as the bacteria are uniformly exposed to the disinfectant.

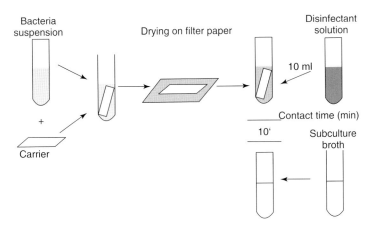

Figure 14.9 Carrier test for liquid disinfectant according to the American Association of Official Analytical Chemists (AOAC).

Table 14.7 DGHM required reduction rates for different bacteria and the yeast *Candida albicans* in quantitative carrier tests.

Method	Carrier	Organism	Reduction
Surface	Ceramic tile	Staphylococcus aureus	$5 \log_{10}$
		Pseudomonas aeruginosa	$5 \log_{10} \leq 60$ min or $4 \log_{10}$ 4 h
		Enterococcus hirae	$5 \log_{10}$
		Candida albicans	$4 \log_{10}$
Instruments	Frosted glass	Staphylococcus aureus	$5 \log_{10}$
		Pseudomonas aeruginosa	$5 \log_{10}$
		Enterococcus hirae	$5 \log_{10}$
		Candida albicans	$4 \log_{10}$

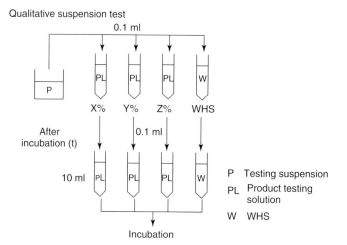

Figure 14.10 General principle of a qualitative suspension test. P = test organism, PL = disinfectant, X, Y, Z are different concentrations of disinfectant, W: water hardness standard and negative control.

There are three principal stages in disinfectant analyses:

- **Phase 1** is the primary testing or screening.
- **Phase 2** includes quantitative suspension and carrier tests.
- **Phase 3** means field trials.

In general a qualitative test is a useful first indicator of activity and can be used as a screening tool to define the range of activity. By enumeration of survivors a basic qualitative test becomes quantitative. Bactericidal tests routinely determine the

Table 14.8 Neutralizing agents for selected biocides.

Disinfectant	Neutralizing agent
Phenol	Dilution
	Tween 80
Chlorine and iodine	Sodium thiosulfate
	Sodium sulfite
	Nutrient broth
Glutaraldehyde	Glycine
Formaldehyde	Dimedone and morpholine

number of survivors in a quantitative suspension test by direct culture. Virucidal activity in a quantitative suspension test is determined by infection of cultured cell monolayers.

Field tests must take practical and real life conditions into consideration for disinfection. They are usually performed after the time–concentration relationship of the disinfectant to certain microorganisms has been determined. Their aim is, simply speaking, to verify whether the proposed use dilution is still adequate.

As mentioned before in this chapter many national microbicidal efficacy criteria and regulations demand **a defined logarithmic reduction for the microbial count**.

> For example, for **high-level disinfection** bacterial spores have to be reduced by a minimum of $6 \log_{10}$, which equals an inactivation of 99.9999%. For bacteria a minimum $5 \log_{10}$ kill is required. Fungi and viruses have to be reduced by $4 \log_{10}$ to classify a biocide as high-level effective.

Moreover, for accurate assessment of surviving organisms following contact with biocides, activity must be arrested at the moment of sampling. This is usually achieved by the addition of an appropriate neutralizing agent – an alternative approach is membrane filtration. Table 14.8 shows examples of suitable neutralizing agents.

14.9
Resistance to Biocide Inactivation – a Growing Concern

The term *"biological agent resistance"* is often applied when disinfection technologies fail and this failure is often due to inadequate use of the disinfectant. Besides these technical failures there is evidence that bacteria and viruses can develop resistance to biocides.

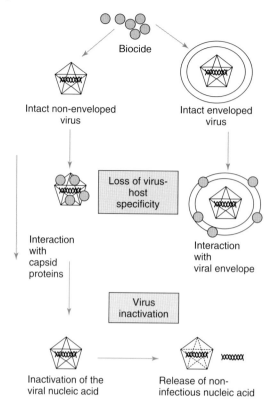

Figure 14.11 Mechanisms of viral deactivation. Adapted from Reference [31].

For deeper insight into the different mechanisms of bacterial and virus resistance against biocides we refer to advanced textbooks and related literature [31]. However, the following subsection should give the reader a first idea of the problem of microbial resistance.

14.9.1
Resistance of Viruses

Three main mechanisms are currently discussed to explain increased viral resistance to disinfection. These mechanisms, illustrated in Figure 14.11, are [31–33]:

- viral aggregation
- adaptation
- multiplicity reactivation.

Viral aggregation is often found on naturally contaminated surfaces or in body fluids and is mainly caused by charge effects. Moreover, viruses located in the center of an aggregate withstand the inactivation effect of biocides more effectively than viruses located at the exterior edge of the aggregate.

In addition, individual virus particles often tend to adhere to other particulate material and surfaces, an effect that can neutralize biocidal activity.

The **adaptation of viruses** to biocides is another point of growing importance. Repeated exposures to inadequate levels of biocides could lead to an adaptation of viruses to these conditions. Genetic mutations could occur and the selective pressure to lower susceptibility to biocides could lead to physical alterations of the macromolecular structure of the virus and altered interaction with the host cell.

The third possibility for increasing viral resistance could be the **multiplicity reactivation** after viral inactivation. This behavior of viral populations was detected with pox as well with influenza virus and describes the phenomenon that UV- or chemical-inactivated viruses can replicate in cells because of complementary reconstruction of genetic material between different member viruses in the inactivated population. Thereby, the virus reobtains its infectivity.

14.9.2
Resistance of Bacteria

As explained above, biocides have multiple target sites on vegetative bacteria and therefore also their biocide resistance could have diverse reasons. We distinguish between **intrinsic** and **acquired resistance**. While the intrinsic insusceptibility to certain biocides is due to an inherent cellular feature, the second type of resistance is acquired after biocide exposure. Acquired resistance frequently results from genetic changes and arises either by selection and mutation or by the acquisition of a resistance plasmid or transposon and is mainly found with Gram-positive and Gram-negative bacteria. They have not been described for mycobacteria and spores.

> Usually, we term bacteria resistant to biocide exposure when they are able to limit their internal biocide concentration to harmless levels.

Resistance could be caused by changes in the bacterial cell wall and the permeability of the cell membrane. Moreover, bacteria could regulate the efflux of chemical agents by up- and down-regulation of transporters. Furthermore, tolerance toward biocides could be obtained by their intrinsic degradation. This phenomenon has been described by Reuter [34] but might be temporary. In addition, bacteria surrounded by extracellular matrices and bacteria that are associated within the three-dimensional structure of biofilms are often less susceptible to disinfectants than planktonic bacteria. In both cases the extracellular matrices or outer layers act as a barrier and usually reduce the penetration of biocides. Well-known examples of these phenomena include *Burkholderia cepacia*, *Staphylococcus aureus*, and *Pseudomonas aeruginosa*.

14.10
New and Emerging Technologies for Disinfection

Although the focus is currently on the refinement and optimization of existing disinfection methods, new sterilization and disinfection technologies may provide substantial benefits. Moreover, they might be necessary because of an increasing microbial resistance to certain biocides and because new pathogens like severe acute respiratory syndrome virus (SARSV) emerge. Alternative disinfection technologies with either less time requirement or reduced need for complex equipment might be the application of plasma sterilization or super-oxidized water. In addition, persistent antimicrobial agents and nanoparticles as well as combined formulations are highly interesting topics for future disinfection. In any event, current data have primarily been generated by the manufacturers and need independent validation and their cost-effectiveness has to be assessed compared to standard technologies.

14.11
"Is Clean Clean Enough" or "How Clean Is Clean Enough"?

To determine the decontamination level or cleanup it is essential to address (i) public perception of risk to health, (ii) political support, (iii) public acceptance of recommendations based on scientific results and criteria, and (iv) economic concerns [35, 36]. As described and defined in Chapter 4 the incubation period, infectious dose, and fatality rate for biological warfare agents differ significantly [37, 38]. Therefore, from the scientific point of view, different threshold (limit) values for biological warfare agents should also exist. Public institutions and politicians often insist on zero living organisms after the decontamination of a public building, but in contrast to these requests are regulations and statements from the US Environmental Protection Agency (US EPA), and the AOAC, concerning the decontamination of medical equipment and the requirement for a 6 $\log_{10}$ reduction. According to the fact that, for example, natural inhalation of anthrax is extremely rare, zero concentration of biological agents is in many cases not a necessity and sometimes also not accessible. Another important point is the need for an overarching national/international concept to answer the questions "Is clean clean enough?" or "How clean is clean enough?" The key methods to verify an effective and successful decontamination are sampling, identification, and verification. If poor methods for validation are used, even decontamination methods with an inadequate inactivation rate of the biological agent could seem to give a safe cleanup [35].

Moreover, a successful decontamination process is only one important part of an efficient protection strategy, but this issue will be discussed in more detail in Chapter 16.

References

1. U.S. Department of Health and Human Services (2007) *NIOSH Pocket Guide to Chemical Hazards*, DHHS (NIOSK), Washington, DC, September 2007, Publication No. 2005-147, http://www.cdc.gov/niosh/docs/2005-149/pdfs/2005-149.pdf (accessed 1 January 2011).
2. Science Applications International Corporation (2005) Compilation of Available Data on Building Decontamination Alternatives, U.S. Environmental Protection Agency, Washington, DC, EPA/600/R-05/036, March 2005, http://www.epa.gov/nhsrc/pubs/600r05036.pdf (accessed 1 January 2011).
3. Fink, R., Liberman, D.F., Murphy, K., Lupo, D., and Israeli, E. (1988) *Am. Ind. Hyg Assoc. J.*, **49**, 277–279.
4. Munro, K. (1999) *Appl. Environ. Microbiol.*, **65**, 873–876.
5. Everall, P.H. (1982) *J. Clin. Pathol.*, **35**, 245–263.
6. Kahnert, A., Seiler, P., Stein, M., Aze, B., McDonnell, G., and Kaufmann, S.H.E. (2005) *Lett. Appl. Microbiol.*, **41**, 1–5.
7. Meszaros, J.E., Antloga, K., Justi, C., Plesnicher, C., and McDonnell, G. (2005) *Appl. Biosaf.*, **10**, 91–100.
8. Russell, A.D., Furr, J.R., and Maillard, J.Y. (1997) *ASM News*, **63**, 481–487.
9. STERIS Corporation (2003) Bio-Dekontamination mit VHP Monographie "Technische Daten" (Europa), 17 pp.
10. Richardt, A. and Russmann, H. (2006) *Biol. Unserer Z.*, **5**, 322–329.
11. Klapes, N.A. and Vesley, D. (1990) *Appl. Environ. Microbiol.*, **56**, 503–506.
12. Meszaros, J.E., Antloga, K., Justi, C., Plesnicher, C., and McDonnell, G. (2005) *Applied Biosafety*, **10**.(2), 91–100. "http://absa.org./abj/abj/ (accessed 16.05.2012).
13. Assadian, O. and Kramer, A. (2008) *Wallhäußers Praxis der Sterilisation, Desinfektion, Antiseptik und Konservierung*, 1st edn, Georg Thieme Verlag, Stuttgart. ISBN: 978-3131411211.
14. Jursch, C.A., Gerlich, W.H., Glebe, D. et al. (2002) *Med. Microbiol. Immunol.*, **190**, 189–1997.
15. Nuanualsuwan, S. and Cliver, D.O. (2003) *Appl. Environ. Microbiol.*, **69** (3), 1629–1632.
16. Maillard, J.Y. (2002) *J. Appl. Microbiol. Symp. Suppl.*, **92**, 16–27.
17. Russel, A.D. (1990) *Clin. Microbiol. Rev.*, **3** (2), 99–119.
18. Montville, T.J., Dengrove, R., De Swiano, T., Bonnet, M., and Schaffner, D.W. (2005) *J. Food Prot.*, **68**, 2362–2236.
19. Spotts Whitney, E.A., Beatty, M.E., Taylor, T.H., Weyant, R., Sobel, J., Arduino, M.J., and Ashford, D.A. (2003) *Emerg. Infect. Dis.*, **9**, 623–627.
20. Francis, A. (1956) *Proc. Pathol. Soc. (Great Britain and Ireland)*, **71**, 351–352.
21. Xu, S., Labuza, P., and Diez-Gonzalez, F. (2008) *Appl. Environ. Microbiol.*, **74** (11), 3336–3341.
22. Urakami, I., Mochizuki, H., Inaba, T., Hayashi, T., Ishizaki, K., and Shinriki, N., (1997) *Biocontrol Science*, **2**, 99–103.
23. White, G.C. (1999) *Handbook of Chlorination and Alternative Disinfectants*, 4th edn, John Wiley & Sons, Inc., New York.
24. Watson, H.E. (1908) *J. Hyg.*, **8**, 536–542.
25. EPA (2003) *LT1ESWTR Disinfection Profiling and Benchmarking: Technical Guidance Manual*, EPA Guidance Manual, US Environmental Protection Agency, Washington, DC.
26. Russell, A.D. (2004) in *Russell, Hugo and Ayliffe's Principles and Practice of Disinfection, Preservation and & Sterilization*, 4th edn (eds A.P. Fraise, P.A. Lamber, and J.Y. Maillard), Blackwell Publishing Ltd, Oxford, ch. 3, pp. 98–127. ISBN: 9781405101998.
27. Hilgren, J. et al. (2007) *Appl. Environ. Microbiol.*, **73** (20), 6370–6377.
28. Ayliffe, G.A.J. (1989) *J. Hosp. Infect.*, **13**, 211–216.
29. Wallhäuser, K.H. (1995) *Praxis der Sterilisation, Desinfektion, Konservierung*, 5th edn, Georg Thieme Verlag, Stuttgart.

30. Mulberry, G.K. (1995) in *Chemical Germicides in Health Care* (ed. W.A. Rutala), Association for Professionals in Infection Control, Washington, DC, pp. 224–235.
31. Maillard, J.Y. (2005) *Ther. Clin. Risk Manage.*, **1** (4), 307–320.
32. Weber, D.J., Barbee, S.L., Sobsey, M.D., and Rutala, A.W. (1999) *Infect. Control Hosp. Epidemiol.*, **20**, 821–827.
33. Roberts, C.G. (2000) in *Disinfection, Sterilization and Antisepsis: Principles and Practice in Healthcare Facilities* (ed. W.A. Rutala), APIC, Minneapolis, pp. 63–69.
34. Reuter, G. (1994) The effectiveness of cleaning and disinfection during meat production and processing Influence factors and use recommendations, Fleischwirtschaft. **74** 808–813.
35. Raber, E., Jin, A., Noonan, K., McGuire, R., and Kirvel, R.D. (2001) *Int. J. Environ. Health Res.*, **11**, 128–148.
36. Raber, E., Carlsen, T., Folks, K., Kirvel, R., Daniels, J., and Bogen, K. (2004) *Int. J. Environ. Health Res.*, **14**, 31–41.
37. Benenson, A.S. (ed.) (1990) *Control of Communicable Disease in Man*, 15th edn, American Health Association.
38. Ingelsby, T., Hendeson, D.A., Bartlett, J.G., Ascher, M.S., Eitzen, E. *et al.* (1999) *J. Am. Med. Assoc.*, **281**, 1735–1745.

15
Radiological/Nuclear Decontamination – Reduce the Risk
Nikolaus Schneider

Technologies for decontamination of nuclear and radiological contaminated environment, equipment, and personnel are well known from the nuclear industry and, with regard to the subject of this book, from the nuclear weapon testing in the middle of the last century. However, due to the possible impacts on persons exposed to radiation, the need for improved decontamination procedures led to intensive research activities. Over the decades, the processes of contamination were investigated and decontamination procedures were elaborated to reduce the radiation exposure. New threats, contaminations from industrial sources or terrorist activities ("dirty bomb") are taken into consideration nowadays. Overall, effective decontamination technologies are available to reduce the contamination and to avoid an uncontrolled spread of the contamination.

15.1
Why Is Radiological/Nuclear Decontamination So Special?

Before we start our tour through radiological/nuclear decontamination technologies it is important to understand the significant difference to biological and chemical decontamination techniques. While chemical or biological decontamination generally means that the agents will be destroyed or transformed into less harmful products there is no practically applicable way to influence radioactive decay. The only procedures for radiological contaminations are based on removal of the radioactive substances from the contaminated surface. The aim of nuclear/radiological decontamination can only be to detach as far as possible the radioactive particles from the surface to reduce the dose rate caused by the residual contamination on the material, thus reducing the external irradiation hazard, minimize contact hazard, and prevent the re-aerosolization of residual particles, which may be an inhalation/ingestion issue.

The term chemical, biological, radiological, and nuclear (CBRN) includes two different aspects, radiological and nuclear:

- Nuclear implies all contaminations resulting from the explosion of a nuclear weapon, that is, radioactive fallout, wash-, and rainout and to some extent neutron activation. Although, fortunately, such an event has become more and more unlikely, the particular characteristics of nuclear contamination will be considered here. The resulting high dose rates and the effects on men and material demand a quick and effective decontamination.
- "Radiological" covers all other aspects and sources of radioactive contamination. This means, on the one hand, an accidental release from industrial facilities and, on the other hand, nuclear terrorism, which has especially come into the focus of interest. The threat of a terrorist attack using an explosive device containing radioactive material, a so-called "dirty bomb," has been discussed intensively and controversially.

The properties of radiological contaminations vary distinctly, and depend on many different factors that affect the means and measures for efficient decontamination. Interaction between the radioactive material and the surface structure is the factor that determines the extent to which contaminations can be removed from the surface and the effort required to do so. Nuclides, chemical compounds, particle size, solubility, and environmental influences like humidity or temperature are some of the factors that affect contamination. The properties of the surface, on which the radioactive material is deposited, determine essentially the decontaminability of the material. Smooth, even surfaces like glass, plastics, some metals, or painted surfaces can be decontaminated quite effectively. The more structured, rough, or porous a surface gets, the less effective decontamination will be. Considering the most likely location for the use of a dirty bomb, an urban environment, a contamination of buildings, roads, or pavements will take place, where a quick and effective decontamination is nearly impossible. Under CBRN conditions the first goal must be a "first response" to reduce quickly the hazard for people/personnel in contaminated areas and to recover contaminated material for further use. RN (radiological and nuclear)-decontaminants and RN-decontamination procedures described in this chapter will focus mainly on typical scenarios for CBRN responders. Clean-up techniques for nuclear industry or expensive and time-consuming procedures will not be dealt with.

RN-decontamination procedures are mainly washing/rinsing processes, supported by detergents, and mechanical means like high pressure, scrubbing, or brushing. Chemicals like desorbents or chelating agents can improve the process by forming chemical complexes with radionuclides that are stronger than binding forces to the surface and will also prevent the adhesion of solved nuclides to the surface during the decontamination process. Figure 15.1 depicts the different types of contaminations, the processes that occur at a surface during contamination, and the implications for decontamination.

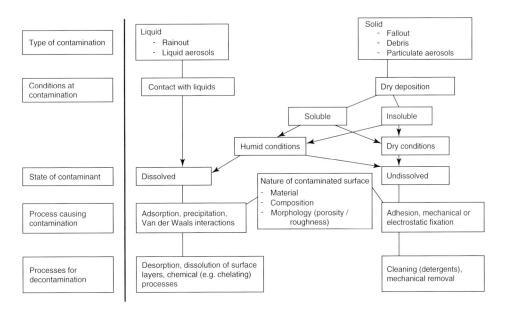

Figure 15.1 Main contamination/decontamination scheme.

15.2
Contamination

We learned in the chapter on RN-detection about the general features of radiation that can be detected. Now we have to discuss the same issue under the aspect of RN-decontamination.

15.2.1
Nuclear Weapons

Nuclear fission [1–3] occurs when the nuclei of certain isotopes of very heavy elements like uranium-235 or plutonium-239 capture neutrons. The nuclei of these isotopes are barely stable and the addition of a small amount of energy to one by an outside neutron will cause it to promptly split into two fragments, with the release of a great deal of energy (180 MeV of immediately available energy) and several new neutrons (an average of 2.52 for ^{235}U and 2.95 for ^{239}Pu):

$$n + {}^{235}U \longrightarrow F1 + F2 + xn$$

These neutrons lead to further fission reactions. If on average more than one neutron from each fission triggers another fission a chain reaction starts as the number of neutrons and the rate of energy production will increase exponentially with time.

The fission does not lead to distinct fragments but to a wide spectrum of mainly radioactive isotopes. The mass partition between the two fragments is asymmetrical, the mass numbers range from about 70 to 160. The yield of the different isotopes produced shows maxima at about 90–110 and 130–150 mass units (Figure 15.2).

It is important to understand that overall about 250 nuclides of 35 different chemical elements can be expected. The total activity of fission products resulting from a 1-kT ^{235}U weapon is estimated 1.5×10^{19} Bq 1 h after the explosion [4]. Most of these nuclides have short half-lives. As a guideline, taking the radiation dose rate 1 h after the explosion as a reference, the dose rate will decrease to 1/10th after 7 h, after 7×7 h (roughly 2 days) to 1/100th, and after $7 \times 7 \times 7$ h (about 2 weeks) to 1/1000th of the initial dose rate. On this basis, it could be concluded that if we wait long enough the problem would solve itself. But with regard to the extremely high initial dose rates, this will take years, as Hiroshima and Nagasaki and the nuclear test sites have shown. Even then, some long-lived nuclides remain.

15.2.1.1 Nuclear Fallout Contaminates the Ground
Nuclear explosions on or near the ground will drag large masses of soil, dust, and debris into the fireball. The radioactive fallout is the dust produced by the nuclear explosion and carried into the air by the mushroom cloud. The fission products of the explosion will condense, adhere or chemically react on or into particles. The particle sizes range from less than 1 µm to several millimeters [5]. The total mass of fission products is marginal compared to the inactive soil. So the properties of

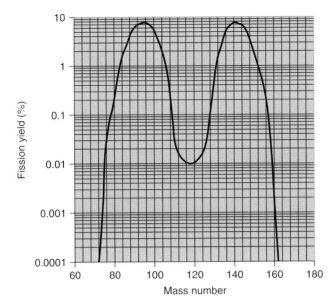

Figure 15.2 Yield of fission products for ^{235}U fission.

fallout will be mainly determined by the inactive parts, and not by the particular chemistry of the nuclides. Particles larger than 300 μm will fall near ground zero (Figure 15.3). Dose rates in this area will be extremely high, and in addition to the high degree of destruction, no decontamination of military relevance will take place.

Small particles, 20 μm or below, will be taken very high into the atmosphere and will mainly remain there for longer periods (worldwide fallout). The fraction that will mainly be an issue for decontamination has a particle size from about 50 to 200 μm. Depending on weapon size, and weather/wind conditions, these will come down within some hours and will contaminate large areas downwind.

These particles are mainly insoluble and are deposited quite loosely on material or equipment (e.g., vehicles). Mechanical brushing or washing with water, optionally under increased pressure, will remove the biggest part of the contamination. To remove conventional dirt, which to some extent will fix the contamination, the addition of surfactants will support decontamination. As the goal of a military decontamination under these conditions can only be to reduce the hazard to the individual using, handling, servicing, and/or transporting the contaminated item, these simple methods are mainly sufficient for the decontamination task.

15.2.1.2 Rainout/Washout

The phenomenon of rainout is widely underrated, although it was observed both in Hiroshima and Nagasaki [6, 7]. At relative air humidities above 70% it is assumed that rain will be induced due to the enormous thermal lifts, temperature gradients, heating, and cooling down processes and the presence of fine particles as

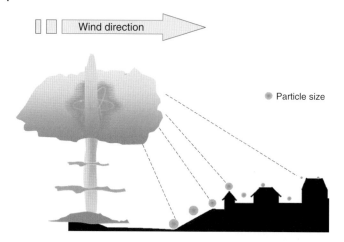

Figure 15.3 Factors affecting the fallout. The fallout pattern can be influenced by the initial altitude and also ground burst in contrast to one from altitude. *Particle size*: under conditions of constant wind, larger particles land closer to ground zero than smaller particles, which land farther away. *Altitude*: the degree of deposition also depends on the altitude the particles are taken up to in the explosion cloud; the higher the altitude the further away the particles land from ground zero. *Wind*: since different altitudes may have different wind directions and speeds, the particle movement and the final destination is the sum of all wind forces.

condensation nuclei. Radionuclides in this rain will be partly solved. It is estimated that up to 20% of the radioactive load of the rain will be present in aqueous solution. Radionuclides in ionic form will reach the surface, and will be bound chemically or by electrostatic forces to the material. Therefore, decontamination of rainout is more difficult than a "dry" fallout, as simple "washing off" methods will not be sufficient to give satisfactory decontamination results – decontaminants are necessary (Figure 15.1). For the decontamination of ionically sorbed contaminants the addition of a strong complexing agent to the aqueous decontamination solution (also containing a surfactant for conventional cleaning) is inevitable.

15.2.2
Radiological Contaminations – Radiological Dispersal Device (RDD)

A nuclear strike and widespread contamination of large areas have become improbable, but the issue of a radioactive contamination for CBRN responders – military or civilian – is still present. Beside the danger of accidental release of radioactivity from industrial facilities, especially the question of nuclear terrorism has come into the focus of interest. As described in chapter 5.4 a radiological dispersal device (RDD), more commonly known as *"dirty bomb,"* is a conventional blasting device containing radioactive material. The probability and the effects of an RDD attack are rated controversially amongst experts, but there is no doubt that radiological attacks constitute a credible threat [8–10]. In addition, it is important to realize that

Figure 15.4 Russian RTGs in scruffy condition. (Image from Finnmark regional government)

guidelines for RDD cleanup have to include the possibility that a RDD incident occurs with a high likelihood in an urban area. The decontamination processes have to be balanced with the affected population in terms of reoccupying the contaminated zone [11].

The effect of an RDD would surely not result in the hundreds of thousands of fatalities that could be caused by a nuclear weapon. Fatal radiation will remain limited to small areas around the blast, but large urban areas could be contaminated with radiation levels that exceed given exposure limits, and evacuation of the zone would become necessary. Furthermore, the psychological effect on the population should not be underestimated, even at very low radiation levels. The efficacy of a RDD and the health and environmental consequences are mainly due to the type and quantity of the radioactive material.

Radioisotope thermoelectric generators (RTG) convert the heat released by a radioactive decay into electrical energy. For example, Russian Radioisotope Thermal Generators (Figure 15.4), contain up to 1.5 PBq 90Sr, one of the most hazardous nuclides when incorporated [12]. Many of these generators are still in use in remote regions in Russia and other states of the former USSR. Safe containment and control of the radioisotope is an increasing challenge.

Dispersal will depend on the physical and chemical properties of the radioactive material used in an RDD. Metallic forms like cobalt, iridium, or polonium would be difficult to disperse, in contrast to a powder of oxides (americium, californium, and plutonium), which could be dispersed fairly readily. Cesium is commonly used as water-soluble CsCl, while strontium may be found in different forms (metallic, oxide, ceramic, or even as water soluble salt). Furthermore, the original material could be chemically or physically altered to enhance dispersal. The explosion of a dirty bomb would also likely physically and chemically alter the materials to produce a mixture that could include oxides as well as nitrates (from the explosives) over a range of particle sizes.

The different properties of contaminations will influence the means and methods to be applied for decontamination and the efficacy of decontamination measures.

We can expect:

- different nuclides;
- large particles, debris;
- very fine particles, aerosols, soluble, and insoluble material;
- liquid contaminations (solved nuclides);
- condensates, precipitates.

However, it is important to note that actions to be taken for decontamination depend on contamination and the properties of the surface.

Military decontaminants and the procedures for nuclear decontamination have been designed only for contaminations by fallout particles following a nuclear strike. They are only of limited suitability for the decontamination tasks after an RDD event.

The diversity in chemical and physical properties of radiological contamination and the low limits for residual contamination make radiological decontamination much more demanding. Although the basic principle and procedures of decontamination are the same, radiological decontamination will require efforts that go beyond the standard military decontamination procedures. This could include the need for more time, more personnel, adapted techniques and equipment, or specified decontaminants.

15.3
Decontamination

Especially for RN-decontamination we need to define some parameters. With these parameters we are able to determine the decontamination efficiency and can give an outlook on possible improvements.

15.3.1
Decontamination Efficiency Calculation

The decontamination efficiency is determined by measurement before and after the decontamination process; the ratio of activities indicates directly the quality of decontamination. This allows an objective evaluation of a decontamination process independent of the initial contamination. The results can be expressed in three ways:

1) percent decontamination efficiency
2) percent residual
3) decontamination factor.

---

1) **Percent decontamination efficiency:**

$$D = \frac{C_0 - C_t}{C_0} \times 100$$

where

D = percent decontamination efficiency;
C_0 = average or individual measurement values of initial contamination;
C_t = average or individual measurement values post-decontamination.

2) **Percent residual:**

$$R = \frac{C_t}{C_0} \times 100$$

where

R = percent residual;
C_0 = average or individual measurement values of initial contamination;
C_t = average or individual measurement values post-decontamination.

3) **Decontamination factor:**
The decon factor (DF) is the ratio between the initial activities divided by the final activity:

$$DF = \frac{C_{initial}}{C_{decon}}$$

where $C_{initial}$ is the activity before decontamination and C_{decon} is the activity after the decontamination step.

Measurement of the residual activity after decontamination can be expressed as dose rate at a given distance to the surface (e.g., $\mu Sv\ h^{-1}$ at 10 cm) or as surface contamination in $Bq\ cm^{-2}$. Notably, these are the units generally used for the threshold values given in legal regulations.

15.3.2
Decontamination Procedures for RN Response

There is a very wide spectrum of RN-decontamination measures and RN-decontamination procedures for a large variety of tasks [14, 15]. Within the frame of this book, we will focus only on procedures applicable for CBRN response forces, be it military or civilian (fire brigade, emergency aid, and companies supporting the authorities). The goal of radiological/nuclear decontamination is to remove the contaminant from the contaminated surface to the lowest reasonably achievable level by quick and easily applicable procedures so as to reduce the hazard to personnel in using, handling, servicing, and/or transporting a contaminated item (from vehicles to personal equipment), entering or staying in a contaminated zone for limited time, and the decontamination of personnel and their protective gear.

In cases when it can be assured that only a dry deposition of radioactive particles has taken place, it is appropriate to start with dry mechanical means to remove the particles from the surface. This may include brushing, sweeping, vacuuming, or blowing off, for example, with compressed air. The use of water under high pressure may be convenient as well, if the kinetic energy of the strong water stream effects an immediate removal of the particles.

Depending on the properties of the contaminating particles a longer contact of water may lead to a leaching process of nuclides out of the particles or solving the particles themselves. In this case, the nuclides will reach the surface in ionic, molecular, or colloidal form and will adhere or be bound to the surface. This will make decontamination much more difficult.

On soiled surfaces, for example, vehicles in use, the soil may act as protection for the surface. The contaminant is deposited on the soil layer and removing the soil means eliminating most of the contamination as well. Thus, conventional surfactant-based cleaners will serve to some extent as decontaminants.

But, generally, a decontamination agent will be needed.

- A RN decontamination agent consists of a solvent or a dispersing medium and one or more "active" substances in the medium that initialize or support the decontamination process. This may be: (i) acids or alkalis (both inorganic or organic), (ii) oxidizing or reducing chemicals, (iii) complexing (chelating) agents, (iv) surfactants (surface active substances), and (v) auxiliary additives (emulsifiers, stabilizers, abrasives, etc.).

Similar to the challenges for C-decontaminants (Figure 13.3) a decontamination agent for use in RN scenarios should principally meet the following requirements (Figure 15.5):

- highly effective at even low concentrations;
- efficiency for a wide spectrum of conceivable contaminants;

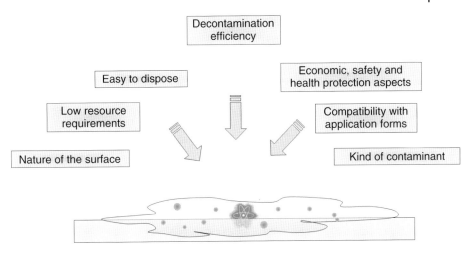

Figure 15.5 Requirements for RN decontaminants.

- not corrosive or aggressive to surfaces;
- quick and easy operational application;
- low resource requirements (material, chemicals, water, personnel, and time);
- compatibility with simple methods of application;
- environmental safety;
- sufficiently stable (storage of stock supply without loss of efficiency);
- allow processing of the resulting radioactive waste.

Obviously, a universally competent decontaminating agent, adequately suitable for all kinds of contaminations and decontamination tasks, that fulfills all the requirements listed above will not be achievable. A decontaminant can only be the best possible compromise, taking into account:

- the nature of the surface to be decontaminated;
- the kind of contaminant and the presumed type of strongest bond between contaminant and surface;
- the decontamination efficiency required;
- economic, safety, and health protection aspects (Figure 15.5).

15.3.3
RN Decontamination Agents

Since the 1960s the decontaminants A1 and A2 have been in service for nuclear decontamination. A1 is a surfactant (alkyl-aryl sulfonate), while A2 is a complexing agent [di-Na-EDTA (ethylenediaminetetraacetic acid)]. Both are water-soluble powders and are used in aqueous solution.

This quite simple decontamination agent was, as tested with simulants, fairly effective for nuclear contaminations. For radiological contaminations it also works to some extent but by far not sufficiently for all types of contamination that might be expected.

Problems in application and, particularly, environmental aspects (non-biodegradability of compounds) forced the search for alternatives [16]. A decontamination agent should be universally applicable for radiological and nuclear decontamination and in addition to the criteria mentioned above, meet the following requirements [17]:

- liquid formulation,
- highly concentrated ingredients,
- components: surfactants, complexing agents, mild oxidizers, and additives,
- enhanced contact time surface/decontaminant,
- biodegradability,
- stability over longer periods (storage).

The main active components for the decontamination agent are the surfactant and complexing (chelating) agent – this has not changed compared to the old solution.

15.3.3.1 Surfactant

Surfactants (surface-active agent) generally consist of a hydrophobic portion, usually a long alkyl chain, attached to hydrophilic or water solubility enhancing functional groups. They can be categorized according to the charge present in the hydrophilic portion of the molecule (after dissociation in aqueous solution) as anionic, cationic, nonionic, or ampholytic surfactants [18].

The surfactant has several tasks in the RN-decontamination process. It reduces the surface tension of the aqueous solution, thus allowing a better wetting of the surface and an improvement of contact between decontaminating compounds and the surface. It also serves as a conventional cleaner, removing dirt from the surface by enclosing the particles in the aqueous phase or stabilizing radioactive contamination deposited in urban areas [19] as shown in the Figure 15.6.

A very important function is the production of stable foam (Figure 15.7), when the decontaminant is applied on the surface. Depending on the surfactant used, the foam sticks to the surface and breaks down slowly, constantly releasing fresh solved decontaminant (chelating agent) to the surface, which reacts with the radionuclides and runs off. The reaction equilibrium is shifted continuously to the side of the chelate product, supporting the decontamination process.

The surfactant chosen for the new decontamination agent is a mixture of sugar-based surfactants (alkyl polyglycosides) (Figure 15.8) from renewable resources [20].

They are non-ionic surfactants: carbohydrates (sugar) are the hydrophilic part of the surfactant, with fatty alcohols, or acids, as the hydrophobic group. They are distinguished by low toxicity, are fully biodegradable, and are rated in the lowest German Water Hazard Class (WGK 1).

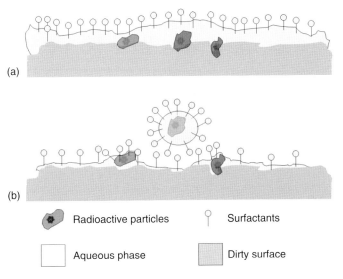

(a)

(b)

- Radioactive particles
- Surfactants
- Aqueous phase
- Dirty surface

Figure 15.6 Removing and solubilizing particles from surfaces. Modified from Reference [19].

Figure 15.7 Application of foam on test surface (a) and after 5 min residence time (b).

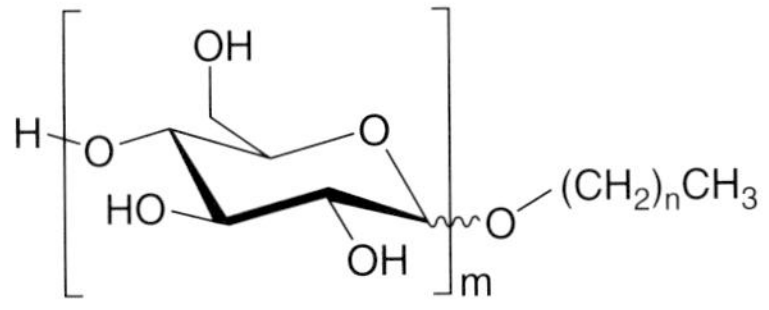

Figure 15.8 Alkyl polyglycosides, which can be used as the basis for surfactants for RN decontamination.

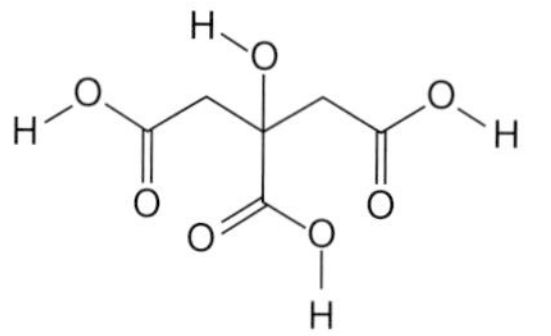

Figure 15.9 Citric acid (2-hydroxypropane-1,2,3-tricarboxylic acid).

The cleaning abilities are excellent and by mixing different compounds (variation of chain lengths) the desired foam properties are achieved.

15.3.3.2 Chelating Agent

Some polyamino carboxylic acids, especially EDTA and derivates, are well known and widely used as chelating agents for many different cations. This makes them ideal for decontamination purposes and the decontamination of fallout and fission products, because the formed complexes are stronger than the binding forces to the surface. Therefore, the di-Na salt of EDTA was chosen many years ago as the widely known NATO decontaminant A2.

But, due to low biodegradability and its ability to mobilize heavy metal ions into the environment, its use and the release to the environment in larger quantities is restricted today and alternatives had to be found [21]. Another strong chelating agent for a wide spectrum of cations is citric acid. Although the complex formation constants for many metals are lower than those of EDTA, it has proved to have even better decontamination abilities in practice. Citric acid is a hydroxy-tricarboxylic acid (2-hydroxypropane-1,2,3-tricarboxylic acid) (Figure 15.9).

The carboxylic groups encapsulate the metal, depending on size and charge, as shown for calcium:

$$Ca^{2+} + 2\ CitH_3 \longrightarrow [Ca(Cit)_2]^{4-} + 6H^+$$

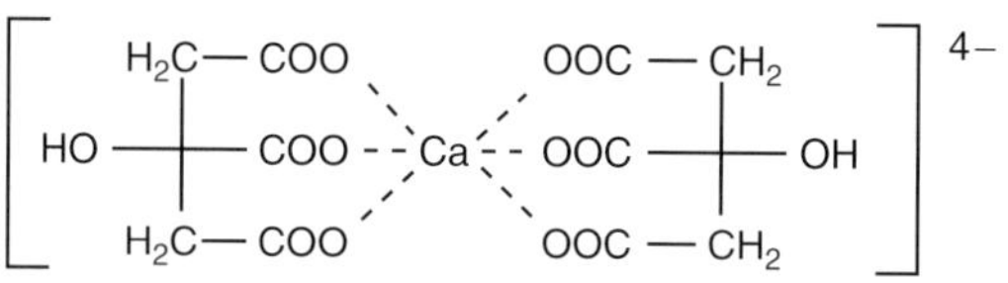

$$\begin{bmatrix} HO - \overset{H_2C-COO}{\underset{H_2C-COO}{\mid}} - COO - - Ca - - OOC - \overset{OOC-CH_2}{\underset{OOC-CH_2}{\mid}} - OH \end{bmatrix}^{4-}$$

Beside the complex formation properties for many of the nuclides to be expected as contaminants in a radiological or nuclear scenario, its low toxicity and high biodegradability make citric acid and its salts a natural choice as decontaminant for outdoor use. In combination with sugar surfactants, a formulation for a decontamination agent was investigated that combines decontamination efficiency with high environmental compatibility (Figure 15.10).

15.3.4
Specific Decon Processes, Alternative Procedures

As stated before, the first responder's goal is a quick and easy procedure with the highest possible efficiency. Obviously, these procedures have their limitations due to the properties of the contaminated surface and binding processes of nuclides to surfaces, for example, buildings (Figure 15.11). However, for large areas or

Figure 15.10 RN decontamination: application on an armored vehicle of the TEP 90 decon system.

Figure 15.11 Decontamination of a building.

items it is the only practical way of handling the task to reduce the risk in a short time.

A more far-reaching decontamination will be at first limited to special points of interest, material, or equipment urgently needed.

15.3.4.1 Spraying/Extraction Systems

The principle of these systems is well known from commercial carpet cleaning. Water or aqueous solutions of cleaners or decontaminants are sprayed onto the surface under elevated pressure and vacuumed off in the same working step. Determinant for the decontamination result is the specifically designed spraying–extraction heads adapted to the particular surface or decontamination task. For military decontamination purposes the technique is used especially for the decontamination of interiors of vehicles, where a spill of water or decontaminant

Figure 15.12 Spray extraction: (a) principle of spray extraction; (b) decontamination of the hatch of a tank; (c) decon shuttle.

must be avoided (Figure 15.12). On a smaller scale it can also be used for sensitive equipment or even for human skin.

15.3.4.2 Surface Ablation Techniques

Radioactive contaminations that remain fixed to surface or – for porous material like concrete – have penetrated into the material cannot be removed by a simple washing/rinsing process. To attack these contaminations, removal of the upper layers of the surface could be the method of choice. Conventional methods like sanding or grinding or even stronger technical appliances like sand blasting are in use to remove dirty or contaminated layers of a surface, but often lead to a certain degree of damage to the material, which, for example, in the decontamination of equipment, should be avoided.

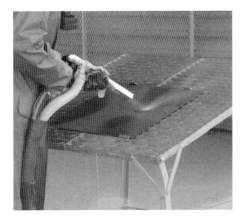

Figure 15.13 Dry-ice cleaning of material samples.

There are two more sophisticated technical solutions that are taken into consideration for a smooth and more controlled removal of surface layers, which we now introduce briefly.

Dry Ice Blasting This method uses dry ice pellets, or dry ice snow produced from liquid CO_2, that are shot onto the surface to be treated by compressed air (Figure 15.13). The process mechanism involves temperature, blasting, and solvent effects. The ice particles cool the surface abruptly, which leads to a separation of surface adhering particles. Owing to the sudden sublimation of the CO_2 ice, small surges in pressure are created. These pressure surges release micro impurities on surfaces, even in the pores. By blasting, the surface impurities are removed from the surface.

Although decon results are, depending on contamination conditions, competitive to other procedures, problems like the uncontrolled spill of the removed radioactive particles into the surrounding atmosphere and the logistics in providing and storing the dry ice material allow only a very limited use.

Laser Cleaning The treatment of surfaces with pulsed laser beams is an acknowledged technique that not only affects the surface soiling but can also be used for the ablation of material itself to a defined depth. It is widely used for paint stripping (e.g., from airplanes) or the removal of graffiti from walls. Applied for decontamination purposes it will be able to remove contaminations that have penetrated into material, where most conventional procedures fail (Figure 15.14). Different types of laser can be used for material treatment, but the high intensities needed can only be supplied by pulsed lasers. At the experimental stage, for a very fine surface ablation of only the upper layers [22] an excimer laser (wavelength 158–351 nm) is used; most commercial laser cleaning system are based on Nd-YAG lasers (wavelength 1064 nm). Generally speaking, powerful, very short (nanoseconds), rapid, moving laser pulses produce micro-plasma bursts, shockwaves, and thermal

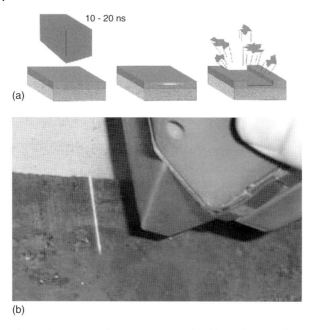

Figure 15.14 Laser cleaning: (a) principle of laser cleaning; (b) removal of rust using a laser cleaning system.

pressure that lead to mechanical tensions, sublimation, and ejection of the target material, coating, or contaminant [23, 24].

Commercial laser cleaning systems are available from small 20 W portable "backpack" devices up to large systems of 1000 W. The laser beam is delivered via a flexible optic fiber cable to a handheld laser optic where different parameters can be adjusted. A vacuum cleaner can be connected to safely collect detached radioactive particles and dust.

The treatable surface area per time depends on the power of the laser and the desired depth of the ablation. To achieve high efficiency, the process is quite time consuming and elaborate, but in the hands of specialists it is a sophisticated method for specific decontamination tasks.

15.4
Conclusions and Outlook

Radiological decontamination always means that the radioactive material has to be removed from the contaminated surface. This is different to chemical or biological decontamination, where the agents can be destroyed by decontaminants.

Many different means, methods, and procedures are known for the elimination of radioactive contaminations from surfaces, especially from the nuclear industry, but within the frame of this book only those procedures applicable for response in a CBRN scenario are considered. Historically, the decontamination of nuclear weapon fallout was the main point of interest. But these mostly insoluble particles were considered not to be strongly bound or fixed to surfaces and therefore could be removed by quite simple means like washing and rinsing, optionally supported by surfactants and/or chelating agents.

New threats, especially nuclear terrorism and the potential use of RDDs (dirty bombs), may lead to contaminations with distinctly different physical and chemical properties that will make decontamination much more demanding. Limits for residual contaminations according to civil legislation have to be achieved – magnitudes lower than those valid in a nuclear scenario.

Decontaminants that are quick and easily applicable, environmentally compatible, and efficient for a wide spectrum of contaminants had to be developed for the use in CBRN response. Although very efficient products are commercially available, it is not, and will not be, possible to have a procedure that covers everything. Especially, the decontamination effectiveness in urban environments is very limited due to the properties of the building materials. For items of particular interest or importance, mechanically enhanced procedures or surface ablating techniques are taken into consideration. Having in mind that a radiological terrorist attack is most likely to happen in an urban environment, an improved decontamination of infrastructure is one of the tasks for the future.

References

1. Glasstone, S. and Dolan, P. (1977) *The Effects of Nuclear Weapons*, 3rd edn, U.S. Department of Defense, U.S. Department of Energy.
2. Bohr, N. (1939) *Phys. Rev.*, **56** (5), 426–450.
3. Duderstadt, J.J. and Hamilton, L.J. (1976) *Nuclear Reactor Analysis*, John Wiley & Sons, Inc., New York.
4. Gut, J. (1971) Ein Modell für den lokalen radioaktiven Ausfall bei nuklearen Explosionen mit Bodensprengpunkt und seine Anwendung. Dissertation, Eidgenössische Technische Hochschule Zürich.
5. Crocker, C.R. et al. (1966) *Health Phys.*, **12**, 1099.
6. Molenkamp, C.R. (1979) An Introduction to Self-Induced Rainout, University of California, Lawrence Livermore Laboratory, United States UCRL 52669.
7. Knox, J.B. and Molenkamp, C.R. (1974) Investigations of the Dose to Man from Wet Deposition of Nuclear Aerosols, University of California Radiation Laboratory, UCRL 76109.
8. Sopko, J.F. (Winter 1996–1997) The changing proliferation threat, *Foreign Policy* (105) Washington Post Newsweek Interactive, LCC, pp. 3–20.
9. Ring, J.P. (2004) *Health Phys.*, **86**, 42–47.
10. Barnett, O.J., Parker, L.L., Blodgett, D.W., Wierzba, R.K., and Links, J.M. (2006) *Radiiol. Nucl. Terror.*, **47** (10), 1653–1661.
11. Eraker, E. (2004) *Non-Prolif. Rev.*, **11**, 167–185.
12. Sneve, M.K. and Reka, V. (2007) Upgrading the Regulatory Framework of the Russian Federation for the Safe Decommissioning and Disposal of Radioisotope Thermoelectric Generators. Strålevern Report 2007:5, Norwegian Radiation Protection Authority, ISSN: 0804-4910.
13. Petersen, J.M., MacDonell, M., Haroun, L., Monette, F., Hildebrand, R.D., and Taboas A. (2007) Radiological and Chemical Fact Sheets to Support Health Risk Analyses for Contaminated Areas, Prepared by Argonne National

Laboratory Environmental Division in collaboration with U.S. Department Energy, Richland Operations Office and Chicago Operation Office (1-133).

14. Severa, J. and Bár, J. (1992) *Handbook of Radioactive Contamination and Decontamination*, Elsevier, Amsterdam, ISBN: 978-0444987570.
15. Simon, A.D. (1980) *Entaktivierung – Dekontamination*, Militärverlag der Deutschen Demographischen Republik, Berlin.
16. Schöberl, P. and Huber, L. (1988) *Tenside Surfact. Detergents*, **25** (2), 99–107.
17. Schneider, N. (2009) Development of a new decontaminant for radioactive contaminations. Presented at Workshop on Response to Chemical, Biological and Radiological/Nuclear Terrorist Attacks, Ottawa 2009.
18. Tadros, T.F. (2005) *Applied Surfactants: Principle and Applications*, Wiley-VCH Verlag GmbH, Weinheim, ISBN: 3527306293.
19. Fox, G.A., Fuchs, J.W., Medina, V.F., and Atapattu, K. (2007) *J. Environ. Eng.*, **133** (3), 255–262.
20. Hill, K.H. and Rohde, O. (1999) *Fett/Lipid*, **101** (1), 25–33.
21. Combes, R., Barratt, M., and Balls, M. (2003) *Altern. Lab. Anim.*, **31**, 7–19.
22. Steglich, K.H. (2007) Reinigung und Strukturierung empfindlicher Oberflächen mittels Excimerlaserstrahlung – Entwicklung und Anwendung eines Applikationsgerätes. Dissertation. Helmut Schmidt Universität, Hamburg.
23. Tam, A.C., Park, H.K., and Grigoropoulos, C.P. (1998) *Appl. Surf. Sci.*, **127–129**, 721–725.
24. Porprawe, R., (2005) *Lasertechnik für die Fertigung*, Springer-Verlag, Berlin, 314 ff.

Part VI
CBRN Risk Management – Are We Prepared to Respond?

copyright by Jörg Pippirs, http://www.artesartwork.de

CBRN Protection: Managing the Threat of Chemical, Biological, Radioactive and Nuclear Weapons,
First Edition. Edited by A. Richardt, B. Hülseweh, B. Niemeyer, and F. Sabath.
© 2013 Wiley-VCH Verlag GmbH & Co. KGaA. Published 2013 by Wiley-VCH Verlag GmbH & Co. KGaA.

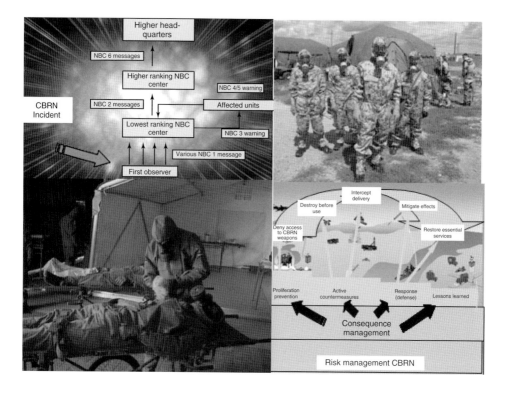

16
Preparedness

Marc-Michael Blum, Andre Richardt, and Kai Kehe

Throughout human history there have been many threats to the security of nations. When these threats become reality they are not only responsible for widespread illness, injury, and large-scale losses of life but also for the destruction of property and devastating economic losses. Over recent decades we have seen new technological advances, economic globalization, and ongoing international political unrest as components of risk to national and international security. Nowadays, trend analysis and technology foresight are predicting new technological breakthroughs in science and therefore new potential threats for human society. For countermeasures, local, national, and international authorities and organizations need to collaborate to be prepared and to establish the best available response strategy to minimize the potential loss of lives, to reduce economic loss, and to avoid new international conflicts. The special aspects of risk management with respect to CBRN threats are discussed in this chapter.

16.1
Introduction to Risk Management

In the previous chapters we have learned a lot about the technical aspects of chemical, biological, radiological, and nuclear weapons of mass destruction (WMD) and the possibilities for technical countermeasures. However, understanding these countermeasures is only one aspect of an effective strategy against WMD. While the application of such countermeasures will be discussed in the next section we will focus on necessary preparations as well as measures to prevent a chemical, biological, radiological, and nuclear (CBRN) event from happening at all. We are not able to cover every factor and eventuality in any imaginable CBRN event and therefore it is our aim to give a general idea of how to approach risk management. Without proper risk management, which preferably covers many aspects of potential CBRN threats, an effective public response cannot be established. But before we go further we have to understand the basics of risk management. What is risk management? Risk management is a process of identifying opportunities and avoiding or mitigating losses.

CBRN Protection: Managing the Threat of Chemical, Biological, Radioactive and Nuclear Weapons,
First Edition. Edited by A. Richardt, B. Hülseweh, B. Niemeyer, and F. Sabath.
© 2013 Wiley-VCH Verlag GmbH & Co. KGaA. Published 2013 by Wiley-VCH Verlag GmbH & Co. KGaA.

- **Risk management**: a process of identifying opportunities and avoiding or mitigating losses. To achieve these ambitious stipulations, it must be a structured and systematic process. In this process it is necessary to establish an understanding of the risks associated in a way that enables an organization to minimize losses and maximize opportunities (Figure 16.1).

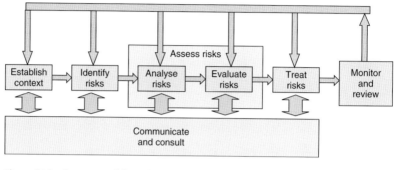

Figure 16.1 Summary of the risk management process [1].

We can state that fear of the impacts of a CBRN incident for population and economy, new asymmetric actors, and the scientific revolution in life science and nanotechnology are the drive to install procedures for effective countermeasures in the case of a CBRN incident. In this general environment of continual change and limited resources, the management of CBRN risk has become a critical issue. It is of vital interest that decision-makers thus know about possible outcomes and take steps to control their impact. Therefore, it is necessary to understand that risk management has to be part of an organization's culture. Only when it is integrated into its philosophy, practices, and business plans does risk management become the business of everyone in the organization. Otherwise, the failure of not managing CBRN risk in an adequate way can lead to fatal consequences such as:

- personal injury,
- public health crises,
- economic losses,
- environmental damage,
- community losses,
- loss of professional or technical reputation and standing,
- criminal charges.

However, this is the "negative" approach, which reflects only on the losses or undesired outcomes that might arise to explain why risk management is necessary. On the other hand, the risk management approach can also be used to identify and

16.1 Introduction to Risk Management

prioritize opportunities ("positive" risks), with little change to the process. What could be achieved with "positive" risk analysis?

- any associated positive consequences (i.e., opportunities/benefits) that would offset the negative consequences of particular events, as identified in a traditional risk analysis;
- and/or the opportunities/benefits that could arise from a set of events selected solely on the basis of their potential for positive outcomes.

Nevertheless, the traditional approach to CBRN risk management is to evaluate the potential losses and possibilities to avoid or to mitigate. Here we are discussing the basics of risk management. But what is risk? Risk can be defined as the chance of something happening that will have an impact upon objectives.

> - **Risk**: Can be defined as the chance of something happening that will have an impact upon objectives. It is measured in terms of consequences and likelihood. Therefore, the three basic elements of risk are: an event, the likelihood of the event, and the consequences of the event.

It is common practice that, for a particular event, likelihood and consequence can be combined to produce a level of risk, expressed either qualitatively or quantitatively. Risk analysis can help public and private organizations to manage and to

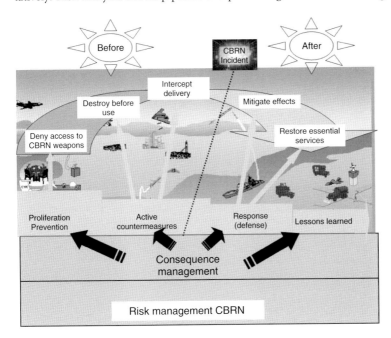

Figure 16.2 Risk management CBRN. Adapted from Figure 1.1 of Reference [5].

communicate the risk posed by uncertain threats [2–4]. Based on these elements we can discuss some general aspects of CBRN risk management (Figure 16.2).

The general idea of CBRN risk management (Figure 16.2) is to have a consequence management (CM) that is predominantly an emergency management function. For CM several definitions are available. In a more general definition CM can be defined as measures to protect public health and safety.

> - **Consequence management** (CM) includes measures to protect public health and safety; to restore essential government services and provide emergency relief to all institutions and individuals affected by a WMD incident.
>
> In general the responsible agencies have to use well-rehearsed structures if they want to deter and to prevent possible WMD incidents, to respond during an incident, and to learn from the whole process to mitigate further incidents. Therefore, is it necessary to further graduate the whole process:
>
> - **Proliferation prevention**:
>   – identify threats and own vulnerabilities;
>   – deploy protective measures to deny access to CBRN material and expertise to third parties;
>   – active countermeasures;
>   – awareness of ways of delivery;
>   – intercept delivery.
> - **Response**:
>   – awareness of actions to take if a CBRN event occurs to mitigate and to restore essential services;
>   – train to respond to a CBRN event.
> - **Lessons learned**:
>   – analyze incident and response;
>   – improve the whole risk management process.

But before an effective CBRN risk management based on the above-mentioned points can be established, we should first be aware of the special key elements influencing a counter-CBRN strategy and of the particular nature of chemical, biological, radiological, and nuclear threats.

16.2
Key Elements Influencing a Counter-CBRN Strategy

A well-known cliché about terrorism states that the defense against terrorism must succeed every time, but that terrorists must succeed only once. This is true only from plot to plot but this logic is reversed within each plot [6]. The terrorist must succeed at every stage of the plot leading to a successful attack while the defense only needs to succeed once to stop the plot and potentially catch the plotters. This insight points towards a strategy of broad defense. This means that a broad number of potential defenses should be put in place. Instead of picking one bullet point from the list in the introduction (one that promises a high rate of success) it is more advisable to use several lines of defense even if each of them has a slightly lower chance of success than the "big thing" on the list [7, 8]. We should also have in mind that costs are an important factor as well. Available resources to counter CBRN threats will not be unlimited. The assumed costs should normally correspond to the amount of residual risk a government/society would see as acceptable. However, in our risk-averse societies this residual risk might be so low that the associated costs are unbearable. One should also keep in mind that human beings can be quite irrational when dealing with risks. An individual is still far more likely to die in a car accident than to be killed in a terrorist attack (even a conventional one) – yet millions of people still commute to their jobs every day and drive their cars without unease.

This leads us to an important point in which CBRN attacks are different from conventional ones using guns or explosives: they are rare but their consequences can be far more devastating and long lasting. In addition, the term weapons of mass destruction that is commonly used for CBRN weapons indicates the amount of fear and insecurity associated with this kind threat. As their use is rare there is only limited information and experience for comparison. The sarin attack on the Tokyo subway and a similar incident in Matsumoto a year before were the only use of nerve agents for a large-scale terrorist plot in history. Even though several attempts were made, the anthrax letters in the USA are basically the only incident of bio-terrorism that remains in public perception. This leaves room for speculations, hysteria, and undefined fear. This adds to the point that even just the potential threat that a terrorist could use CBRN weapons (and not the actual use of them) could lead to the effects a terrorist would like to achieve with his actions. This makes CBRN preparedness somewhat different to preparedness for conventional attacks (Figure 16.3, Table 16.1). It should also be kept in mind that response to a CBRN incident might turn out to be far more difficult than response to, for example, a car bomb. Weapon effects might lead to widespread destruction (nuclear) or long-term contamination (chemical and biological and radiological). Cleanup of the Brentwood US postal facility that was contaminated with anthrax in 2001 took 26 months and cost $130 million [9]. This illustrates that certain types of CBRN weapons can also inflict severe economical impact besides their threat to human lives.

A final point to be made at this stage is to point out the differences between countering a CBRN threat from a non-state actor like a terrorist and one from states. In recent history the strategy employed by the two superpowers during the cold war to avoid being attacked by nuclear weapons was deterrence. Even though nuclear strategy can be highly complex one could express this kind of deterrence as the ability of an actor to counter a nuclear attack by making sure (or to let it appear as likely to the adversary) that such an attack would be retaliated in kind (also with nuclear weapons) so that this would result in no benefit for an

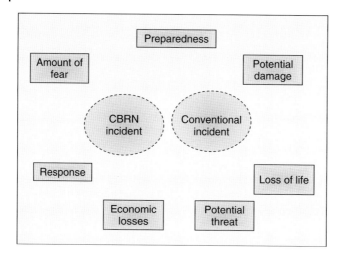

Figure 16.3 CBRN incident versus conventional incident (e.g., suicide bomber).

Table 16.1 CBRN incident versus conventional incident.

	CBRN incident	**Conventional incident**
Amount of fear	High	Middle–high
Potential threat	Low	High
Potential damage	High–middle	Low–middle
Response	Difficult and complex	Low–middle
Potential for loss of life	High	Middle
Preparedness	Difficult and complex	Difficult
Economic losses	High	Middle–low

actual attack as it would result in one's own defeat. However, this is not possible with terrorist groups. Here deterrence fails. First of all there are no targets for retaliation. There is no enemy capital or industrial centers. In addition, as terrorist actors might remain unidentified or disappear after an attack there is no personal target. Finally, a terrorist might accept death as a consequence for his actions (like suicide bombers). As a conclusion, one has to accept that a concept based only upon deterring terrorist actors from using CBRN weapons will in all likelihood fail.

16.3
A Special Strategy for CBRN

Law enforcement in several countries can draw from long experience with respect to anti-terrorism initiatives and actions (in Europe think about the UK in Northern Ireland or Germany, France, and Italy in their fight against the Red Army Faction, Action Direct, or the Red Brigades). Even though the largest threat today is regularly considered to be Islamist terrorism, including the new phenomenon of suicide bombers, "classic" anti-terrorist strategies can still be applied even though in a modified form. These kinds of strategies will not be discussed in detail here and the reader if referred to the literature on this topic. Instead we will focus on the special elements that can be, or have to be, added to an anti-terrorism strategy with regard to CBRN [10]. Two key elements can be identified that make CBRN different from conventional means of terrorist actions:

1) it is significantly more difficult to obtain CBRN materials (including precursors) compared to guns or normal explosives;
2) it is also more difficult to recruit specialists with knowledge in handling and production of CBRN agents compared to people with experience in handling guns and explosives.

Therefore, a counter-CBRN strategy (Figure 16.4) should put special emphasis on both aspects and try to develop effective means of denial of access to CBRN material and specialized personnel. As far as we can see current measures seem to be at least somewhat effective in this regard. If we examine attacks from insurgents or Al-Qaida in Iraq and Afghanistan the classical means of attacks are still car bombs, suicide bombers, and remotely detonated improvised explosive devices (IEDs). The materials required are easy to obtain, there is widespread knowledge about the design of bombs and IEDs, and there is experience in the use and effects of these weapons. On the other hand, several Al-Qaida sources stress the organizations desire to obtain some CBRN capabilities and it should be clear that a successful CBRN attack will be highly prestigious for the organization that carries it out. But there are also possibilities for an improvised use of CBRN materials. In recent years there have been several attacks on trucks in Iraq carrying toxic industrial chemicals (TICs) like chlorine (Chapter 1). However, there was little effect from the chlorine release itself as the detonation and the subsequent fires created a chimney effect that carried the chlorine into the atmosphere – a clear indication of a lack of chemical and physical knowledge. On the other hand, these incidents are also clear indicators that chemical attacks can also be carried out using industrial

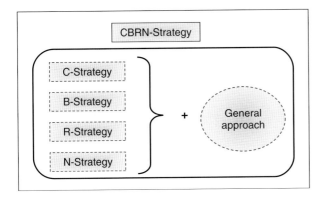

Figure 16.4 CBRN strategy to counter CBRN threats.

chemicals that are produced and consumed in bulk amounts. Therefore, chemical preparedness must also focus on TICs [sometimes also termed toxic industrial materials (TIMs)].

Now we have heard a little about chemical threats, but what about biological, nuclear, and radiological threats? In fact this reveals one of the major problems when discussing CBRN issues. All four threats that are combined in the term *CBRN* differ fundamentally with respect to possible weapon effects, the required knowledge and expertise for a successful attack, the difficulty in obtaining raw materials, and the chances of interception by law enforcement during the planning stage. Even though the title of this section talks about "a" strategy for CBRN, the best approach will be to examine the four different threats individually and identify key elements important for countering them. In the end these four individual areas should be examined for redundancies, synergies, and possible approaches to combine them in a general anti-CBRN approach (Figure 16.4). Keeping in mind the two key points identified in this section we will examine each threat on its own with regard to obtaining the required materials and the "human factor," which includes knowledge, expertise, and experience.

16.3.1
Chemical Threats

According to the Chemical Weapons Convention (CWC) (see also Chapter 2) any toxic chemical or precursor is considered a chemical weapon "except where intended for purposes not prohibited under this Convention, as long as the types and quantities are consistent with such purposes." This means that TICs could also be used as a chemical weapon even though they have a legitimate use. If we look at chemical threats it is therefore advisable to separate toxic chemicals in general and those specifically designed as chemical warfare agents. Not every toxic substance can easily be used as a chemical weapon. Preferably, the substance should be volatile or easily form an aerosol to be able to enter the body via the respiratory tract or alternatively it should be able to penetrate human skin. In addition, the substance must be reasonably stable to air, moisture, and sunlight. Otherwise the substance might decompose before a potential victim can ingest it. This excludes a lot of toxic chemicals from the list of potential chemical weapons. When large scale chemical warfare was considered for the first time in history during World War I several industrial chemicals such as chlorine or cyanide were used because of existing manufacturing plants and available raw materials. Later in the war new chemicals like sulfur mustard were introduced specifically as chemical warfare agents (Chapter 1). Would a terrorist group aim for more readily available TICs or try to make special agents like sulfur mustard or even nerve agents? Aum Shinrikyo carried out the attack on the Tokyo subway using the nerve agent sarin (quite impure and with a far from perfect dispersal method). The attack resulted in 12 fatalities, hundreds of poisonings, and thousands of people seeking medical attention. This means that at least one terrorist group has successfully used nerve agents in an attack. The terror effect and fear caused among the population are certainly enhanced if "real" chemical warfare agents are used. While the group of organophosphorus nerve agents contains the most toxic agents, other agents like sulfur mustard can also be effective terror weapons as they cause very painful wounds that are difficult to treat medically and which would result in disfigured patients generating just the kind of TV pictures required for a maximal terror effect. However, TICs are attractive as well. As they are produced in large quantities they can be good targets for theft or attacks against storage tanks and containers. In 2007 Iraqi insurgents used chlorine gas cylinders together with conventional high explosives as a kind of improvised chemical weapon. The attacks turned out to be quite ineffective because much of the gas either reacted due to the heat of the explosives or it was lifted into the atmosphere due to a chimney effect caused by the explosion and subsequent fires. But even if the attacks can be considered as a failure they show that industrial chemicals can be a source for chemical weapons if specific warfare agents are not available and their synthesis is beyond the abilities of the terrorist group. Even though not the result of an attack, the release of multiple tons of the highly toxic and volatile chemical methyl isocyanate (MIC) at a Union Carbide plant in Bhopal, India caused more than 3000 immediate fatalities.

Handling and storing large amounts of highly toxic substances is also a significant problem for the chemical industry with respect to operational and work safety. In many cases reactive and toxic intermediates are produced for immediate use for another reaction in the same facility. An example is the handling of phosgene, which was used as a chemical warfare agent in World War I but is also an important building block for polyurethane and polycarbonate polymers. Production of phosgene for direct use avoids storage and long distance transport. Where toxic and volatile chemical are stored in bulk amounts near populated areas a change of this practice should be initiated as a defensive step. Notably, however, a high tonnage of toxic chemicals is still transported around the world. Especially problematic is transport by rail or road as routes regularly pass through heavily populated areas.

What is the situation with "real" chemical warfare agents? If we look at the most dangerous group, the nerve agents, we find that all are organophosphorus compounds. The CWC not only lists the agents themselves on the list of regulated chemicals but also important precursor chemicals. An example is pinacolyl alcohol, which is required for the synthesis of soman and has very little other commercial use. Other precursors include reactive intermediates containing phosphorus that are required to attach other relevant groups to the molecule's core. Effective control of these compounds can in fact make it much more difficult for a terrorist group to obtain nerve agents. On the other hand, the amount of agent required for an attack is relatively small. Instead of tons only a few kilograms are required. Instead of relying on direct precursor chemicals one might also start with more basic chemicals and go through more reaction steps toward the final compound. Even though national legal regulation on obtaining potential precursor chemicals and international agreements and export control regimes are important building blocks in a strategy for chemical defense they are by far not sufficient in themselves.

If we examine the situation with synthetic drugs, it seems that regulating precursor chemicals does not effectively stop drug makers from running clandestine laboratories.

If control of materials is difficult, can we effectively control the human factor? Probably not – or at least not effectively. Even though the synthesis of nerve agents cannot be found via common databases like SciFinder or Crossfire, the synthetic routes are by no means a secret. "Cooking recipes" can be found in the open literature. However, the production of nerve agents is not trivial because of the toxicity of the chemicals involved and those finally produced. The "wannabe" chemical terrorist might easily end up dead even with a chemical education. However, an experienced chemist with caution and the required protective gear (fume hood, ventilation, mask, gloves) might be able to manufacture these agents without a great personal risk. An "experienced chemist" in the field of organophosphorus chemistry is not required and their numbers are simply too large to monitor this group with respect to chemical warfare agents. However, if someone with chemical expertise turns up in an investigation against a potential terrorist group, one should take a close look at him and his professional past. It might well be that the person, if actually involved in a terrorist plot, might be assigned to a job that has nothing to do with chemistry. In addition, he might be involved in the manufacturing of high explosives or other chemicals. But if he is involved in a chemical plot his specialization, his interest in certain parts of the literature, or his ordering pattern when buying chemicals might give hints. In the end, the field of chemical threats is most likely the one where controlling the human factor is most difficult and least rewarding. However, there is still a group of specialists that should be monitored. These are those chemists who have been involved in actual chemical weapons programs. The number of these specialists will most likely decline in the future as only a few countries still have ongoing offensive weapons programs due to the success of the CWC, but the defensive programs that are still allowed under the CWC and have a legitimate purpose and generate several people with in-depth knowledge about agents, their chemical and physical behavior, as well as their toxicity and other special properties.

To formulate the chemical threat part of our counter-CBRN strategy we should therefore focus on the following points:

- Improve the control of precursors and improve export control for equipment.
- Strengthen the Chemical Weapons Convention. While only a few counties are non-signatories to the conventions, universality is still the final goal and current world events might open a window of opportunity to include further countries (like Egypt, Syria and Israel) to the CWC.
- Improve backup checks of people working in laboratory environments with highly toxic compounds. Keep track of people with former occupations in chemical weapons programs.
- Continue efforts to improve detection technologies, including sampling and sample preparation techniques.
- Improve international co-operation in the field of evaluating methods for chemical analysis (Figure 16.5).

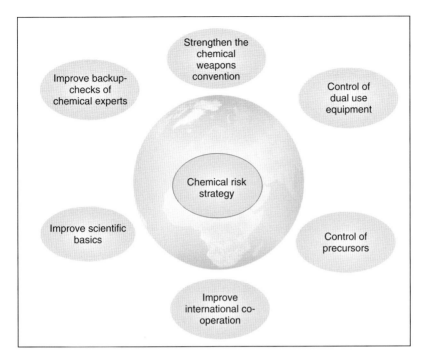

Figure 16.5 Strategy for prevention of chemical incidents.

16.3.2
Biological Threats

Biological threats are fundamentally different from chemical, nuclear, or radiological threats. As no radioactivity is involved a major possibility for detection is not available. In addition, biological weapons do not lead to destruction as in the nuclear case but specifically target humans (or animals or plants). Probably the most pronounced characteristic is that biological agents can be self-replicating weapons (Chapter 4). This can add to the spread of effects and could even lead to a pandemic (Chapter 1). Owing to these possible effects, certain biological agents can lead to uncontrolled outbreaks of disease [27]. Other biological agents like anthrax are basically noncontiguous but if used as spores they can contaminate large areas for prolonged times and are difficult to decontaminate. As biological agents encompass bacteria, viruses, and toxins, this subgroup of CBRN threats is probably the most diverse. This diversity also has consequences when we look at the control of material and the importance of human factors.

Equipment to grow microorganisms on a small scale is readily available worldwide and delivered daily to hundreds of companies, universities, and other research institutions. Essentially, this means that all laboratory equipment used in microbiology or biotechnology is effectively dual-use. This problem is well known but difficult to tackle. If a terrorist group has enough financial resources it would be only a small problem to carry out a basic bioweapon program by using a fake business façade. However, it becomes much more complicated when terrorists want to handle highly pathogenic organisms or viruses that require laboratories with a safety level of 3 or even 4 (Chapter 4). Such laboratories would have to be built and operated in a clandestine fashion as they will arouse suspicion and are normally under strict government control. As all equipment required to produce biological agents also has legitimate uses a strategy to control materials must be different from those systems used in the nuclear/radiological area. Two questions have to be asked:

- **First**: What kind of equipment and consumables are the best indicators for bioweapon production?
- **Second**: Does equipment or consumables exist that are suspicious when ordered in high amounts or with specific purchase patterns?

Even if we can give a satisfactory answer to these questions for the present situation the problem in the biological field lies in the rapid advancement of science and technology. In no other threat area does present knowledge change so rapidly and so profoundly. An example is the emergence of the new field of synthetic biology. In contrast to gene technology where genes are transferred from one organism to another, synthetic biology aims at creating new and artificial biological systems. Arguments were raised that this could be used to create new pathogenic organisms or to turn harmless microbes into toxin producers. This means that another crucial part of bio-preparedness is the monitoring of emerging new technologies and scientific breakthroughs. As technology monitoring is also relevant for other threat-areas it will be discussed in more detail below. The dual-use dilemma in the biological field also became visible when several persons were arrested who obtained pathogens for an alleged bio-attack from organism deposition banks, which without doubt are important for science and industry. Rules on how to obtain pathogens were changed after these incidents. The problem is also seen by most companies that offer synthetic genes to customers. Orders are now regularly checked for sequences associated with toxins or virulent factors, but as pointed out above the fast pace of scientific progress makes it difficult to screen for all possible relevant sequences.

If it is difficult to control dangerous biological materials and organisms, would it at least be possible for law enforcement and customs to detect them when they are shipped? Here the situation is also difficult. Important technologies for bio-detection rely on the use of specific antibodies that bind specifically to a bacterium or virus, sequencing of DNA, or the use of the polymerase chain reaction (PCR) employing specific oligo-nucleotides directed against characteristic genes of pathogens. All of these techniques are not trivial and require an experienced operator. This holds true not only for the detection itself but is even more important for sampling and sample preparation as this can lead to false-positive or false-negative results. Manufacturers of detection equipment have tried hard to make products available that could be handled by people with little training and some progress has been made, but of all threats discussed in this chapter the biological is the most difficult when it comes to detection.

While an international framework of arms control treaties exists in the nuclear field, the situation is less encouraging for biological agents. In the 1970s an international treaty entered into force, which was signed by the USA and the Soviet Union right at the beginning. The Convention on Biological and Toxin Weapons contains a strict ban on biological weapons but it completely lacks verification mechanisms. Owing to an accidental release of anthrax in the city of Sverdlovsk in 1979 it became apparent that the Soviet Union had an ongoing bioweapons project despite being a signatory of the convention. During regular review conferences of the treaty it was proposed by several countries to introduce verification mechanisms but as the treaty was negotiated by the Conference on Disarmament (CD) in Geneva, which requires unanimous decisions, is has been impossible to reach consensus on this issue up to the present date.

As control of biological materials, genes, and organisms remains problematic, we have to look at whether it is possible to exercise any control on the human component of the biological threat. If we assume that every student of biology or biochemistry could build a working bioweapon, probably with low efficiency, the situation would not be very encouraging. But the handling of pathogenic organisms requires some expertise and experience. The number of people with such experience (like working in laboratories with safety levels 3 or 4) is only a tiny fraction of those active in the life sciences. In addition, having a pathogen or toxin does not automatically mean having a bioweapon. The Aum-Shinrikyo cult in Japan, responsible for the Sarin attack on the Tokyo subway, tried to use botulinum toxin. While the toxin is one of the most toxic compounds known to man it is also very difficult to handle, as it is a quite unstable protein, which easily loses activity by exposure to light or air. Absatz crucial knowledge about biological weapons is available from the open literature. If a terrorist group has the required financial and organizational means to support a small number of scientists over a couple of years it should well be possible to produce pathogens in the required batches and quality. However, it would still be necessary to weaponize these pathogens. This includes proper dispersion of the biological material, which is not trivial as it is, for example, sensitive to the heat developed by an explosive charge. The ideal particle size is also crucial. Anthrax spores as used in the infamous anthrax letters in the USA in 2001 are an example of this. If the spore particles are too large they will rapidly settle on the ground due to gravity. If the spores have the optimal size, they will stay in the air for much longer times and they will also more easily enter the smallest alveoli in the lung when inhaled. Only a few people have expertise in this area and a good number of them were active in former bioweapon projects. Consequently, is it possible for someone with just basic training in microbiology to become a bio-terrorist, as argued by some experts in the public debate, or is more profound training, experience, and expertise necessary? Both views are probably correct and it depends on the kind of bio-attack a terrorist would like to carry out. Several organisms and viruses that can be handled in a laboratory with safety level 2 (which could be built much easier as a clandestine lab than a level 3 laboratory) could be

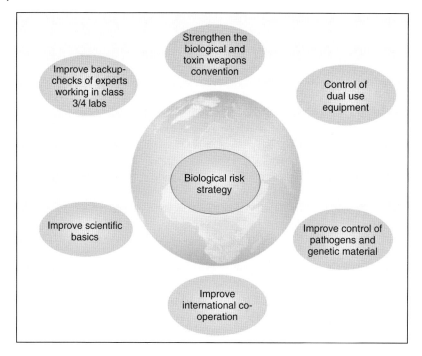

Figure 16.6 Strategy for prevention of biological incidents.

used in an attack, which would result in only a few fatalities but a large number of ill people. This could be sophisticated enough for a terrorist group although the effects and casualties would be rather restricted compared to an attack using a high-end bioweapon with highly dangerous pathogens and advanced methods of dispersion.

It should be clear by now that none of the possible defenses to stop a bio-attack from happening has a very high chance of success on its own. If an effective defense is to be established all available means have to be used in a combined effort.

To formulate the biological threat part of our counter-CBRN strategy we should therefore focus on the following points:

- Improve control of pathogens and of genetic material encoding for virulence factors and toxins.
- Strengthen the Biological and Toxin Weapons Convention. If the inclusion of a functioning verification mechanism turns out to be unrealistic it should at least be used as a formal meeting point for state parties to discuss improved measures for biosecurity and biosafety.
- Improve backup-checks of people working in laboratory environments with safety levels 3 and 4. Keep track of people with former occupations in biological weapons programs.
- Continue efforts to improve detection technologies, including sampling and sample preparation techniques (Figure 16.6).

The key message here can be summed up as follows: as rapid field-detection of pathogens, toxins, and genes encoding virulence factors or toxin products is very difficult, and at the moment it might be practically impossible, one should focus on denying access of terrorists to these materials. Scientists who could gain access to laboratories with safety levels 3 or 4 should be properly screened and people with explicit weapon knowledge should be tracked. Offering jobs to these people should be considered so as to limit the scientific recruiting ground for new terrorists.

16.3.3
Radiological Threats

Even though a nuclear explosion also releases radiation, a radiological threat is one where radiation is released without any form of nuclear yield. The most well-known and discussed issue is that of a so-called "dirty bomb." A dirty bomb uses conventional explosives to disperse radioactive material over a certain area. Alternative dispersion methods can be envisaged: dispersion of light particles with the wind, contamination of drinking water, and so on. Depending on the means of dissemination of the radioactive material and the way it will be brought into contact with humans, different kinds of radiation might pose different risks. While alpha rays have a short free travel path in air and are effectively shielded by a piece of paper they can pose a significant threat if incorporated into the human body. An alpha emitter in drinking water will be highly dangerous as alpha particles may cause huge damage while an alpha emitter not incorporated in the body poses a very small risk. Probably the most effective materials for dispersal are gamma emitters as gamma radiation is very penetrating and difficult to shield. Many radioactive materials are used outside of the nuclear weapons programs or civilian installations like power plants or research reactors. The radioactive cobalt isotope ^{60}Co, for example, is widely used for food sterilization and in medical applications as a radiation source. ^{137}Cs also has some medical applications and is still the primary radiation source around the Chernobyl reactor (together with ^{90}Sr and ^{131}I). It was also responsible for several severe radiation poisonings, including four fatalities, during the Goiânia accident. In 1987, an old nuclear medicine source was scavenged from an abandoned hospital in Goiânia, in central Brazil. It was partially broken and subsequently handled by many people, resulting in four deaths and serious radioactive contamination of 249 other people. *Time* magazine has identified the accident as one of the world's "worst nuclear disasters." Even though intensive cleanup of the contaminated sites took place after the incident became known, about 7 TBq of radioactive activity remained in the environment (compared to an activity of about 40 kBq of ^{241}Am that is found in a normal smoke detector). The original container for the radioactive material was 51 × 48 mm. This should illustrate the extreme problems associated with clean-up procedures, as radioactivity cannot be destroyed. For the special problems of nuclear and radiological decontamination the reader is pointed to Part Five on cleanup technologies.

Dirty bombs might be an attractive weapon for terrorists that are able to obtain radioactive material but in insufficient quantities to produce a critical mass for a nuclear bomb or radioactive materials that are unsuitable for the construction of a nuclear bomb. It is hard to classify a radiological weapon as a weapon of mass destruction. A real destructive effect would only be produced by conventional explosives used for dispersal and the only "unconventional" effect would be radiation. It is also highly unlikely that a dirty bomb would be a weapon of mass killing. This depends on how much radioactive activity per area would be dispersed and in what kind of way people would come in contact with this radiation. A more likely scenario would result in several people suffering from more or less severe cases of radiation sickness, although a certain number of fatalities is also possible. But these numbers would not be comparable to the consequences of a nuclear explosion. On the other hand, a dirty bomb can be a very effective terror weapon and can also produce significant economic impact. The existing fear of people with respect to radioactivity and the imperfectness of decontamination methods would leave an affected area with a certain amount of residual radiation. Even if this radiation were under the accepted thresholds the value of land and property would fall dramatically and people (if they have a choice) would choose to live in other areas.

When it comes to preparedness against radiological threats there are certain parallels to the nuclear case but also fundamental differences. To start with the differences, it would be clear that the required expertise and knowledge to build a dirty bomb is by magnitudes smaller than in the nuclear case. Some handling experience with radioactive substances is certainly a benefit but this kind of information is in the public domain and a terrorist group with interest in such weapons would probably train a few people themselves. The optimal amount of explosives for optimal dispersion (not too dilute over a large area and not too concentrated over a very small area) might not be calculated correctly the very first time, but as the experience with IEDs shows the user learns from experience and becomes more sophisticated. Therefore, control of specialist knowledge will fail as an efficient instrument in the radiological case.

What remains is the effective control of radioactive material. While highly enriched uranium (HEU) and plutonium are normally heavily guarded, this is not necessarily the case for other radioactive materials. Strong radioactive sources have many civilian uses such as in medicine or for industrial radiography and are used in less strictly controlled areas compared to dedicated nuclear facilities. "Lost sources" are a severe problem as – among others – the Goiânia accident has shown. Other incidents with so-called "orphan sources" (a self-contained radioactive source that is no longer under proper regulatory control) have occurred. Protection, control, and accounting are again essential elements of preparedness, as in the nuclear case, but are more difficult to achieve for small sources in civilian use. On the other hand, a frequently reported fear is that terrorists might be able to get hold of some highly radioactive waste, like spent fuel rods from a nuclear power plant. This is probably much more unlikely than the theft of a strong radiography source. One also has to keep in mind that the handling of spent fuel rods or similar highly radioactive material requires special installation for handling and it is highly unlikely that even a sophisticated terrorist group would be able to work with this material. As in the nuclear case radiation monitoring at critical entry points into countries is also a required measure. Apart from the classic work of law enforcement and intelligence agencies it is one of the only chances to detect and reclaim stolen material. If the radioactive material is properly shielded it might turn out to be very hard to detect, but if the original shielding container was broken or the material not properly repacked and if traces of the material can be found on the outside of the container there is a fairly good chance of interception.

Even though the consequences of a "dirty bomb" attack would be less severe than a nuclear explosion the more limited options by which to impede a radiological attack from happening should be a reason for severe concern. For this reason, the Nuclear Security Summit that took place in Washington in April 2010 also highlighted the fear of terrorist use of a dirty bomb as one of the prime objectives of nuclear security.

To formulate the radiological threat part of our counter-CBRN strategy we should therefore focus on the following points:

- Improve control and accounting of radioactive material used in the civilian domain, especially material used in sources that are easy to gain access to.
- Improve international norms dealing with radioactive materials and nuclear security.
- Improve radiation detection at critical entry points to a country (ports, border control stations, airports, rail fright terminals) (Figure 16.7).

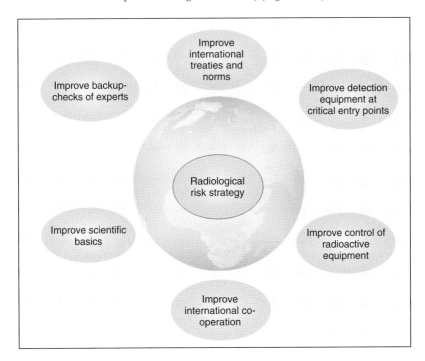

Figure 16.7 Strategy for prevention of radiological incidents.

The key message of this section can be summed up as follows: deny terrorist access to significant amounts of highly radioactive material and if this fails try to find the material with the support of radiation detection.

16.3.4
Nuclear Threats

Let us now look at nuclear threats because the possible use of a nuclear device by terrorists against a military, or even worse civilian, target is seen by many as one of the most frightening scenarios. As pointed out before, such an attack could hardly be deterred by one's own nuclear weapons. The question remains whether terrorist would directly use such a weapon or use it for nuclear blackmail. To be an effective mean for blackmail a terrorist organization would necessarily have to demonstrate that they are in the possession of a nuclear device and that there is at least a credible chance that the device will work as desired. Therefore, we consider it as highly likely that a nuclear device would be used and if an organization is in the possession of more than one weapon the others could be used for nuclear blackmail or also used directly.

Even though the scenario of a nuclear device exploded by terrorists is terrifying because of its consequences it is also undeniably the most difficult CBRN weapon to obtain. States in the possession of nuclear weapons will not share one of their own weapons and give it to a terrorist group that they cannot control completely. It is also possible to track down the origin of the fissile material because of characteristic impurities. Given the special detonation characteristics one would probably be able assign a weapon to a certain country with all the political consequences associated with this. An alternative to obtaining a nuclear weapon from a nuclear weapon state would be the theft of a functional weapon. In addition, this scenario seems to be unlikely to lead to success for the terrorists. Nuclear weapons normally are heavily guarded and even if terrorists should be able to steal a weapon they would still be unable to detonate it due to safeguard mechanisms inside the nuclear device that would make an unauthorized detonation impossible. However, such a theft would bring the terrorists into possession of weapons grade fissile material. As weapons grade uranium (highly enriched ^{235}U also known as HEU) or plutonium cannot be obtained from nature (as in case of certain biological agents and toxins) or be easily made or purified from other raw materials (as for certain chemical agents) terrorists have to rely on already existing stockpiles of fissile material for their purposes. These can in principle be obtained from existing weapons, weapons production facilities, and other nuclear facilities working with weapons grade material. Therefore, protection of this material concentrates on three aspects:

1) protection of facilities against unauthorized access;
2) control of fissile materials so that even if unauthorized access occurs the material does not leave the facility;
3) accounting of fissile material so that a potential theft could be immediately verified and detected.

Not all countries have implemented these measures to the highest possible standard and nuclear proliferation is associated with the risk of an uncontrolled spread of nuclear material and nuclear know-how. Therefore, several international treaties, agreements, and export control regimes exist (Chapter 2) and we will discuss some special aspects from recent history. The International Atomic Energy Agency (IAEA) was assigned to oversee the provisions of the nuclear nonproliferation treaty (NPT) and to be responsible for setting up reliable safeguards for civilian nuclear installations like nuclear power plants or research reactors [11]. This includes the implementation of a high-level structured approach to integrate nonproliferation objectives, international and national safeguards, and physical security by means of a safeguards-by-design (SBD) process (Figure 16.8) into the overall design and construction process for a nuclear facility. This starts from initial planning, through design and construction, and also includes operation [11].

Safeguard activities are carried out at over 900 facilities in more than 70 countries [12]. An international export control regime covering nuclear technology was established in 1974. The Nuclear Suppliers Group (NSG) is a multinational body concerned with reducing nuclear proliferation by controlling the export and re-transfer of materials that may be applicable to nuclear weapon development and by improving safeguards and protection of existing materials. The purpose of the

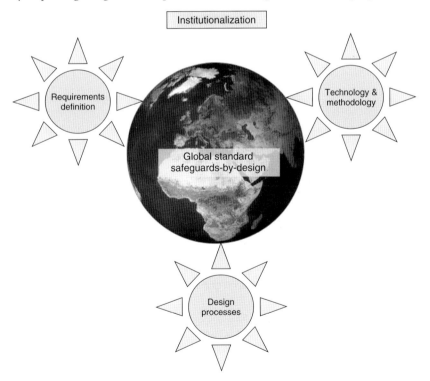

Figure 16.8 High-level framework to institutionalize safeguards-by-design (SBD). Adapted from Reference [11].

group therefore supplements the IAEA's activities. Nuclear weapons free zones (NWFZs) need to be mentioned as well as they cover large areas of the world's land surface, including almost the complete southern hemisphere.

Current NWFZs include Africa (Pelindaba Treaty), Latin America and the Caribbean (Treaty of Tlatelolco), Antarctica (Antarctic Treaty), South Pacific including Australia and New Zealand (Rarotonga Treaty), ASEAN states (Bangkok Treaty), Mongolia (self-declared NWFZ), Central Asia (Kazakhstan, Kyrgyzstan, Tajikistan, Turkmenistan, and Uzbekistan – Treaty of Semei). In addition, the sea bed and outer space are NWFZs (Seabed Treaty and Outer Space treaty) (Chapter 2).

Another fairly new initiative is the Global Initiative to Combat Nuclear Terrorism (GICNT) that started work in 2006 by agreeing on a Statement of Principles. The goals of the initiative are to bring together experience and expertise from the nonproliferation, counter proliferation, and counterterrorism disciplines, integrate collective capabilities and resources to strengthen the overall global architecture to combat nuclear terrorism, and to provide the opportunity for nations to share information and expertise in a legally non-binding environment. Finally, an important future development would be the negotiation of a Fissile Material Cut-Off (FMCT) treaty. Such an agreement might prohibit the production of fissile material for nuclear explosives and the production of such material outside of international safeguards. An FMCT might extend verification measures to fissile material production facilities (enrichment and reprocessing) that are not currently subject to international monitoring. An FMCT's ban on unsafeguarded production of fissile material would place a quantitative constraint on the amount of fissile material available for use in nuclear weapons.

Obama's 2010 Nuclear Security Summit and the International Non-proliferation Regime

After almost a decade characterized by ad hoc initiatives aimed at curbing the illegal trafficking of nuclear related material, the US administration of President Barack Obama seems to be eager to work in order to coordinate the different bilateral and multilateral programs and achieve some sort of institutionalization of the different initiatives. The Nuclear Security Summit might be the first building block of a new international nuclear non-proliferation regime but in order to reach a successful conclusion the US government will have to weigh the advantages of institutionalization against the need for flexibility and wider participation [13].

Forty-six countries were invited by the White House to attend the Summit:

- the four nuclear weapons NPT (Non-proliferation Treaty) member states,
- three nuclear states not party to the NPT (Indian, Israel, and Pakistan),
- several non-nuclear states that are NPT parties and members of the Non-Aligned Movement.

What Was the Aim of This Summit?

Nuclear terrorism is declared as "the most immediate and extreme threat to global security." However, the chances of terrorists acquiring a complete nuclear device from some illegal source are quite unlikely. A more realistic option could be to work on a dirty nuclear device using radioactive material available from different non-military sources. Although such bombs will be less destructive than a complete atomic warhead, the psychological effect on the civilian population would definitely be strong. This possible threat posed by the conjunction of loose nuclear weapons and materials around the world and radical terrorism seems to be the drive for the discussion during the summit. The aim of this summit was to think about transforming nuclear security into the fourth pillar of the nuclear nonproliferation regime.

What Are the Summit Milestones?

The primary focus throughout the two-day event remained the threat from terrorist acquisition of unsecured nuclear material. The Nuclear Security Summit:

- served as the staging point for the announcement of many on-going and new efforts to mitigate nuclear material dangers;
- included a short communique setting out broad goals released on 13 April;
- included an accompanying Work Plan with more specific, if voluntary, steps toward the goals [14].

If we discuss measures to deny access to fissile material we have to address the question of how much material would be required for a nuclear device. Modern nuclear weapons are highly sophisticated developments, including the experience of earlier devices and of numerous nuclear tests carried out in past decades. If we assume that a terrorist group with the necessary know-how was to build a nuclear bomb, much more material would be required than in today's most modern weapons. While a sophisticated implosion-type device based on plutonium could be built with as little as 8 kg (some even claim 4 kg) of plutonium, a terrorist-built bomb would require much more. A gun-type device relying on HEU (similar to "Little Boy," the bomb detonated over Hiroshima), if built as a rather crude design, would require at least 20 kg of HEU and could easily need more than 100 kg. A plutonium-based implosion-type design would require less material but is by magnitudes more complicated than the simple gun-type bomb based on HEU. It requires detailed knowledge of hydrodynamics involved in the shockwave that will implode the hollow plutonium core and expertise in the design of the right arrangement of explosives to generate the shockwave (so-called explosive lenses). Plutonium is also highly toxic and difficult to machine. Small pieces of plutonium metal also ignite spontaneously in air so the work has to be carried out using an inert atmosphere. As a result one would aim for HEU, but as pointed out above a lot of it would be required.

An advantage for law enforcement is that plutonium and HEU can in principle be detected by the radiation (neutrons and gamma rays) they emit. But things are more complicated than they look at first glance. Even though the radiation of the two elements (several isotopes of U and Pu emit different radiation upon decay or spontaneous fission) is characteristic, this radiation can be shielded using other materials like lead. One would also try to position the material as far away from a possible external detector as possible (like trying to arrange it in the very center of a container). Shielding and distance will make detection harder. This can only be countered by prolonged exposure times at a screening station or larger detectors. Both approaches are limited by practical limitations. Another option is the use of active neutron or X-ray interrogation of cargo containers or the use of new Muon based detection technologies. For details the reader is referred to Chapter 9.

After taking a look at the problems of stopping terrorists from obtaining fissile material in the required quantities and the detection if illicit shipping of these materials we have to draw our attention to the "human factor." Even with the required amounts of fissile material it is still a large step toward a working nuclear weapon. Therefore, a terrorist group would have to recruit at least one specialist as hire for money or even by ideological appeal. More than one specialist would be needed if the group aimed at an implosion-type device as several fields of expertise would be required (explosives and fuses, hydrodynamics and weapon design, plutonium machining, and so on). Obviously, such specialists would come out of the weapons programs of the official five nuclear weapons states (USA, Russia, France, UK, China), the "unofficial" nuclear weapons states (Israel, India, Pakistan, North Korea), and finally the former nuclear weapons states (non-Russian parts of the former Soviet Union, South Africa). The collapse of the Soviet Union and the inability of Russia to properly pay nuclear scientists led to a lot of unease in the West. The fear existed of a massive brain drain to states, and perhaps also to non-state actors, seeking a nuclear weapons capability. Therefore, several NATO countries implemented programs to give these specialists jobs in the West. It still remains unknown how many nuclear specialists are currently "on the loose" and where they have found new employers.

One special case, because of its scale, impact, and public prominence, where the "human factor" played an important role for nuclear proliferation should be discussed here. This case is linked to A.Q. Khan, also known as the father of the Pakistani nuclear bomb. In early February 2004, the Government of Pakistan reported that Khan had signed a confession indicating that he had provided Iran, Libya, and North Korea with designs and technology to aid in nuclear weapons programs, and said that the government had not been complicit in the proliferation activities. The Pakistani official who made the announcement said that Khan had admitted to transferring technology and information to Iran between 1989 and 1991, to North Korea and Libya between 1991 and 1997, and additional technology to North Korea up until 2000. A few days later Khan appeared on national television and confessed to running a proliferation ring. He was pardoned the next day by the Pakistani President Musharraf but held under house arrest. However, in an interview in 2008 he blamed President Musharraf and the Pakistani Army for the transfer of nuclear technology. He claimed that Musharraf was aware of all the deals and he and the army were actually finally responsible for them. Regardless of what the truth really is, and a lot of arguments speak in favor of the involvement of the state, the example of Khan shows that experts can ship knowledge and in this case also equipment to countries considered to be proliferation risks. It is also proof that nuclear smuggling is in fact taking place.

To formulate the nuclear threat part of our counter-CBRN strategy we should therefore focus on the following points:

- try to improve measures that make it more difficult to (i) obtain fissile material (even in small quantities), (ii) evade accounting of fissile material, and (iii) transport and ship fissile material without detection;
- improve the international treaty regime created around the nonproliferation of nuclear weapons (NPT);
- start negotiating a FMCT (Fissile Material Cut-Off) treaty;
- improve radiation detection at critical entry points to a country (ports, border control stations, airports, rail fright terminals);
- keep track of nuclear scientists no longer active in official weapons laboratories; make further efforts to offer jobs to those scientists who might be vulnerable to hiring efforts by proliferators (Figure 16.9).

Essentially, this can be broken down into two main messages: stop or at least slow-down nuclear proliferation and keep those fissile materials that exist under the best possible control.

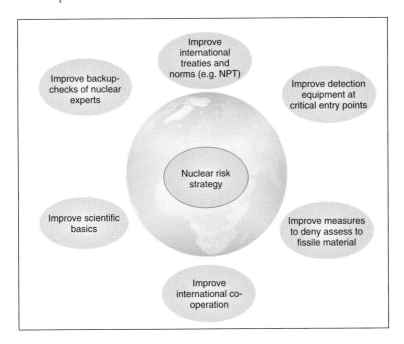

Figure 16.9 Strategy for prevention of nuclear incidents.

16.4
Proliferation Prevention

Based on the above learned special details for chemical, biological, radiological, and nuclear aspects we are now able to discuss and summarize some aspects of an overall CBRN risk management (Figure 16.2). Proliferation prevention is the basis for successful CBRN risk management. In addition, in a world with growing technological innovations, rapid economic globalization, and ongoing international political unrest one nation alone is not able to establish an effective strategy against the proliferation of WMD. Only in international partnerships, where trust is the keyword, can effective strategies be implemented and observed. New global players are arising at present and arrogate their right of codetermination.

- **Proliferation prevention**: can be defined as restricting the spread of CBRN weapons and preventing adversary acquisition of CBRN materials and technology. These restriction measures have to support political, economic, intelligence, military, and diplomatic efforts. This is the basis to discourage actors from the pursuit and acquisition of WMD and to deny proliferation to terrorists and unsecure states by enabling interdiction of dangerous materials (Figure 16.10). If diplomatic efforts have no effect, military and intelligence activities based on international treaties and laws can be used to support the implementation of treaties, agreements, sanctions, and export control procedures.

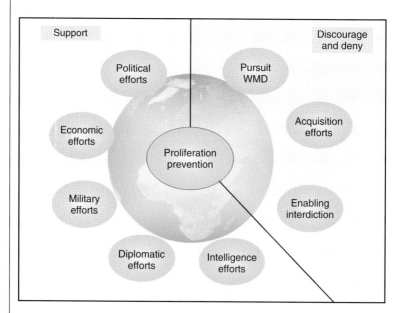

Figure 16.10 Elements of proliferation prevention.

We will discuss some points without covering every detail. The aim of this discussion is to raise awareness that proliferation prevention is more a political and diplomatic issue than one based on a scientific and technological discussion. We can state that intelligence directly supports strategy, planning, and decision-making. Without collected information by the intelligence community the operational or tactical detection of a CBRN threat is nearly impossible. In addition, early detection of precursors or any signs of an attack based on intelligence enables us to make an accurate characterization of the CBRN threat possible and an informed prediction of an impending attack. With this information effective countermeasures to neutralize the threat can be established. Interoperability strengthens CBRN-countermeasure capabilities by cooperation between different international security organizations.

Another important point is treaties and agreements. Without treaties based on international law and accepted by the majority of the international community it is nearly impossible to avoid proliferation of WMD effectively. Owing to the importance of these international treaties we discussed this issue in more detail in Chapter 2. Based on the treaties, single nations are able to build international partnerships bilaterally and multilaterally. One additional example is the Proliferation Security Initiative (PSI) (see below).

Proliferation Security Initiative

In May 2003, President George W. Bush initiated a voluntary partnership of countries working together to develop broad measures to prevent the proliferation of WMD. The partnership became known as the *Proliferation Security Initiative*. Today, the PSI is composed of core member states – Australia, Canada, France, Germany, Italy, Japan, The Netherlands, Norway, Poland, Portugal, Russia, Singapore, Spain, the United Kingdom, and the United States – and carries the signatures of more than 60 states. The PSI creates a global partnership in the prevention of WMD, and is an international mechanism to interdict shipping vessels, cargo trucks, and aircraft suspected of transporting WMD. On 4 October 2003, German and Italian security forces operating under the auspices of the PSI intercepted a marine vessel in the Mediterranean Sea on its way from Malaysia to Libya. The ship was carrying components for a "turn-key" uranium enrichment centrifuge factory that appeared to have been produced using the expertise of Pakistani nuclear specialists. The ship contained the production equipment of nuclear weapons technology that carried the distinct trademark of the Pakistani scientist A.Q. Khan, who was later revealed to have been selling nuclear secrets to rogue states around the world. Following the successful interdiction operation in October 2003, Libya's then leader, Colonel Muammar Gaddafi, agreed in December 2003 to permanently dismantle Libya's WMD programs and allow weapons inspectors into the country to start the dismantling process:

> Colonel Gaddafi made the right decision, and the world will be safer once his commitment is fulfilled. We expect other regimes to follow his example. Abandoning the pursuit of illegal weapons can lead to better relations with the United States and other free nations.
>
> President Bush, National Defense University, 2004.

In addition, strategic communications shape perceptions at the global, regional, and national levels.

16.5
Active Countermeasures

While counterforce operational capabilities are proactive in nature, active defense operational capabilities contain reactive characteristics. Active defense operations attempt to intercept CBRN weapons *en route* to their targets. Successful active defense operations can complement counterforce activities by forcing an adversary to alter attack strategies and expose CBRN assets. If counterforce operational capabilities are unsuccessful or unavailable, successful active defense operations can reduce the threat, lessen the number of attacks, thwart an attack, allow more effective passive defense and CM responses, and enhance operational capability following a CBRN attack.

Effective active defense measures take into account active defense capability sets (detect, divert, and destroy) with various planning considerations for each of the layered-defense domains (i.e., space, air, and surface). This layered defense approach incorporates networked space, air, and surface systems, and employs both kinetic and non-kinetic means of defeat. Measures include, but are not limited to, missile defense (ballistic and cruise), offensive counter-air/defensive counter-air (OCA/DCA), special operations, and force protection operations to defend against conventionally and unconventionally delivered WMD.

Active defense capability sets include actions to detect, divert, or destroy an adversary's WMD while *en route* while minimizing collateral damage. Detection includes space, air, and surface sensors to locate, characterize, track, and monitor an *en route* CBRN threat. Diversion capabilities cause the CBRN threat to redirect, modify, or miss the intended target set. Capabilities to destroy prevent the function of or neutralize/eliminate the CBRN threat.

However, the main question is not answered yet. Why should we be prepared? Preparedness can reduce fear, anxiety, and chaos that accompany WMD disasters. Furthermore, authorities can reduce the impact of attacks and sometimes mitigate the threat completely if they are able to gather information to stop a CBRN attack before it occurs. To be able to respond to CBRN events in the most effective and appropriate way it is inevitable to make provisions with respect to doctrine and deployment planning:

- personnel (first responders, medical, and so on),
- training,
- equipment (including communications),
- consumable supplies (like decontamination material, antidotes, and so on).

While the above-mentioned points focus on the preparedness for a situation where a CBRN event actually has happened, the whole complex of CBRN preparedness also includes measures to either prevent the incident from happening or to lessen the immediate effects of an incident. This kind of preparedness encompasses a rather wide range of possible options. The following list is therefore incomplete and should rather give an idea of the range of options available:
- political initiatives (CBRN nonproliferation, strengthening treaty regimes, and so on);
- legal regulations (including international treaties);
- intelligence gathering;
- police activities;
- customs activities (against trafficking of illicit CBRN material);
- hardening and surveillance of critical infrastructure;
- adding elements of redundancy to critical infrastructure;
- adding safety elements where accidental release is possible (chemical plants, nuclear power plants, and so on).

16.6
If Things Get Real: Responding to a CBRN Event

A release of CBRN-related substances will cause a primary response of local authorities. Local preparedness is the key to dealing with the early stages of CBRN incidents. The next level is national response, and if their resources are exhausted an international effort is needed. In the case of chemical attacks, member states of the organisation for the prohibition of chemical weapons (OPCW) have the right to call for international aid.

The goals of a coordinated response of a community to counteract the release of or attack with CBRN-related substances are to:
- save lives (search and rescue),
- preserve or restore environment and infrastructure,
- keep public order and enforce legal investigations,
- mitigate effects,
- risk assessment,
- reduce economical impact,
- implement long-term surveillance of affected people, animals, and environment.

16.6.1
Fundamentals of Installation of a Response

Each installation of a CBRN response is unique and will depend on the special circumstances of the incident and on the preparedness of the affected population

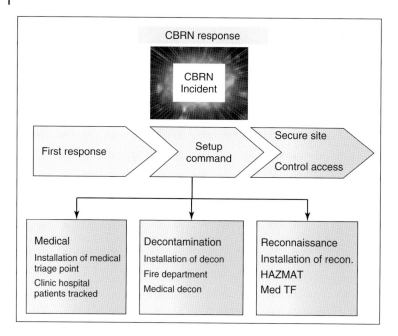

Figure 16.11 Installation of a response after a CBRN attack or release. HAZMAT, hazardous material; Med TF, medical task force.

as well as available resources. Nevertheless, Figure 16.11 depicts some general principles of the flow of response:

1) **Triggering response**: CBRN response actions may be initiated by several events. Overt terrorist attacks or military use of CBRN weapons will need confirmation of whether it was a CBRN attack and will immediately prompt a response. More difficult are detector warnings that produce possible false positive or false negative readings. Medical surveillance may be the first indicator of a hidden release of CBRN material. Some nations have established a highly sophisticated medical surveillance network. Intelligence from several sources (military, national institutions) may provide information of a significant CBRN attack in advance and trigger a response.

2) **Setup of a command structure**: It is necessary to establish a command structure as soon as possible. This will be first on a local level and will seek further assistance until sufficient resources are available. Heavy attacks with or massive release of CBRN material will require an international response and implementation of a civilian headquarter [e.g., by United Nations Disaster Assessment and Coordination (UNDAC)] and/or military command structure (e.g., UN, NATO).

3) **Reconnaissance**: It is crucial to identify the released agent as soon as possible and to predict the affected area. To achieve this goal, simple handheld detectors and highly specialized analytical task forces should sample and report as early

as possible to the command structure to generate a map with hot zones. The reconnaissance teams should wear appropriate personal protective equipment (PPE). The possible presence of explosives may warrant additional specialists [e.g., explosive ordnance disposal (EOD) teams].

4) **Secure site**: To prevent a spread of CBRN material or to protect the affected population it may be necessary either to restrict movement or to evacuate a certain area. In any case it is necessary to control access by police forces and other law enforcement agencies. A secure perimeter should be defined.
5) **Decontamination**. These facilities can be established as soon as the agent and the affected area are identified. Decontamination facilities are necessary to establish a medical triage point.

16.6.2
Detection, Reconnaissance, and Surveillance

Obviously, a nuclear explosion gives a direct indication that a CBRN attack has occurred. Other types of attack might only be recognized after typical injuries or symptoms are visible among the affected population. Especially, biological and radiological attacks are prime candidates for a delayed identification. In the case of a chemical attack it depends highly on the type of chemical agent used as to whether the event is recognized rapidly. While this might be the case for the nerve agents because of their rapid action or with gases like chlorine because of its characteristic odor, the effects of other agents like sulfur mustard are delayed.

In any case first responders need as much information about the attack as possible. This includes the type of CBRN material, the scale of the attack, the risk for downwind drift, and further contamination of downwind areas (depending on the actual weather situation), and the number of potentially affected people. Effective detection capabilities are a prerequisite to accomplish this non-trivial task. These capabilities must include a wide array of techniques, from simple detection paper via handheld analytical devices up to high-end laboratory equipment. Even preliminary detection results can be of tremendous help to first responders. If a nerve agent was spread it is sufficient for first responders to know this and in addition it might be an asset to know if it is a volatile or persistent agent. This information enables the specialists in warning and reporting to make a first prediction for the spread of contamination or the risk for areas in a downwind direction. In a later stage the exact identity of the agent can be determined as this kind of information is important for medical treatment beyond the first use of antidotes. However, the wrong identification of the CBRN material can also have fatal consequences. The nerve agent tabun, for example, can contain substantial amounts of cyanide, which is a degradation product. If a first detection only "sniffs" cyanide but fails to detect the nerve agent the first responders lives might be at risk. Therefore, detection should be carried out very carefully. Proper collection of samples for laboratory analysis is also important.

In a typical CBRN event the available information for the local commanders and specialists will be incomplete and fragmented. After first detection results flow in, the situation becomes clearer. However, decision-making must always rely on available data and information. If certain types of information are unavailable at a certain point this is a fact one has to live with. One might then issue orders to obtain further information or not (because of lack of time and resources at that moment). The important message here is that incomplete data will be the rule rather than the exception. Continuing efforts to gain further data are an essential part of command and control.

But even if sufficient data about the event is available, ongoing surveillance of the affected and nearby areas should be carried out. One experience with IED attacks in Iraq and Afghanistan is that quite often additional IEDs were used to either hit arriving first responders and medics or EOD personnel. The same is possible in a CBRN scenario. The "secondary" attack might be carried out either using CBRN material but also using conventional explosives. This tactic would lead to much slower progress made by first responders and would also require EOD personnel on the scene. Apart from this risk surveillance is also important to monitor movement of people. Many potentially contaminated people would not wait for first responders to arrive on the scene but would seek medical assistance by themselves. This was clearly demonstrated during the Tokyo sarin attacks. This might lead to contamination spread and cause problems by overloading medical capacities. In short, while detection and reconnaissance give answers to what has happened and the state of the current situation, surveillance helps in monitoring how a situation is developing over time.

16.6.3
Risk Assessment

CBRN risks are characterized as potentially harming affected people or environment. Once a CBRN risk has been identified, it is necessary to assess its present impact and future consequences. All further actions should be guided by the risks posed by the CBRN incident. Thus, proper risk assessment is a crucial task to plan further actions. This should replace the "intuitive" risk assessment: fear. In developed countries a lead federal agency has implemented guidelines for risk assessment. For example, in the United States, the Environmental Protection Agency (EPA) provides a good collection of relevant documents on its homepage. A review of risk assessment guidance may be found at the EPA's National Center for Environmental Assessment website (*http://epa.gov/risk*, accessed 9 June 2012). The evaluation of health and environmental risk is a highly developed technical discipline. Nevertheless, decision makers in CBRN scenarios need a rapid risk assessment to resolve the problems in a timely manner.

A human health and environmental risk assessment is a continuous process to integrate radiological, microbiological and toxicological, environmental, and statistical information. The EPA approach defines four steps to achieve this goal (Figure 16.12):

- **Step 1 – hazard identification** Prompt identification of the CBRN hazard is crucial for the initiation of further appropriate preventive and medical countermeasures.

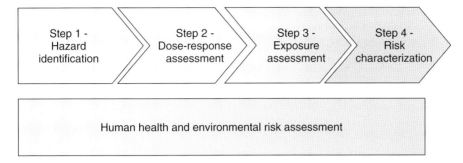

Figure 16.12 Human health and environmental risk assessment.

Data from detection, identification, and surveillance teams are summarized and evaluated to decide under which conditions stressors have the potential to cause harm to humans and, if so, under what circumstances. It is important to identify a potential terroristic background of the CBRN release and to include information from the intelligence community. An epidemiologic surveillance system may detect a delayed onset of symptoms and a covert release of toxic or radiological substances. Step 1 should include the extensive use of internet databases to gather information. Specialized information is collected at national poison control centers, industrial facilities, centers of disease control, specialized research facilities, and universities. Most nations have highly trained experts in the police and armed forces, which should be contacted immediately. The extent of technical, medical, and analytical support available in this phase relies on the pre-incident preparation.

Sometimes a rapid laboratory confirmation cannot be achieved in a timely manner. In this case a decision based on the observed clinical infections or toxidromes is necessary to start risk-reduction strategies or to rule out hoaxes.

- **Step 2 – dose-response assessment** This involves analysis of the quantitative relationship between the amount of exposure to a radiation, infectious agent, toxin or toxic substance, and the produced injury, disease, or intoxication. In the case of inhalation exposure it is necessary to refer to the evaluated acute lethality guideline levels (AEGL) if possible. AEGLs are highly useful in emergency planning, prevention, and response. They define threshold exposure limits for mild effects, serious adverse effects, and lethality (Chapter 6). Hereby, toxicity data from human studies are the most applicable, and are preferred over data from animal studies.
- **Step 3 – exposure assessment** This entails estimation of the magnitude, frequency, and duration of an exposure on the affected population. If the incident involves a release of either an aerosol or a chemical cloud, computer modeling may help to predict the affected area on the basis of gathered weather data. In the case of a biological outbreak with significant person-to-person transmission, standard epidemiological methods apply.

- **Step 4 – risk characterization** Integration of all information and data from steps 1–3 to determine the likelihood and severity of CBRN-related injuries, diseases, or intoxications.

16.6.4
CBRN Warning and Reporting

A critical component of CBRN response is the timely information of responders and affected populations of CBRN hazards that might affect them. This means keeping track of the spatial locations of CBRN events, the type and scale of the event as well as predicting the downwind drifts of chemical vapors, biological aerosols, fallout, and similar volatile risks. The CBRN warning and reporting service of the NATO armies is highly standardized and is described in NATO publication ATP-45(C) [15]. As it also serves as a guideline for civilian CBRN warning and reporting in many countries the NATO system will be explained to give an idea of the setup of such a system.

The system uses a set of standardized message types (nuclear, biological, chemical) NBC 1–6 exchanged between the source level (NBC observation posts, survey and reconnaissance teams, sites, formations, units, and sub units) and a hierarchical system of NBC centers (Figure 16.13). To rapidly calculate affected

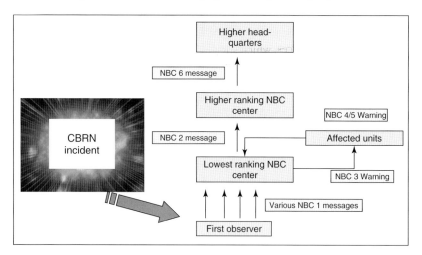

Figure 16.13 Way of passing NBC 1–6 messages. NBC 1: observer's report, giving basic data. NBC 2: report for passing the evaluated data from collected NBC 1 reports. NBC 3: report for immediate warning of predicted contamination and hazard areas. NBC 4: report for reporting detection data and passing monitoring and survey results. This report is used for two cases. Case one: used if an attack is not observed, and the first indication of contamination is by detection. Case two: used to report measured contamination as a part of a survey or monitoring team. NBC 5: report for passing information on areas of actual contamination. This report can include areas of possible contamination, but only if actual contamination coordinates are included in the report. NBC 6: report for passing detailed information on NBC events (intelligence report).

areas a set of special NBC weather forecasts are used [Chemical Downwind Message (CDM), Effective Downwind Message (EDM), and Basic Wind Message (BWM)].

Observations at the source level (like an NBC observations post) are collected and preprocessed at a low level of command (typically company level). This information is then transferred to the lowest ranking NBC center in the form of a NBC 1 message (initial report with data from the unit level). The NBC center will probably collect several NBC 1 messages from different units. It consolidates this information and creates a NBC 2 message (report for passing evaluated data), which is then passed on to higher-ranking centers and subordinate adjacent units. The center also creates a warning message for affected units based on the information of the NBC 2 message and NBC weather information. This message (NBC 3) is then passed down to possibly affected units. Units will use the NBC 3 message to determine if they are affected by a hazard and at what time a drift cloud could reach the unit at the earliest time. Instead of complicated computer models a rather simple system with tabulated values is used that contains "worst case" values. Affected units will then order appropriate countermeasures (like moving the unit and/or issue the order to wear protective masks and clothing). If available, monitoring equipment and detectors will be used. NBC 4 messages are used to report actual contamination and NBC 5 messages contain coordinates that describe a contaminated area and are usually prepared from data received in NBC 4 messages. Finally, a NBC 6 message summarizes information concerning a CBRN event and is prepared at a lower level, but only if requested by higher headquarters. It is used as an intelligence tool. The NBC 6 Report is submitted to higher headquarters (Figure 16.13).

In summary, the NBC centers fulfill a dual role. They collect the data coming in from the source level and relay back warning messages. They also communicate upward to higher-ranking centers, which will then relay it to adjacent units but also up to the highest command structures. They also track the types and numbers of attacks or incidents. In general, CBRN warning and reporting assures a standardized and easy to handle flow of relevant data with respect to CBRN hazards. It achieves this task by evaluating and condensing data from the source level and generating relevant warnings. While the calculation and preparation of NBC 3 messages according to ATP-45(C) can be easily carried out by hand with just a few simple tools, the availability of specific analysis software like the widely used Hazard Prediction and Assessment Capability (HPAC) in combination with local reconnaissance data can lead to refined and more precise warnings.

16.6.5
Command and Control, Communication

Command and Control is a crucial element of CBRN response. While personnel on the scene will conduct detection, sampling, decontamination, and medical response these actions and efforts must be coordinated. There must also be a central authority that gathers information, processes data if necessary, and redistributes it according to the needs of the individual actors. The process of CBRN warning and reporting described in the previous section is an example for such a task. But as warning and reporting is a very specific task there must also be a higher center of command that is able to effectively control operations. This can be a very demanding and difficult task. First of all we have to remind ourselves that large-scale CBRN events are rare and it is difficult to gain experience from practice. Response to a CBRN event will also require the cooperation of different actors and agencies. This is probably less of a problem for those responders that are used to working together like fire departments, police, and medical emergency services. But things become more difficult if new actors join in like the military. However, there are countries in which the military is also an integrated actor in disaster relief so it depends greatly on the specific location where a CBRN event happens.

Communication is the major tool of a command and control cell. While a nuclear explosion will cause disruption of the common means of communications (phone lines, cell phones, computers) due to the electromagnetic pulse (EMP) of the explosion other attacks will affect communication networks either only to a small extend or not at all. However, communicating between different actors can be a problem if different types of equipment are used. Using radio transmitters with different frequency bands or combinations of analog and digital transmission can be a significant problem.

In the case of an event two things must happen as soon as possible: (i) establish a line of command and (ii) establish stable lines of communication. This can only happen effectively if proper preparations were made. If it is unclear who the "chief in command" is, or how to contact important sources of data and information, response will be ineffective or even end up in chaos. Therefore, proper planning and training exercises are required. Equally important is to learn lessons from such training.

16.6.6
Technical Response

The prime objective of an effective response is to save lives, avoid casualties, and mitigate the CBRN effect. To be able to rescue people from the affected area and to give them appropriate medical treatment the responder must be able to enter an affected area and operate in it without risks to his own life. While personal protective equipment can be a first solution to this problem, decontamination of personnel, equipment, and infrastructure is equally important if the CBRN material is persistent and there is a risk of contamination spread.

The technical means for effective decontamination are highly dependent on the type of CBRN material dispersed. Notably, radioactive material will stay radioactive. Decontamination in this case means the removal of this material from a surface (which is indeed the definition of decontamination). In the case of chemical or biological material it is generally not only a physical removal that should be achieved but also a detoxification or disinfection. Therefore, CB decontamination tries to chemically alter toxic chemical or toxins into non-toxic reaction products or to destroy the biological agent (virus, bacteria including spores, and other type of organisms) so that it is no longer infectious.

The equipment (machines but also decontamination fluids) will also depend on what will be decontaminated. For personnel there will be a differentiation between injured/ill persons and those not showing any kinds of symptoms. While the first will be treated by the medical first responders (see next section) "normal" decontamination of people will be carried out by normal CBRN response units. This includes the decontamination (and/or removal) of clothing followed by some kind of shower with or without scrubbing. Decontamination of equipment is the next important area. This includes the equipment of the responders, including their personal protective equipment like suits or masks, as it must be assumed that this equipment will be contaminated after operations in an affected area. Equipment also means machines and vehicles used in a contaminated area. Finally, there is decontamination of infrastructure, which is also often associated with post-disaster recovery. An example of this kind of decontamination is the treatment of affected buildings after the anthrax letter attack in the USA in 2001. Two mail processing facilities of the US Postal Service that were affected had to be decontaminated in a lengthy and costly process. This also brings us to the question of what kind of residual contamination can be accepted after decon. This depends on the circumstances and on what is technically achievable. A useful differentiation that is used by the US military, gives three levels of decontamination:

- **Immediate decontamination**: Minimize casualties, save lives, and limit the spread of contamination. Immediate decon is carried out by individuals upon becoming contaminated. It is conducted using supplies and equipment they carry.
- **Operational decontamination**: Conducted by teams or squads using decontamination equipment, organic to battalion-size units. If this equipment is not available, units will request vehicle wash-down through command channels. This mission will normally be tasked to the supporting chemical unit. These procedures limit spread of contamination and allow temporary relief from mission oriented protective posture (MOPP) level 4 (all protection including overgarment, mask, boots, and gloves worn).
- **Thorough decontamination**: Thorough decon is the most resource-intensive type of decontamination. These operations reduce contamination to negligible risk levels. The use of these procedures restores operational power by removing nearly all contamination from units and individual equipment. Troops can operate equipment safely for extended periods at reduced MOPP levels. A small risk from residual contamination remains, so periodic contamination checks must be made after this operation.

As these categories are valid in a war-like scenario it makes sense to add another category: *Thorough decontamination in peacetime*. To properly restore a CBRN affected area to a pre-event status the contamination has to be reduced to a level where extensive exposure and contact are possible without the need for protective equipment. This is not always possible, especially if the area was heavily contaminated with radioactive material. An example is the exclusion zone around the nuclear reactor of Chernobyl (core meltdown in 1986), where no people are allowed to live permanently. While radiation levels regarded as "safe" are well established, things are more difficult in the case of chemical or biological contaminations. "How clean is clean?" is an important question to answer and closely linked to the even more crucial question "How clean is safe?" Especially, biological agents that are highly infectious and can therefore spread between humans require proper decontamination and sterilization efforts. In general, the required thresholds allowed by legal regulations should not be reached, but it can also be a political question as to whether further clean-up is required. This is the case if a civilian population wants to return to an affected area for prolonged times (residence). The inflicted costs of this clean-up might be one of the major effects of radiological attacks given the widespread fear of people with respect to radioactivity.

16.6.7
Medical Response

First of all it is essential that technical and medical responders have access to PPE. This should include appropriate respiratory protection as well as autoinjectors with antidotes (e.g., atropine, obidoxime, or pralidoxime). Collective protection of relevant infrastructure (hospitals) may be available in highly industrialized countries. Vaccination and protective antibiotic/antidote treatment of those who are involved in the response is highly desirable but depends on the preparedness of the response force [16]. It is advisable to have special packages available with medical supplies for special medical tasks (e.g., intubation) or for treatment of intoxications with antidotes (e.g., against organophosphorous poisoning) [17]. It is crucial to have a stockpile of decontamination equipment that is applicable to severely wounded patients [e.g., reactive skin decontamination lotion (RSDL)] [18].

First responders, paramedics, and other high-risk personnel may need some sort of prophylaxis (e.g., pyridostigmine) or post-exposure prophylaxis to minimize exposure effects. Normally, such drugs are only available for soldiers in wartime.

It is crucial to establish a *hot zone* to ensure a proper contamination control. This will establish a controlled entry and exit with decontamination of all patients and personnel. However, it is very likely that contaminated patients will leave the hot zone and try to gain access to medical treatment in hospitals. Therefore, all hospitals have to be secured immediately to control entry into their facility. For proper protection of medical facilities a decontamination line should also be established at the entry. It is very likely that more people will try to get medical help than were actually exposed (so-called "worried well"). This will cause the need for proper risk communication and the presence of psychological intervention teams. In addition, it is important to sort injured people into groups with different signs of injuries and contamination effects (triage).

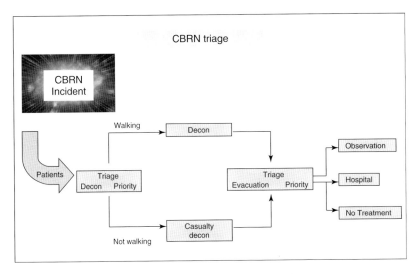

Figure 16.14 Triage is a continuous process at every stage of the CBRN patient management. After extraction from the hot zone a first triage determines for each patient their decon priority. After cleaning and stabilization of the patient a second triage is necessary to determine the evacuation to further specialized medical facilities.

- **Triage**: A process for sorting injured people into groups based on their need for or likely benefit from immediate medical treatment. Triage is used in hospital emergency rooms, on battlefields, and at disaster sites when limited medical resources must be allocated.

Triage (Figure 16.14) is a dynamic process and necessary at each stage of the rescue chain [19]. The special kind of CBRN exposed casualties may change classical triage criteria because every patient has to pass the decontamination line. This bottleneck with limited possibilities for advanced medical support will make triage decisions difficult. The most experienced medical doctors are needed to perform triage to direct the limited resources to those who are most likely to benefit from them.

According to the US military the triage categories are:

- **Immediate**: The casualty requires immediate medical attention and will not survive if not seen soon. Any compromise to the casualty's respiration, hemorrhage control, or shock control could be fatal.
- **Delayed**: The casualty requires medical attention within 6 h. Injuries are potentially life-threatening, but can wait until the immediate casualties are stabilized and evacuated.
- **Minimal**: "Walking wounded;" the casualty requires medical attention when all higher priority patients have been evacuated, and may not require stabilization or monitoring.

- **Expectant**: The casualty is expected not to reach higher medical support alive without compromising the treatment of higher priority patients. Care should not be abandoned – spare any remaining time and resources after immediate and delayed patients have been treated.

After entering the hot zone, some kind of basic life support (BLS) is needed to rapidly evacuate the casualties. However, intubation or infusion poses the risk of introducing CBRN related substances into the body. Thus, any *advanced* life support potentially harms the patient and should be restricted to otherwise dying patients. Beside this, if the agent is a known or a typical toxidrome is present, specific antidotes are applicable intramuscular (i.m.) to ensure survival of the patient.

Specific diagnostic aids should be available for detecting and confirming exposure to CBRN agents. It is important that physicians are aware of typical symptoms (e.g., toxidromes) to initiate first tests and treatment. Further (radiological) measurement or testing for CB-agents should be established after the decon line. Handheld tests for relevant biological agents should be brought to the decon line. Transportation of probes to the next specialized laboratory to confirm a biological attack has to be organized at the earliest stage. Some nations have the capability of mobile microbiological laboratories, which can be deployed at short notice [20].

For chemical attacks the measurement of acetylcholinesterase activity (after nerve agent exposure) by newer advanced techniques, such as the detection of specific DNA adducts (after mustard gas exposure), should be established after the decon line [17]. It is necessary to establish a close connection to the nearest poison control center as well as all hospitals.

It may be difficult to evaluate or confirm CBRN hoaxes in a timely manner. If available a specialized "fact finding team" deployable at short notice may be helpful. If any CBRN-related agent is identified, all necessary antidotes, vaccines, and antibiotics should be transported to the medical treatment facilities. Not every patient will go through a decon line at the incident. The Tokyo incident showed that *most* patients found their way directly to surrounding hospitals. As all of those patients are a potential risk to contaminate the facility, it is necessary to decontaminate them before entering the hospital. Thus, hospitals should be equipped with decontamination facilities.

16.6.8
Risk Communication

We have learned a lot about the special considerations that we should have in mind if we want to counter a CBRN incident effectively. However, in the whole process of risk management we should not forget the concerned public and, therefore, risk communication plays a vital role in bringing all relevant information to the affected population. The media's role is crucial due to its tremendous power to influence the view of citizens [21]. Without an effective risk communication strategy the risk is high for ineffective, fear-driven, and potentially damaging public responses (Figure 16.15) to a serious crisis such as unusual disease outbreaks and bioterrorism [22].

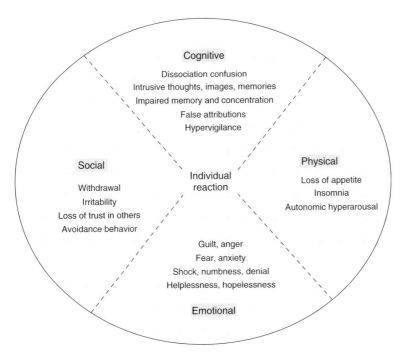

Figure 16.15 Normal individual reactions include (i) cognitive, (ii) physical, (iii) emotional, and (iv) social reactions.

Here we have to understand one key question, which could be crucial: "If you are not certain, how can we know we are being protected?" [23] This is not a question about data it is about personal and family safety. That is the issue to be addressed in risk communication [23]. Owing to the importance of risk communication we will cover this point of risk management in a little more detail.

- **Risk communication**: An interactive process of exchange of information and opinion among individuals, groups, and institutions. It often involves multiple messages about the nature of risk or expressing concerns, opinions, or reactions to risk messages, or to legal and institutional arrangements for management. Risk communication is the process of providing the public with information that serves to reduce anxiety and fear as well as provide suggestions for planning that will assist the public in responding appropriately to some crisis (or impending crisis) situation. Typically, the crisis situation has the potential to impact large groups of people.

The provision of clear advice, reliable information, and communication of potential risks is compulsory. One other important point is that appropriate risk communication procedures foster the trust and confidence that are vital in a crisis situation [24, 25]. To achieve an effective risk communication public officials have to know the steps needed to respond to the challenges of managing such crises as CBRN incidents and to understand some general rules for risk communication:

- People perceive risks differently, and people do not believe that all risk is of the same type, size, or importance. Therefore, be honest and open. Never mislead the public by lying or failing to provide information that is important in their understanding of issues.
- The cardinal rule of risk communication is: first do no harm. A threatening or actual crisis often poses a volatile equation of public action and reaction.
- Know the crucial information to convey in initial messages to prompt appropriate public responses after a crisis situation.
- Separate unnecessary messages and fundamental messages to be delivered prior to, during, and after an incident.
- Know the obstacles that hamper effective communications and how they can be minimized.
- Anticipate the possible questions from the public.
- Understand the news media's responsibilities and support news media's agents to meet them [23].

Therefore, we have to accept that, for example, the information on how people should avoid contact or minimize infection is by far not enough. The population has to be informed that medical countermeasures are available and where to obtain them. Any releasable information should be provided in simple and clear language. This information must be consistent and not contradictory. It is important to report back every information and detection result to decision makers and centers of medical expertise (e.g., poison control centers). This is important to avoid unproven reports in the media.

We have to understand that a CBRN incident could lead to a highly unstable public situation. It is essential to note that every crisis, especially a CBRN incident in combination with heightened public emotions and limited access to facts, leads to rumor, gossip, speculation, assumption, and inference, and in the end to an unstable information environment. To avoid this situation, it is vital for spokepersons in a crisis to understand what the key elements are for an effective CBRN risk communication (Figure 16.16)?

For each of these key elements it is necessary to jot down all pieces of information and to develop the key messages. Review your ideas and background material for information that provides support to your key messages. It is important to train for the whole communication process.

16.6.9
Medical Support and Post-disaster Recovery

It is necessary to build up sufficient stockpiles of drugs and vaccines as well as gloves, masks, and so on to ensure ongoing medical treatment. If necessary it is essential to establish mechanisms for rapid licensing procedures of drugs and

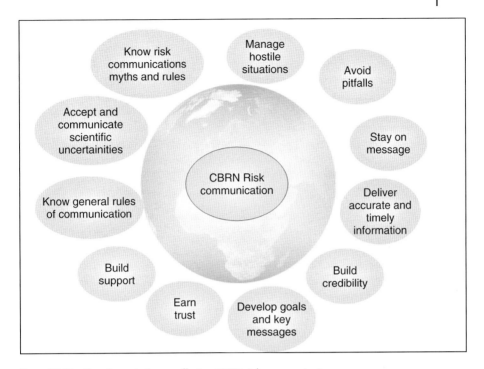

Figure 16.16 Key elements for an effective CBRN risk communication.

vaccines in crisis situations. The stress and fear during a CBRN attack may cause post-traumatic stress disorder (PTSD) syndromes to the professional responders and the affected population. Thus, mental health professionals need to be deployed to the incident. The environment has to be decontaminated and restored as soon as possible. In some cases it will not be possible to achieve this goal. The area has then to be secured from public access. An example of such a restricted area was Gruinard Island, which was the site of a biological warfare test (anthrax) by British scientists to test British vulnerability against an attack with anthrax spores.

16.7 Research

Proper security research is necessary to meet the challenge of a CBRN attack. Unfortunately, funding of such research is highly dependent on risk perception. On the other hand it, takes a long time to educate and train CBRN scientists. For medical doctors, to specialize in these areas of expertise is difficult, as these professionals are a limited human resource and are well paid elsewhere. Thus, it is recommended to integrate skills to manage CBRN patients into current curricula of medical education. Nevertheless, a pool of highly specialized experts is need.

It is further desirable to fund research activities that are related or similar to CBRN caused health problems. A lot of research areas have to be covered:

- development of antidotes and vaccines;
- development of an integrated concept to run a decontamination facility for critically ill CBRN patients;
- development of decontamination solutions that are applicable to wounds and mucous membranes;
- evaluation of analytical capabilities to detect highly toxic or infectious agents;
- development of software that integrates medical data, characteristics of CBRN agents, and models of damage propagation.

16.8
Aftermath Action – Lessons Learned

The final analysis of lessons learned from past CBRN disasters crosscut each of the disaster management components: the need for effective organization and communication and the need to follow a systematic approach:

- risk analysis,
- risk assessment for humans and environment,
- damage and risk control,
- assessment of the remaining risk,
- monitoring and adaption of risk management.

- **Risk analysis**: It is necessary to identify hazards and to adopt legal regulations to reduce production or to prohibit recognized sources if necessary. New hazardous materials must be gathered early with risk potential, protection, diagnosis, and therapy. The collection is made by the appropriate national law, hazardous substance regulations, and transport regulations (including hazardous impurities and secondary products in case of fire or decomposition).
- **Risk assessment**: The risk assessment correlates with the possible vulnerability of a population. This allows security necessary resources in comparison with existing capacity to identify the existing weaknesses. Only on this basis, is a rational resource management of teams possible.
- **Damage and risk control**: The best risk management is to avoid the risk. If possible, hazardous materials should be replaced by less toxic substances or the required quantities should be restricted.
- **Assessment of the reamaining risk**: An "acceptable risk" should be defined for all hazardous materials and necessary safety limits defined.
- **Monitoring and adaption of risk management**: It is important to perform a continuous evaluation and adjustment of risk management. This should ensure a strategy to counter a hazardous challenge with state of the art methods.

In addition, we have to accept that after every disaster experiences result in new knowledge, and that previous lessons learned may not have been fully integrated into emergency management plans. We have to take into account new and old lessons to improve the risk management of future CBRN incidents. Some obvious lessons learned are more national and international interaction between various leaders and encouraging the shedding of distrust and territoriality. Furthermore, joint training exercises and conferences are helpful in this regard. Other lessons, such as managing convergence and emergence, may require new ways of thinking and new approaches to CBRN disaster management. In addition, we can highlight several understudied disaster topics that apply to the threat of terrorism:

- vulnerability
- values
- victimization and visibility
- symbolism
- landscapes of hazard
- metropolitan contexts
- gaps in existing terrorism research [26].

To understand these topics and the reliance on new technologies is of high importance to improve CBRN risk management. Therefore, research into all scales of terrorism is of vital interest. Without the knowledge gained from this research and its implementation, preparedness, detection, response, and recovery are much more difficult processes when dealing not only with CBRN incidents but overall with all kinds of natural, technological, and terrorist-induced disasters.

16.9
Conclusions and Outlook

In this chapter we have learned the basics of CBRN risk management. Now, we understand more about the specific elements for an effective CBRN counter strategy. However, natural, technological, and terrorist-induced disasters are analogous in many important ways before, during, and after impact. Thus, lessons can be drawn and applied among all three. We can state that a common approach to disaster management can be conceived for natural, technological, and terrorist-related disasters involving preparedness, detection, response, and recovery. One of the most important points in all types of disasters is risk communication. So that all groups can work together efficiently and successfully during emergency situations, communication is needed among:

- policymakers,
- first responders,
- public health workers and public service officials,
- practitioners,
- researchers and the public.

Furthermore, an incident command structure is needed to effectively manage disaster situations. Extensive planning and preventive measures are needed for all disasters, but equally important is the ability to improvise solutions for unforeseen problems that inevitably develop.

References

1. AS/NZS (2004) AS/NZS 4360 *Australian and New Zealand Standard on Risk Management*, 3rd edn, Standards Association of Australia and New Zealand, Sydney. ISBN: 0 7337 5904 1.
2. Parnell, G.S., Smith, C.M., and Moxley, F.I. (2010) *Intell. Adversary Risk Anal.: Risk Anal.*, **30**, 32–48.
3. Haimes, Y.Y. (2009) *Risk Modeling, Assessment, and Management*, John Wiley & Sons, Inc., Hoboken. ISBN-10:0-470-28237-1.
4. Ayyub, B.M., McGill, W.L., and Kaminskiy, M. (2007) *Risk Anal.*, **27** (4), 789–801.
5. United State Air Force (2007) Counter-Chemical, Biological, Radiological, and Nuclear Operations, Air Force Doctrine Document 2-1.8. Available at http://www.fas.org/irp/doddir/usaf/afdd2-1-8.pdf (accessed 6 March 2012).
6. Levi, M. (2007) *On Nuclear Terrorism*, Harvard University Press. ISBN-10:0-674-02649-7.
7. Hoffman, B. (2009) *Terror. Pol. Violence*, **21**, 349–377.
8. Alexander, D.A. and Klein, S. (2006) *J. Postgrad. Med.*, **52**, 126–131.
9. Wein, L.M., Liu, Y., and Leighton, T.J. (2005) *Emerg. Infect. Dis.*, **11**, 69–75.
10. Maurer, S.M. (2009) *WMD Terrorism: Science and Policy Choices*, MIT Press. ISBN-10: 0-262-51285-8.
11. Bjornard, T. et al. (2009) Safeguards-by-design: early integration of physical protection and safeguardability into design of nuclear facilities. Proceedings of Global 2009, Paris, France, September 6–11, 2009, Paper 9518.
12. International Atomic Energy Agency (IAEA) (2002) IAEA Safeguards Glossary– 2001, Edition, Vienna, Austria. Available at http://www-pub.iaea.org/MTCD/publications/PDF/nvs-3-cd/PDF/NVS3_scr.pdf (accessed 6 March 2012).
13. Sorvillo, E. (2009) Obama's 2010 Nuclear Security Summit and the International Non-proliferation Regime, ISPI Policy brief, N. 163: 8 pp. Available at http://www.ispionline.it/it/documents/PB_163_2009.pdf (accessed 6 March 2012).
14. Diehl, S.J. (2010) The April 2010 Nuclear Security Summit: One More Step Toward the Mountaintop, Issue Brief, NTI.
15. NATO ATP-45(C) (2005) Reporting Nuclear Detonations, Biological and Chemical Attacks, and Predicting and Warning of Associated Hazards and Hazard Areas (Operators Manual), North Atlantic Treaty Organization.
16. Macintyre, A.G., Christopher, G.W., Eitzen, E. Jr., Gum, R., Weir, S., DeAtley, C., Tonat, K., and Barbera, J.A. (2000) *J. Am. Med. Assoc.*, **283** (2), 242–249.
17. Marss, T.C., Maynard, R.L., and Sidell, F.R. (2007) *Chemical Warfare Agent: Toxicology and Treatment*, 2nd edn, John Wiley & Sons, Ltd, Chichester.
18. Walters, T.J., Kauvar, David S., Reeder J., and Baer D.G. (2007) *Mil. Med.*, **172** (3), 318–321.
19. Ramesh, A.C. and Kumar, S. (2010) *J. Pharm. Bioallied Sci.*, **2**, 239–247. http://www.jpbsonline.org/text.asp?2010/2/3/239/68506.
20. Burkle, F.M. (2002) *Emerg. Med. Clin. N. Am.*, **20** (2), 409–436.
21. Mauroni, A. (2006) *Where Are the Wmds?: The Reality of Chem-Bio Threats on the Home Front and the Battlefront*, US Naval Institute Press. ISBN-10:1-59114-486-8.
22. Brooke, M.R., Amlôt, R., Rubin, H.J., Wessely, S., and Krieger, K. (2009) *Int. Rev. Psychiatry*, **19** (3), 279–288.
23. U.S. Department of Health and Human Services (2002) *Communicating in a Crisis: Risk Communication Guidelines for Public Officials*, Department of Health and Human Services, Washington, DC.

24. Covello, V., Peters, R.G., Wojtecki, J.G., and Hyde, R.C. (2001) *J. Urban Health: Bull. N.Y. Acad. Med.*, **78**, 382–391.
25. Maxwell, R. (1999) in *Risk Communication and Public Health* (eds P. Bennett and K. Calman), Oxford University Press, London, pp. 95–107.
26. Demuth, J.E. *et al.* (2002) *Countering Terrorism: Lessons Learned from Natural and Technological Disasters*, National Academy of Sciences, Washington, DC, A Summary to the Natural Disasters Roundtable (1-36).
27. O'Toole, T. (2001) *J. Urban Health: Bull. N.Y. Acad. Med.*, **78**, 396–402.

Index

a

absorbed dose   145, 249
acetylcholine   96–97
achieved protection factor (APF)   308–309
acquired resistance   407
activated carbon   299, 319
active detectors   218–219
acute exposure guideline levels (AEGLs)   176, 177, 182, 183–184, 310, 463
acute whole-body exposure, radiation effects of   146
Adamsite (DM)   92, 93
aerogels   368
aerosol
– detection   217
– – aerosol particle sizer, flame photometry, and fluorescence aerosol particle sizers (FLAPS)   220–222
– – cloud detection   217–219
– – detector layout topology, sensitivity, and response   222–223
– – radio detecting and ranging (RADAR) and light detection and ranging (LIDAR)   219–220
– sampling   224
– – surface sampling   227–228
– – techniques   226
aerosolization process   163–164
aerosol particle sizer (APS)   220
African Nuclear Weapon Free Zone Treaty. See Pelindaba Treaty (1996)
agent containing particles per liter of air (ACPLA)   222
agent vapor filtration mechanisms   285
Agreement on Reconciliation, Non-aggression, Exchanges, and Cooperation   62

air-filtration unit (AFU) and auxiliary equipment   342–344
airlock   346–348
air-purifying escape respirator (APER) with CBRN protection   313
air-purifying respirators (APRs) with canisters for ambient air   314–315
– powered air-purifying respirator (PAPR)   315–316
– self-contained breathing apparatus (SCBA)   316–317
air regeneration unit (ARU)   348
alkylating agents   78, 386
alpha radiation   164, 247
altitude   245
American Association of Official Analytical Chemists (AOAC)   402, 403
analytical chemistry   180–182
Antarctic Treaty (1961)   61
anthrax   121, 408, 437, 445
Anti-Ballistic Missile Treaty (ABMT)   46, 50–51, 56
Antonine Plague   15
aptamers   233
aqueous-based decontaminants   358, 377
– water   358–359
– water-soluble decontamination chemicals   359
arms control and international controlling bodies   42
arms reduction and prohibition of use   41
arsenicals   83–85
arsine   88
Asian Flu Pandemic   16
Aspergillus niger   391
asphyxiants   89

Index

atomic absorption spectrometry (AAS) 190–192, *193*, 221
Avian Flu Threat 16

b

Bacillus anthracis 106, 114, 119, 121, 213, 384, 396, 399
Bacillus atrophaeus 384
Bacillus spores 237, 389, 396
– chemicals and gases for inactivation of 397–398
Bacillus stearothermophilus 392
Bacillus subtilis 392
bacteria 105–107
– resistance of 407–408
– target sites 394–395
Bangkok Treaty (1995) 62
bellows effect 321
beta radiation 164, 247
Bethe, Hans Albrecht 24
bioaerosols 216–217
biocide 384, 385
– action mechanism against microorganisms 385–386
– – chemicals for disinfection 386–387
– – fumigation 388–390
– biological target sites 393
– – bacterial target sites 394–395
– – viral target sites 393–394
– inactivation resistance 405–406
– – of bacteria 407–408
– – of viruses 406–407
– neutralizing agents for selected 405
biological agent real-time sensor (BARTS) *217*
biological agents (B-agents) 211. See also biological warfare agents
– aerosol detection 217
– – aerosol particle sizer, flame photometry, and fluorescence aerosol particle sizers (FLAPS) 220–222
– – cloud detection 217–219
– – detector layout topology, sensitivity, and response 222–223
– – radio detecting and ranging (RADAR) and light detection and ranging (LIDAR) 219–220
– bioaerosols 216–217
– developing and upcoming technologies 238–239
– generalized analysis process of *214*
– ideal detection and identification platform *215*
– identification 229
– – chemical and physical identification 236–238
– – immunological methods based on ELISA 229–233
– – molecular methods 233–235
– sampling 223–224
– – aerosol sampling 224–228
– sensor system challenges *215*
biological and chemical weapons 51–52
biological dosimetry 271
biologically active particles per liter of air (BAPLA) 223
biologically active units per liter of air (BAULA) 223
biological threats 443–447
biological warfare agents 13, 103, 383
– in ancient times BC 16–18
– antimicrobial efficiency evaluation 401–402
– biocide action mechanism against microorganisms 385–386
– – chemicals for disinfection 386–387
– – fumigation 388–390
– biocide biological target sites 393
– – bacterial target sites 394–395
– – viral target sites 393–394
– biocide inactivation resistance 405–406
– – of bacteria 407–408
– – of viruses 406–407
– and borderline to pandemics, endemics, and epidemics 121
– carrier tests versus suspension tests 403–405
– cleanup 408
– disinfection and decontamination principles 384
– – chemical methods 385
– – definition of terms 384–385
– – physical methods 385
– disinfection
– – levels 390–392
– – new and emerging technologies for 408
– genetically engineered pathogens and 121–123
– harmful pandemics in history 14–16
– inactivation, as kinetic process 399–401
– in Middle Ages to World War I 18
– from 1980 to today 20–21
– origin, spreading, and availability 118–119
– – delivery methods 120
– risk classification of 110–111
– – potential 111–114, *115–116*
– routes of entry of 114–118
– specialty of 104

– spores problem   395–399
– types of   104–105
– – bacteria   105–107
– – fungi   109–110
– – toxins   108–109
– – viruses   107–108
– warning list   113
– from World War I to World War II   18–20
– from World War II to   1980, 20
biological weapon (B-weapon). See biological warfare agents
Biological Weapons Convention (BWC)   46
– implications   53–54
Biopreparat   121
bioterrorism   20. *See also individual entries*
– examples of emergence of   21
Black Death   15
blister agents (vesicants)   75, 78–82
– physicochemical characterization of   80, 81
– toxicity data of   82
blood agents   75, 85–89
– interference with oxygen transport   88–89
– physicochemical characterization of   88
– toxicity data of   89
breakaway   136
breakthrough concentration   284
breakthrough time   284
bremsstrahlung   164
British Standards Institution (BSI)   402
Brucella melitensis   106
Burkholderia cepacia   408
Burkholderia mallei   119
Bush, George W.   457

c

Candida albicans   403, *404*
canister
– air-purifying respirators (APRs) with, for ambient air   314–317
– connection   311–312
– number of   311
– position   310–311
carbon monoxide (CO)   87, 89
casualty protection   325–326
Central Asian Nuclear-Weapon-Free Zone (CANWFZ) treaty. See Semei Treaty (2006)
central burster technology, in artillery shells   100
challenge concentration   284
chelating agent   424
chemical impregnates   275, 285
– and chemical contaminants and   *286*
chemical threats   440–443

chemical warfare agents   4–14, 69, 179–180
– analytical chemistry   180–182
– in ancient times   5
– classification   71
– – NATO code identification   78
– – organs to be affected   75–76
– – physicochemical behavior   72–74
– – physiological agents on humans   76–77
– – route of entry into body   74
– during cold war   10–11
– detectors, testing of   203–206
– dissemination system of   99–100
– false alarm rate and sensitivity limit   184–185
– properties of   70, 78
– – arsenicals   83–85
– – blister agents (vesicants)   78–82
– – blood agents   85–89
– – choking and irritant agents   97–99
– – incapacitating agents   99
– – nerve agents   94–97
– – tear agents (lachrymators)   89–92
– – vomiting agents (sternutators)   92–93
– requirements for   69–71
– sensor system standards and deployment criteria   182
– – acute exposure guideline levels (AEGLs)   183–184
– – recommended concentration and requirement   182–183
– sensor system, technologies for   185–186
– – atomic absorption spectrometry (AAS)   190–192, *193*
– – colorimetric technology   197
– – electrochemical technologies   199–200
– – infrared (IR) spectroscopy   200–203, *202*
– – ion mobility spectrometry (IMS)   192, 194–196
– – mass spectrometry   187–190
– – photoionization technology   198–199
– used in terrorism   11–12
– between two world wars   8–9
– use in World War I   5–8
– in World War II   9–10
chemical warfare weapon (CWA)   4
Chemical Weapons Convention (CWC)   44, 46, 52, 440, 441, 442
– and Biological Weapons Convention (BWC) implications   53–54
– and Organization for the Prohibition of Chemical Weapons (OPCWs)   52–53
Chick–Watson law   400
Chlorine attacks, in Iraq   12
chlorine dioxide gas, fumigation with   388

chloroacetophenone (CN) 90
chloro-alkyl groups 78
chlorobenzylidene-malononitrile (CS) 90, 91
choking and irritant agents 97–99
– physicochemical characterization of 98
– toxicity data of 99
cleaning 384
closed-circuit system 316
Clostridium botulinum 109
clothing and protective clothing 378
cloud detection 217–219
cloud shine 163
Coccidioides immitis 110
collective protection (COLPRO) 275, 291, 331–337
– basic design 341–342
– – air-filtration unit (AFU) and auxiliary equipment 342–344
– – airlock 346–348
– – contamination control area (CCA) 345–346
– – environmental control unit (ECU) 344–345
– – toxic-free area 348
– equipment types *339*
– systems 337–340
colony-forming unit (CFU) 222
colorimetric technology 197
Committee for European Standardization (CEN) 402
Comprehensive Nuclear-Test-Ban Treaty (CTBT) 60
Comprehensive Nuclear-Test-Ban Treaty Organization (CTBTO) 60
Compton effect 152, 153, 154
Conference on Disarmament (CD) 62, 445
consequence management 436
contamination control area (CCA) 345–346
contamination monitors 253
Convention on Biological and Toxin Weapons 445
Coxiella burnetii 106
critical bed depth. See mass transfer zone
cyanogen bromide (CB) 86
cyanogen chloride (CK) 86

d

decontamination, of chemical warfare agents 353–354. See also biological warfare agents; radiological and nuclear decontamination
– catalysis in 373–375
– dispersal and fate 354–356
– media 356, 358
– – aqueous-based decontaminants 358–359
– – heterogeneous liquid media 362–369
– – non-aqueous decontaminants 359, *361*, 362
– procedures 375
– – clothing and protective clothing 378
– – dry procedures 378
– – generalities 376
– – personnel decontamination 379
– – rapid decontamination of personnel and personal gear 379–380
– – thorough personnel decontamination 380
– – wet procedures 376–378
– reaction schemes 369–370
– – sarin 370–371
– – soman 372
– – sulfur mustard 370
– – VX 372–373
delayed triage 469
deployment system for weapons 50–51
detect-to-protect detection 214–215
diphenylchloroarsine (DA) 92
diphenylcyanoarsine (DC) 92
direct ion storage (DIS) detector 264
dirty bomb 160, 416, 447–449. See also radiological dispersal device (RDD)
disarmament 56–58
dose equivalent 145
drinking device 312
dry decontamination procedures 378
dry ice blasting 427
dual use 44
dynamic pressure 138

e

early time waveform *155*
effective dose equivalent 250
electrical erasable programmable read only memory (EEPROM) cell 264
electrochemical technologies 199–200
electrochemiluminescence (ECL) assays 231
emulsions, decontamination 365
endospore structure 106
energy-to-effect categories, general distribution of *132*
environmental control unit (ECU) 344–345
enzyme-linked immunosorbent assay (ELISA), immunological methods based on 229–233
equivalent dose 250
ethyldichloroarsine (ED) 83
ethylene oxide, fumigation with 388

European standards, for disinfectant testing 402
exhalation valve 311
expectant triage 470
explosive radiological dispersal device (eRDD) 160, 162, 163
eye-sight correction 311

f
false alarm rate 184
"Fat Man" 27, 28, 30
filter technology 275–276
– and air cleaning 278–279
– – gas-phase air cleaning 283–286
– – particulate filtration 279–283
– general considerations 276–278
– selection process for CBRN filters 290–292
– test methods 286–287
– – gas filter tests 289–290
– – particle filter testing methods 288–289
first nerve agents 9
Fissile Material Cut-Off (FMCT) treaty 452
fission products 126–127
fission reaction 126
flame photometry. See atomic absorption spectrometry (AAS)
flame photometry detectors (FPDs) 190
fluorescence aerosol particle sizers (FLAPS) 221
foams 367–368
formaldehyde gas, fumigation with 389
formalin 387
Fourier-transform infrared (FTIR) spectromet 201
Francisella tularensis 106, 117
full-facepiece respirators 307
fumigation 388
– with chlorine dioxide gas 388
– with ethylene oxide 388
– with formaldehyde gas 389
– vaporized hydrogen-peroxide (VHP) 389–390
fungi 109–110
fusion-boosted fission weapon 130

g
gamma dose rate and gamma radiation detection 266
– energy response of dose-rate detector 267–268
– metrological dose rate measurements 266–267
– quantitative detection 268–271
gamma radiation 162, 247
gamma spectroscopy 253
gas-filled detectors 256–259
gas filter tests 289–290
gas-phase air cleaning 283–286
Geiger-Müller counter 258
gels 368–369
gene chips 234
genetically engineered pathogens 121–123
Geneva Conventions 49–50
geographical latitude 246
German Emulsion 365
German Society for Hygiene and Microbiology (DGHM) 402, 403, *404*
germicide 384–385, 390
Global Initiative to Combat Nuclear Terrorism (GICNT) 452
Goiânia accident 447, 448
Gram-staining method 105–106, 407
ground shine 163
gun method 129

h
Halabja, death people in 11
half-face piece masks 308
halogens 387
Hansen solubility parameters 364
head harness 312
helmets, CBRN 307–308
heterogeneous liquid media 362
– gels 368–369
– foams 367–368
– macroemulsions 362–366
– microemulsions 366–367
high altitude burst 132, 153–154
high efficiency particulate air (HEPA) *280, 282, 292,* 344, *345*
high efficient gas adsorber (HEGA) 344, *345*
highly enriched uranium (HEU) 448, 450, 453–454
high test hypochlorite (HTH) 365
Hildebrand solubility parameter 363
Histoplasma capsulatum 110
HL-mixtures 84
Hong Kong Flu 16
hot zone 468, 469, 470
HPGe detectors 263
HQ-mixtures 80
hydrodynamic enhancement 144
hydrodynamic front 137
hydrodynamic phase 136
hydrodynamic separation 135
hydrogen cyanide (AC) 86, 89

i

immediate decontamination   467
immediate triage   469
immunochip protein arrays   232
immunochromatographic assays   230
impinging samplers   225
implosion   27
improvised explosive devices (IEDs)   439, 448, 462
incapacitating agents   99
individual protective equipment (IPE)   170, 295–296
– basics   296
– challenges, identification of   296–297
– donning and doffing   305–306
– ergonomics   301–305
– function   299–301
– items, overview of   306–307
– – body protection   317–318
– – canisters   317
– – casualty protection   325–326
– – ponchos   324
– – pouches   323–324
– – protective footwear   323
– – protective gloves   322–323
– – protective suits   318–322
– – respirator design   310–317
– – respiratory protection   307–310
– – self-aid kit   324–325
– quality assurance   326
– way of designing   298–299
– workplace safety   327
indoor air quality   333
infrared (IR) spectroscopy   200–203, *202*
intercontinental ballistic missiles (ICBMs)   43, 50, 56
Intermediate-Range Nuclear Forces Treaty (INFT)   51
international arms control, conflict areas in   43
International Atomic Energy Agency (IAEA)   56, 64, 451
International Monitoring System (IMS)   60
international society of air quality and climate (ISIAQ)   333
International Task Force on the Prevention of Terrorism   36
international treaties   39
– difficulty in international regulations implementation   42–43
– – dual use   44
– – negotiation   43–44
– – technological momentum   45–46
– – trust   43
– – verification   44–45
– as global network   47–49
– – biological and chemical weapons   51–52
– – Chemical Weapons Convention (1993) and Organization for the Prohibition of Chemical Weapons (OPCWs)   52–53
– – Chemical Weapons Convention (CWC) and Biological Weapons Convention (BWC) implications   53–54
– – deployment system for weapons   50–51
– – Geneva Conventions   49–50
– historical development   46–47
– negotiations and   39–41
– – arms control and international controlling bodies   42
– – arms reduction and prohibition of use   41
– – nonproliferation   42
– nuclear weapons and   54–55
– – disarmament   56–58
– – nonproliferation   55–56
– – nuclear-weapon-free zones   60–62
– – test-ban and civil use   58–60
– organizations   63–64
intrinsic resistance   407
iodine   387
iodophors   387
ionization chamber   258
ionizing radiation   141
– prompt   142
ionizing radiation measurement   243
– detectors   256
– – gas-filled   256–259
– – luminescence   259
– – neutron detectors   265–266
– – photo-emulsion   260
– – scintillators   260–262
– – semiconductor detectors   262–264
– gamma dose rate and gamma radiation detection   266
– – energy response of dose-rate detector   267–268
– – metrological dose rate measurements   266–267
– – quantitative detection   268–271
– importance of detection of   244–248
– physical quantities to describe radioactivity and   248
– – absorbed dose   249
– – activity   248–249
– – effective dose equivalent   250
– – equivalent dose   250
– – operational dose quantities   250–251
– tasks   251
– – ambient dose rate measurement   252

– – measurement of activity   254–255
– – nuclide identification   253–254
– – personal dosimetry   252
– – radioactive aerosols detection   255–256
– – searching for gamma and neutron sources   252–253
– – surface contamination measurements   253
ion mobility spectrometry (IMS)   192, 194–196
isothermal sphere   135
Italian–Ethiopian War   8–9

j

Japanese invasion, of China   9
Joint Chemical Agent Detector (JCAD)   204
Joint Declaration on the Denuclearization of the Korean Peninsula   62
Joint Services Operational Requirements (JSOR)   183

k

Khan, A. Q.   454–455, 457
Korean Treaties (1991/1992)   62

l

lab-on-a-chip LOC) technologies   239
Lambert–Beer law   181, 201
laser cleaning   427–428
lewisites   83–85
– physicochemical characterization of   84
light detection and ranging (LIDAR)   219–220
liquid scintillation counting (LSC)   254, 261
"Little Boy"   26, 27, 453
– impact of   28
Loligo vulgaris   375
loose fitting facepieces and visors   308
low air burst   132
luminescence detectors   259
lyogels   368

m

macroemulsions   362–366
magnetic immunocapture assays   231
Manhattan Project   25–28
mask breaking action   92
mass spectrometry   187–190, 236
mass transfer zone   285
matrix-assisted laser desorption/ionization time-of-flight (MALDI-TOF)   236
MAUD report   25
microbiological agents   103
microemulsions   366–367

minimal triage   469
mission oriented protective postures (MOPPs)   305, *306*, 348, 467
monoclonal antibodies   229
Moscow Treaty. See Strategic Offensive Reductions (SORT) agreement (2003)
most penetrating particle size (MPPS)   280–281
multilayer laminates   321
multiple independent re-entry vehicles (MIRVs)   45, 56
mustard gas   79–80, 317, 370
– hydrolysis   370
Mutually Assured Destruction (MAD)   29
myths and facts   3
– biological warfare agents   13
– – in ancient times BC   16–18
– – harmful pandemics in history   14–16
– – in Middle Ages to World War I   18
– – from 1980 to today   20–21
– – from World War I to World War II   18–20
– – from World War II to   1980, 20
– chemical warfare agents   4–14
– – in ancient times   5
– – during cold war   10–11
– – used in terrorism   11–12
– – between two world wars   8–9
– – use in World War I   5–8
– – in World War II   9–10
– radiological and nuclear warfare   22–23
– – Manhattan Project   25–28
– – nuclear arms race   29–34
– – nuclear fission discovery   23–25
– – and nuclear terrorism   35–37
– – world nuclear forces and status   35

n

nanopores   239
National Commission on Terrorist Attacks (2004)   36
NATO code
– identification   78
– for pure chemical warfare agents   76
natural background rejection (NBR)   268
nerve agents   94–97
– physicochemical characterization of   95
– toxicity data of   95
neutron detectors   265–266
neutron induced gamma activity (NIGA)   145
neutron radiation   247
New Strategic Arms Reduction Treaty (NEW START) (2010)   57
NH-mustard (HN1)   81

nitrogen mustard   79–80
– physicochemical characterization of blister agents of   81
– toxicity data of   82
non-aqueous decontamination   377–378
non-explosive radiological dispersal device (neRDD)   160
nonproliferation   42, 55–56
Non-Proliferation of Nuclear Weapons Treaty (NPT)   32, 55, 451, 452
nosecup   312
nuclear age, timeline of   22
nuclear arms race   29–34
nuclear arsenal, time of states with   29
nuclear detonation effects   132
nuclear electromagnetic pulse (NEMP)   152
– early component of   154–156
– electric field generation   152–153
– in high-altitude burst   153–154
– intermediate component of   156–157
– late time component of   157–159
nuclear explosions   126
– direct effects   133
– – blast and shock   137–140, *141*
– – initial nuclear radiation   140–145
– – residual nuclear radiation   145, 147–148
– – thermal radiation   133–137
– indirect effects   149
– – nuclear electromagnetic pulse (NEMP)   152–159
– – transient radiation effects n electronics (TREE)   149–152
– monitoring technologies to detect   58
– nuclear fission   126–128
– nuclear fusion   128–129
– weapon design   129
– – fusion-boosted fission weapon   130
– – pure fission weapon   129–130
– – thermonuclear weapons   130–131
nuclear fission   23, 126–127, 414
– critical mass for fission chain and   127–128
– discovery   23–25
– energy of   *127*
– and ground contamination   414–415
– rainout and washout   415–416
nuclear fusion   23, 128–129
Nuclear Security Summit (2010)   449, 452–453
Nuclear Suppliers Group (NSG)   451
nuclear terrorism   35–37, 453
nuclear threats   450–456

nuclear warfare   22
nuclear weapons   22–23, 54–55, 414–416
– disarmament   56–58
– nonproliferation   55–56
– nuclear-weapon-free zones   60–62
– strategic   145
– tactical   145
– test-ban and civil use   58–60
– thermo   130–131
Nuclear weapons free zones (NWFZs)   60–62, 452

o
Obama, Barack   452
Occupational Safety and Health Administration (OSHA)   87
oil in water (O/W) emulsions   365
open-circuit system   316
Open Skies treaty (1992)   43, 51
operational decontamination   467
operational dose quantities   250–251
operational risk management and low-level exposure   175–177
Oppenheimer, J. Robert   24
organisation for the prohibition of chemical weapons (OPCW)   459
Organization for Security and Co-operation in Europe (OSCE)   51, 64
Organization for the Prohibition of Chemical Weapons (OPCWs)   52–53
Outer Space Treaty (1967)   61
oxidizing agents   386
oxygen mustard   82

p
pandemics   14
panoramic lens   311
Partial Test Ban Treaty (PTBT) (1963)   44, 58, 59
particle filter testing methods   288–289
particles per liter of air (PPL)   222
particulate filtration   279–283
passive detectors   218, 219
Peaceful Nuclear Explosions Treaty (PNET)   60
Pelindaba Treaty (1996)   62
peptide ligands   233
permeable barriers   299–300
– impermeable   300
– selectively   300
personal dosimetry   252
personnel decontamination   379–380
petroleum, oil, and lubricants (POLs)   320, 321, 322

phenols and alcohol 387
phenyldichloroarsine (PD) 83
phosgene 97
photocatalysis 375
– chemical warfare agents photooxidation by 376
photo-emulsion 260
photoionization detectors (PIDs) 199
photoionization technology 198–199
photomultiplier tube (PMT) 191
photons 162, 263
physical protection 275, 276, 290
PIN diodes 263
PIPs detectors 263
Plague of Justinian 15
plaque forming unit (PFU) 222
plutonium 448, 450, 453
point detectors 173, 222
Polaris Sales Agreement 30
polyclonal antibodies 229
polymerase chain reaction (PCR) 233
ponchos 324
pouches 323–324
powered air-purifying respirator (PAPR) 315–316
preparedness 433
– active countermeasures 458–459
– elements influencing counter-CBRN strategy 436–438
– lessons 474–475
– proliferation prevention 456–458
– research 473–474
– responding to CBRN event 459
– – CBRN warning and reporting 464–465
– – command and control, and communication 465–466
– – detection, reconnaissance, and surveillance 461–462
– – installation response fundamentals 459–461
– – medical response 468–470
– – medical support and post-disaster recovery 472–473
– – risk assessment 462–464
– – risk communication 470–472
– – technical response 466–468
– risk management and 433–436
– special strategy for CBRN 438–440
– – biological threats 443–447
– – chemical threats 440–443
– – nuclear threats 450–456
– – radiological threats 447–450
Proliferation Security Initiative 457
proportional counter 258

protection factor (PF), minimum required 308–309
protective footwear 323
protective gloves 322–323
protective suits 318–319
– impermeable 320–322
– permeable 319–320
Pseudomonas aeruginosa 408
Pseudomonas diminuta 375
pure fission weapon 129–130
Pyricularia grisea 110

q
quality assurance 177
3-Quinuclidinyl benzilate (BZ) 99

r
radiation dose 250
radioactive aerosols detection 255–256
radioactivity, physical quantities to describe 248
– activity 248–249
– absorbed dose 249
– effective dose equivalent 250
– equivalent dose 250
– operational dose quantities 250–251
radio detecting and ranging (RADAR) 219
radioisotope thermoelectric generators (RTG) 417
radiological and nuclear decontamination 411–412
– agents 421–422
– – chelating agent 424
– – specific processes and alternative procedures 424–428
– – surfactants 422–424
– contamination and 414
– – nuclear weapons 414–416
– – radiological dispersal device (RDD) 416–418
– efficiency calculation 418–419
– procedures, for response 420–421
– specialty of 412–413
radiological and nuclear warfare 22–23
– Manhattan Project 25–28
– nuclear arms race 29–34
– nuclear fission discovery 23–25
– and nuclear terrorism 35–37
– world nuclear forces and status 35
radiological dispersal device (RDD) 36–37, 159, 160, 161, 162–165, 416–418
radiological exposure device (RED) 159, 160
radiological terrorism 36
radiological threats 447–450

radiological weapons 23, 159–160
– impacts of
– – radiological dispersal device (RDD) 162–165
– – radiological exposure device (RED) 162
– most likely radionuclides for misuse in 161
– radioactive material and 160–161
radiophoto-luminescence detectors (RPL) 259
Raman spectroscopy 238
Rarotonga Treaty (1985) 62
real time polymerase chain reaction (rtPCR) 234
red telephone 43
Registration, Evaluation, Authorization, and Restriction of Chemicals (REACH) 54
remote detector 174
replicate organism detection and counting (RODAC) 227, 228
residence time 284
respirator design 310–313
– air-purifying escape respirator (APER) with CBRN protection 313
– air-purifying respirators (APRs) with canisters for ambient air 314–315
– – powered air-purifying respirator (PAPR) 315–316
– – self-contained breathing apparatus (SCBA) 316–317
respiratory quotient 336
Ricinus communis 118
RSDL 380
Russian Flu Threat 16

s

safeguards-by-design (SBD) process 451
Sampling Identification Biological Chemical Radiological Agents (SIBCRA) 174–175, 223
sarin 95, 96, 370–371
scintillators 260–262
Seabed Arms Control Treaty (1971) 62
self-aid kit 324–325
self-contained breathing apparatus (SCBA) 316–317
Semei Treaty (2006) 62
semiconductor detectors 262–264
sensitivity limit 185
sensor technologies 170
sesquimustard 79, 82
severe acute respiratory syndrome virus (SARSV) 408
solar activity 246
solar proton events (SPEs) 246

soman 372
Southeast Asian Nuclear-Weapon-Free Zone Treaty (SEANWFZ). See Bangkok Treaty (1995)
South Pacific Nuclear Free Zone Treaty. See Rarotonga Treaty (1985)
Spanish Flu 15
special nuclear material 36
speech intelligibility 312
Sphingomonas 375
spider chart *174*
spores, problem of 395–399
spraying and extraction systems 425–426
stand-off detector 173, 222
Staphylococcus aureus 109, 408
Starfish Prime test 153
static overpressure 138
sterilization 384, *390*
STERIS sterilization process 390
sternutators 75
Strasbourg Agreement (1675) 46
Strategic Arms Limitation Talks/Treaty (SALT) (1972) 45, 56–57
Strategic Arms Reduction Treaty (START) 57
strategic nuclear weapons 145
Strategic Offensive Reductions (SORT) agreement (2003) 57, 58
substance tracking, CBRN 170–175
– identification levels 172–173
subsurface burst 132
sulfur mustard. See mustard gas
surface ablation techniques 426–428
surface barrier detectors 263
surface burst 132
surface-enhanced Raman scattering (SERS) 238
surfactants 422–424
swab sampling 227, *228*
Swine Flu Threat 16
Szilárd, Leo 24

t

tabun 94, 95
tactical nuclear weapons 145
tandem mass spectrometry 189
tear agents (lachrymators) 89–92
– physicochemical characterization of *91*
– toxicity data of *91*
Teller–Ulam diagram *131*
terrorist activities 412, 429
test-ban and civil use 58–60
testing procedures, for chemical warfare agent detector 206, *207*

test methodology development (TMD), for chemical warfare agent sensors   204
tetrachloroethylene   365
thermal radiation   133–137
thermo-luminescence detectors (TLDs)   259
thermonuclear weapons   23, 130–131
"Thin Man"   27
thorough decontamination   467
Threshold Test Ban Treaty (TTBT)   59
Tlatelolco Treaty   61–62
total ionizing dose. See absorbed dose
toxic chemicals, eighteenth-and nineteenth-century discoveries of   6
toxic-free area   348
toxic industrial chemicals (TIC)   75, 439, 440–441
toxic industrial materials (TIMs)   296
toxins, biological   108–109, 211, 217
transient radiation effects n electronics (TREE)   149–152
triage   469
Trinity   26, *27*
tritium   265
Tsar Bomba   31–32
Tube Alloys Project   25

u

ultra-low penetration air (ULPA) filters   282
United Nations Children's Fund (UNICEF)   63
United Nations Disaster Assessment and Coordination (UNDAC)   460
United States Army Center for Health Promotion and Preventive Medicine   91
United States Atomic Energy Act   28, 30
United States Atomic Energy Commission   28
United States Environmental Protection Agency (US EPA)   408, 462
US-UK Mutual Defense Agreement (1958)   30

v

vacuum spray extraction sampling   227–228, *228*
vaporized hydrogen-peroxide (VHP)   389–390
Veepox   121
vesicants. See blister agents (vesicants)
Vienna Document (1999)   51
viruses   107–108
– adaptation of   407
– aggregation   406–407
– multiplicity reactivation   407
– target sites   393–394
volatility   72
vomiting agents (sternutators)   92–93
– physicochemical characterization of   93
– toxicity data of   93
VX (ethyl ({2-[bis(propan-2-yl)amino]ethyl}sulfanyl) (methyl)phosphinate)   69, 94, 372–373

w

water-in-oil emulsion   363–365
weapons of mass destruction (WMD)   41, 42, 44, 47, 50, 169, 295, 433, 456–458
weapons treaties, timeline and classification of   *49*
wet decontamination procedures   376–378
wipe sampling   227, *228*
World Disarmament Conference (1932–1934)   47
World Health Organization (WHO)   63, 87
world nuclear forces and status   35

x

Xylene Emulsion   365

y

Yersinia pestis   106, 114